Risk Measurement, Econometrics and Neural Networks

Contributions to Economics

Christoph M. Schneider
Research and Development Management: From the Soviet Union to Russia
1994. ISBN 3-7908-0757-5

Bernhard Böhm/Lionello F. Punzo (Eds.)
Economic Performance
1994. ISBN 3-7908-0811-3

Lars Olof Persson/Ulf Wiberg
Microregional Fragmentation
1995. ISBN 3-7908-0855-5

Ernesto Felli/Furio C. Rosati/Giovanni Tria (Eds.)
The Service Sector: Productivity and Growth
1995. ISBN 3-7908-0875-X

Giuseppe Munda
Multicriteria Evaluation in Fuzzy Environment
1995. ISBN 3-7908-0892-X

Giovanni Galizzi/Luciano Venturini (Eds.)
Economics of Innovation: The Case of Food Industry
1996. ISBN 3-7908-0911-X

David T. Johnson
Poverty, Inequality and Social Welfare in Australia
1996. ISBN 3-7908-0942-X

Rongxing Guo
Border-Regional Economics
1996. ISBN 3-7908-0943-8

Oliver Fratzscher
The Political Economy of Trade Integration
1996. ISBN 3-7908-0945-4

Ulrich Landwehr
Industrial Mobility and Public Policy
1996. ISBN 3-7908-0949-7

Arnold Picot/Ekkehard Schlicht (Eds.)
Firms, Markets, and Contracts
1996. Corr. 2nd printing 1997.
ISBN 3-7908-0947-0

Thorsten Wichmann
Agricultural Technical Progress and the Development of a Dual Economy
1997. ISBN 3-7908-0960-8

Ulrich Woitek
Business Cycles
1997. ISBN 3-7908-0997-7

Michael Carlberg
International Economic Growth
1997. ISBN 3-7908-0995-0

Massimo Filippini
Elements of the Swiss Market for Electricity
1997. ISBN 3-7908-0996-9

Giuseppe Gaburro (Ed.)
Ethics and Economics
1997. ISBN 3-7908-0986-1

Frank Hoster/Heinz Welsch/Christoph Böhringer
CO_2 Abatement and Economic Structural Change in the European Internal Market
1997. ISBN 3-7908-1020-7

Christian M. Hafner
Nonlinear Time Series Analysis with Applications to Foreign Exchange Rate Volatility
1997. ISBN 3-7908-1041-X

Sardar M.N. Islam
Mathematical Economics of Multi-Level Optimisation
1998. ISBN 3-7908-1050-9

Sven-Morten Mentzel
Real Exchange Rate Movements
1998. ISBN 3-7908-1081-9

Lei Delsen/Eelke de Jong (Eds.)
The German and Dutch Economies
1998. ISBN 3-7908-1064-9

Mark Weder
Business Cycle Models with Indeterminacy
1998. ISBN 3-7908-1078-9

Tor Rødseth (Ed.)
Models for Multispecies Management
1998. ISBN 3-7908-1001-0

Michael Carlberg
Intertemporal Macroeconomics
1998. ISBN 3-7908-1096-7

Sabine Spangenberg
The Institutionalised Transformation of the East German Economy
1998. ISBN 3-7908-1103-3

continued on page 307

Georg Bol, Gholamreza Nakhaeizadeh,
Karl-Heinz Vollmer (Eds.)

Risk Measurement, Econometrics and Neural Networks

Selected Articles
of the 6th Econometric-Workshop
in Karlsruhe, Germany

With Contributions of

K. Abberger, Y. Feng, S. Heiler; J. Baetge, C. Uthoff; C. Breitner; W. Härdle, S. Sperlich; W. Holt, P. Refenes; R. Kruse, S. Siekmann, R. Neuneier, H. Zimmermann; T. Severin, W. Schmid; M. Steiner, S. Schneider, J. Wolf; R. Matthes, M. Schröder; T. Ridder; R. Davé, G. Stahl; S. Huschens; H. Schulte-Mattler; T. Wilson

With 115 Figures
and 28 Tables

Physica-Verlag
A Springer-Verlag Company

Series Editors
Werner A. Müller
Martina Bihn

Editors
Prof. Dr. Georg Bol
Universität Karlsruhe
Institut für Statistik und
Mathematische Wirtschaftstheorie
Neuer Zirkel 3
D-76128 Karlsruhe, Germany

Prof. Dr. Gholamreza Nakhaeizadeh
Daimler Benz AG
Postfach 23 60
D-89013 Ulm, Germany

Prof. Dr. Karl-Heinz Vollmer
Südwestdeutsche Genossenschaftszentralbank AG
Karl-Friedrich-Straße 23
D-76049 Karlsruhe, Germany

ISBN 3-7908-1152-1 Physica-Verlag Heidelberg New York

Cataloging-in-Publication Data applied for
Die Deutsche Bibliothek – CIP-Einheitsaufnahme
Risk measurement, econometrics and neural networks: selected articles of the 6th Econometric Workshop in Karlsruhe, Germany; with 28 tables / Georg Bol ... (ed.). With contributions of K. Abberger ... – Heidelberg; New York: Physica-Verl., 1998
(Contributions to economics)
ISBN 3-7908-1152-1

Softcover Design: Erich Kirchner, Heidelberg

SPIN 10691405 88/2202-5 4 3 2 1 0 – Printed on acid-free paper

Preface

On March 19th - 21th 1997, the 6th Econometric Workshop in Karlsruhe, Germany took place at the University of Karlsruhe. The workshop was organized jointly from the Institute for Statistics and Mathematical Economics and the Südwestdeutsche Genossenschaftszentralbank AG (SGZ-Bank AG).

Following the tradition of the Econometric Workshops in Karlsruhe, theoretical as well as empirical contributions in the fields of time series analysis, machine learning and econometrics were presented in the first part. In addition - due to the actuality of the topic - the articles of the second part discuss concepts on Value-at-Risk (VaR).

Klaus Abberger, Yuanhua Feng, and Siegfried Heiler discuss two particular aspects of nonparametric time series analysis. As can be observed from stock market returns, the upper and lower quantiles behave differently, indicating an asymmetrical underlying distribution. To account for this behavior the authors estimate the quantiles of a conditional distribution with kernel methods and give an application to daily series of German stock returns.

In the second part of their paper, the authors use locally weighted regression with polynomials and trigonometric functions to decompose a time series into unobserved components. Bandwidth and polynomial degree is estimated using a mean averaged squared error criterion with a bootstrap variance estimator. An application to the quarterly GDP series demonstrates the behavior of the proposed method.

A central technical problem not only in risk management but also in decision making in general is the retrieval of essential information. Today, data-base management systems are used for day-to-day operation and control purposes. However, to obtain an integrated view on all business-relevant data a data-warehouse has to be designed. Christoph Breitner discusses the necessary requirements and concepts of data-warehouses as well as the retrieval of data using online analytical processing (OLAP).

Debtor analysis using neural nets is studied in the article of Jörg Baetge and Carsten Uthoff. In their study the authors examine financial and non-financial indicators and model a classifier based on neural nets. In their approach feature selection and data preprocessing is emphasized as vital stages in their modeling. In their empirical investigation the performance of classifiers based on neural nets is compared with traditional methods based on balance sheet analysis.

Wolfgang Härdle and Stefan Sperlich introduce an interactive statistical computing environment, which offers modules such as time series analysis or finance-related topics such as option price evaluation. Based on short examples they show main features of the software package which can be accessed via internet.

Diagnostic testing in neural network models is the topic of William Holt and Paul Refenes. They propose a generalisation of the Durbin-Watson test for neural network regression models to detect the presence of first order correlation. The distribution of the statistic is estimated using approximation and exact methods.

Rudolf Kruse et al. show an application of neuro-fuzzy methods in forecasting financial time series. Neuro-fuzzy methods combine properties of fuzzy systems (transparency through the use of rules) and neural nets (adaptation and nonlinear modeling). In their case study the authors use a neuro-fuzzy model to predict daily returns of the German stock index. They compare the results of the neuro-fuzzy model with linear and nonlinear models (i.e. neural nets) and report promising results.

Thomas Severin and Wolfgang Schmid use methods of statistical process control to detect structural changes in financial time series. Assuming that financial time series is subject to a GARCH-model, they apply several control schemes such as Shewart, EWMA, and CUSUM. The authors propose some new residual charts for GARCH-models. In two case studies the properties of these charts applied on German companies' stocks are compared.

Manfred Steiner, Sebastian Schneider, and Benedict Wolf examine the managerial and the pecking order hypothesis. In contrary to the pecking order hypothesis, the managerial hypothesis predicts the preference of internal to external financing if the manager's utility function diverges from the owners', to avoid additional monitoring. In the empirical analysis the authors apply discriminant analysis and several forms of neural networks to analyze the financial behavior of German corporations.

Rakhal Davé and Gerhard Stahl present an empirical study of several candidate models for forecasting losses in relation to positions held against individual risk factors as well as losses in relation to a portfolio of risk factors. Based on eight major currencies and the prices for gold and silver denominated in US-Dollar the behavior of various VaR-models are tested. In their article various measures for evaluating model performance are discussed. Finally, the authors propose a testing framework for models.

The admission of in-house models for estimating the Value-at-Risk (VAR) leads to a variety of propositions used in the banking sector. Stefan Huschens discusses the accuracy of Value-at-Risk estimators. From a statistical point of view VaR is a statistical point estimation of an unknown parameter of the loss distribution. The author takes sampling errors into consideration and derives exact and asymptotic confidence intervals in a parametric context with linear portfolio and normal distributed returns.

Volatility may not be an appropriate measure for risk especially when we look at a portfolio which contains options. Rainer Matthes and Michael Schröder discuss the shortfall concept as an alternative for measuring risk. Their article gives a survey of the major characteristics of shortfall measures. In the second part the authors describe an investment strategy based on a dynamic benchmark adjustment process. In an example based on a portfolio of German assets the behavior of the proposed method is demonstrated.

Quantifying risk exposure led to a revival of interest in standard techniques in statistical quantile estimation. Thomas Ridder gives a survey of parametric and nonparametric approaches to analyze risk exposure by VaR and studies basic statistical properties of various VaR-estimators. In a case study using a foreign exchange rate the accuracy of those VaR-strategies are examined.

Hermann Schulte-Mattler provides a detailed overview of the changing structure of capital requirements for credit and market risk over the past two decades and gives an account of the historical development of the regularity framework for German credit institutions. Finally, the author discusses the requirements of internal risk models and the backtesting procedure to evaluate the accuracy of VaR.

Thomas Wilson examines risk from credit exposure at a portfolio level. The proposed model estimates the expected losses of a portfolio of credit exposures by modeling the marginal and absolute loss distribution conditional on the current state of the economy. The model consists of two components. The first is a multi-factor, systematic risk model for credit risk uncertainty. The second is a method for explicitly tabulating the discrete loss distributions arising from a portfolio of credit exposures.

Many people have contributed to the success of the workshop. Matthias Behnke, Tae Horn Hann, Robert Lechler, Xun Lin, Françoise Mouden, Petra Weth proved indispensable for the organization of the workshop. Julia Branke helped with the layout of the book through several revisions. Her help is much appreciated. We wish to thank the faculty, the managing director Dr. Volker Binder and the Dean, Prof. Dr. Waldmann for their cooperation. Finally, we thank Daimler-Benz AG and SGZ-Bank AG for their sponsorship of the workshop.

Karlsruhe, May 1998

Georg Bol
Gholamreza Nakhaeizadeh
Karl-Heinz Vollmer

Table of Contents

List of Contributors

Dr. Klaus Abberger	University of Konstanz
Prof. Dr. Dr. h. c. Jörg Baetge	Westfälische Wilhelms-Universität, Münster
Christoph A. Breitner	Universität Karlsruhe
Rakhal D. Davé (PhD)	Olsen & Associates Research Institute for Applied Economics
Yuanhua Feng	University of Konstanz
Prof. Dr. Wolfgang Härdle	Humboldt-Universität zu Berlin
Prof. Dr. Siegfried Heiler	University of Konstanz
Prof. Dr. Stefan Huschens	TU Dresden
William T. Holt	London Business School
Prof. Dr. Rudolf Kruse	Otto-von-Guericke University, Magdeburg
Dr. Rainer Matthes	Metzler Asset Management
Ralph Neuneier	Siemens AG, München
Prof. Apostolo-Paul N. Refenes	London Business School
Dr. Thomas Ridder	SGZ-Bank AG, Frankfurt
Prof. Dr. Wolfgang Schmid	Europe-University Viadrina, Frankfurt (Oder)
Sebastian Schneider	University of Augsburg
Dr. Michael Schöder	Zentrum für Europäische Wirschaftsforschung (ZEW)
Prof. Dr. Hermann Schulte-Mattler	Fachhochschule Dortmund
Thomas Severin	Europe-University Viadrina
Stefan Siekmann	Siemens AG, München
Stefan Sperlich	Humboldt-Universität zu Berlin
Gerhard Stahl	Federal Banking Supervisory Office, Berlin
Prof. Dr. Manfred Steiner	University of Augsburg
Dr. Carsten Uthoff	Westfälische Wilhelms-Universität, Münster
Dr. Thomas C. Wilson (PhD)	McKinsey & Company, Zürich
J. Benedict Wolf	University of Augsburg
Dr. Hans Georg Zimmermann	Siemens AG, München

Nonparametric Smoothing and Quantile Estimation in Time Series

Klaus Abberger, Yuanhua Feng, and Siegfried Heiler

University of Konstanz, Fak. für Wirtschaftswissenschaften und Statistik, Postfach 5560, D129, D-78434 Konstanz

Abstract. Two nonparametric approaches to the analysis of time series are discussed in this paper. The first one is quantile regression based on kernel estimation, which may be interpreted as a locally weighted regression with a polynomial of degree zero. If the regressor is the time index, then we arrive at a quantile smoothing. If the regressor is composed of some past time series values, then the approach leads to a quantile autoregression. Both aspects are applied to daily series of German stock returns.

In the second part locally weighted regression with polynomials and trigonometric functions is used for decomposing a time series into unobserved components. For the estimation of bandwidths and polynomial degrees a mean averaged squared error criterion is used with a bootstrap variance estimator. The data-driven procedure works like a moving average in the certral part of the time series. An application to the quarterly German GDP series is presented as illustration.

1 Introduction

In a recent paper, Härdle, Lütkepohl and Chen (1997) give a review of nonparametric time series analysis. In this contribution we focus on two particular aspects of nonparametric methods in time series not covered by the before mentioned review. These are estimation of the quantiles of a conditional distribution $F(y|X = x)$ with kernel methods and the application of locally weighted regression to the problem of decomposing a time series into unobservable components such as the trend-cyclical and the seasonal.

If in the above conditional distribution function X denotes the time index, then the aim of the analysis is to produce a smooth curve of given quantiles in the course of time. Applications to stock returns reveal a pronounced asymmetric behaviour of upper and lower quantiles and manifest that in these and related cases in the financial market volatility is expressed by movements in the lower tails of the distributions, whereas the upper tails show only small changes in the course of time. This is in contrast to many models of the ARCH-, GARCH- and related types, which frequently are based on the assumption of an underlying distribution being symmetrical.

If the variable X in the conditional distribution is composed of past time series values $Y_{i-1}, \ldots, Y_{i-p}$, then the underlying model is a nonparametric quantile autoregression and the estimates are forecasts of conditional quantiles of the distribution at time i. A measure for the goodness of fit of these quantile forecasts is being proposed and applications to stock market data show that past values contain some information about the tails of the conditional distribution, whereas in the central part the gain of taking past values into account is very close to zero.

Locally weighted regression with polynomials and trigonometric functions as local regressors yields a procedure which works like a moving average and which makes it possible to estimate the course of the "smooth" component and the seasonal in one closed approach and that renders many traditional iterative procedures superfluous. Much ink has been spilled regarding bandwidth selection in kernel density estimation and nonparametric regression. These ideas can be exploited for a data-driven selection of local neighbourhood and polynomial degree in the time series decomposition problem, and this holds not only for the central part of the time series but also for each instant in the boundary area.

In contrast to procedures being based on parametric ARMA- or ARIMA models, as proposed e.g. by Harvey and Todd (1983), Kitagawa and Gersch (1984), Hillmer and Tiao (1982) and Maravall and Pierce (1984) (see also the computer programme by Maravall and Goméz, 1994), the proposed procedure will not lead to artificial waves in the smooth course which may happen if the data structure does not correspond to the underlying model. See Stier (1996) for a comparative study. An application to the quarterly German GDP series demonstrates the behaviour of the proposed method.

2 Nonparametric Quantile Regression

Various methods for nonparametric quantile regression have been discussed in the past. Koenker, Portnoy and Ng (1992) propose a spline smoothing, Stute (1986) a nearest-neighbour estimator and Jones and Hall (1990) use locally weighted constant regression. In this paper kernel estimation is considered, which can be shown to be similar to a locally weighted constant regression. Our particular focus is on applications to time series.

2.1 Nonparametric Quantile Estimation

The procedure is based on the conditional distribution function $F_x(y) := F_Y(y|X = x)$ for given values of a regressor variable (vector) X and the corresponding (theoretical) α-quantile $q_\alpha(x) := \inf\{y \in \mathbb{R} \mid F_x(y) \geq \alpha\}$, $0 < \alpha < 1$. For a given data set $\{Y_i, X_i\}_{i=1}^n$, a statistical model can be formulated as

$$Y_i = q_\alpha(X_i) + \xi_i \,, \tag{1}$$

with $P(\xi_i \leq 0) = \alpha$. A kernel estimator for F_x is given by

$$F_{n,x}(y) = \sum_{i=1}^{n} \mathbb{1}_{(Y_i \leq y]}(Y_i) K\left(\frac{x - X_i}{h_n}\right) / \sum_{i=1}^{n} K\left(\frac{x - X_i}{h_n}\right), \tag{2}$$

and the α-quantile of this estimate yields the conditional α-quantile estimator

$$q_{n,\alpha}(x) = \inf\{y \in \mathbb{R} \mid F_{n,x}(y) \geq \alpha\}. \tag{3}$$

For $\alpha = 1/2$ this is the conditional median as a robust alternative to nonparametric kernel regression. But we are particularly interested in α-quantiles in the tails of the distributions.
If the conditioning variable is the time index, then a quantile smooting is achieved. If X is composed of some past time series values Y_{i-1}, to Y_{i-p}, then we arrive at a nonparametric quantile autoregression.
Both models are investigated and applied to financial data by Abberger (1996). He derives expressions for the MSE of $F_{n,x}$ and $q_{n,\alpha}$ and gives conditions for consistency and asymptotic normality of these estimators. For dependent data $\{Y_i, X_i\}$ the concept of strong mixing is used.
In the case of quantile smoothing cross-validation is proposed for bandwidth selection. That is, with the loss function

$$\rho(u) = \alpha \mathbb{1}_{\{u \geq 0\}} \cdot u + (\alpha - 1) \mathbb{1}_{\{u < 0\}} \cdot u \tag{4}$$

introduced by Koenker and Basset (1978) for the parametric quantile regression, the bandwidth is selected according to

$$h_{cv} = \underset{h}{\operatorname{argmin}} \left\{ \sum_{i=1}^{n} \rho\left(Y_i - q_{n,\alpha}^{(-i)}(X_i)\right) \right\}, \tag{5}$$

where $q_{n,\alpha}^{(-i)}(X_i)$ is the leave-one-out estimator, that is an estimator without observation (Y_i, X_i). The cross-validation technique requires independent data. A block-cross-validation may be used to check the sensitivity of the resulting bandwidth for dependent data $\{Y_i\}$. Instead of omitting the i-th observation only, there are whole data blocks $\{Y_j\}_{j=i-b_1}^{i+b_2}$ not taken into account for estimating the conditional distribution and $q_\alpha(X_i)$.

Nonparametric quantile autoregression can be used to obtain quantile forecasts. But forecast measures such as the mean squared forecast error, the relative mean squared forecast error or Theil's coefficient cannot be used to evaluate quantile forecasts, since the actual quantiles are not observable. As a goodness of fit measure for forecasting conditional quantiles

$$\bar{\xi}_\alpha = 1 - \sum_{i=1}^{n} \rho(Y_i - q_\alpha(X_i)) / \sum_{i=1}^{n} \rho(Y_i - q_\alpha). \tag{6}$$

is proposed, where ρ is the above loss function and q_α the unconditional α-quantile. It holds $0 \leq \bar{\xi}_\alpha \leq 1$ and $\bar{\xi}_\alpha = 0$ if $q_\alpha(X_i) = q_\alpha \; \forall i$. Further $\bar{\xi}_\alpha = 1$ if $Y_i = q_\alpha(X_i) \; \forall i$ and all α, i.e. if the conditional distribution of $\{Y|X = x\}$ is a one-point distribution. Figure 1 shows a conical simulated data set of 500 observations. The observations are heteroscedastic and have mean zero. The Bravais-Pearson correlation coefficient is -0.002. Table 1 includes the empirical ξ_α-values for different α. The ξ-value for the median is close to

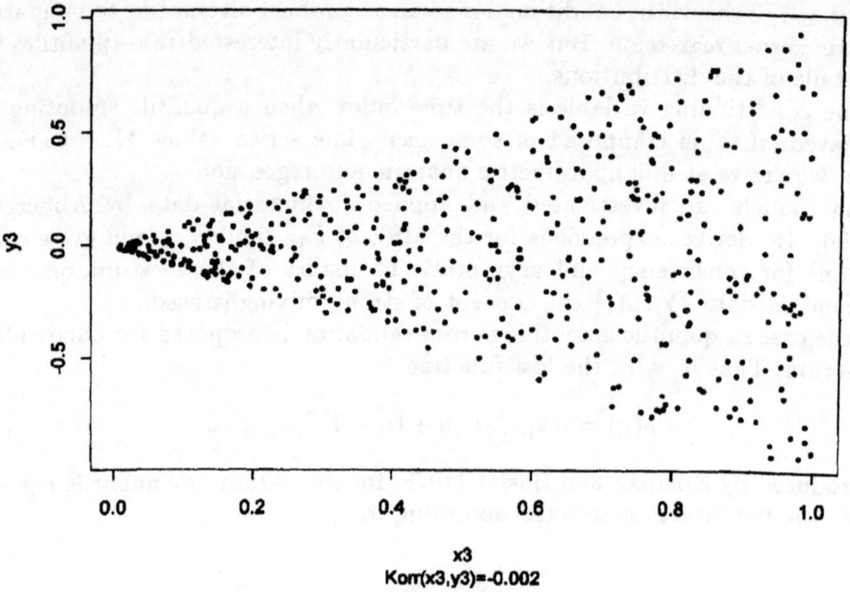

Fig. 1. Simulated heteroscedastic data, $n = 500$

Table 1. ξ_α-values for the data in Fig. 1

α	0.01	0.05	0.10	0.25	0.50	0.75	0.90	0.95	0.99
ξ_α	0.43	0.36	0.27	0.10	0.01	0.11	0.26	0.34	0.41

zero, so there is no information gain by X compared to the unconditional median. But towards the boundaries of the distribution the ξ_α-value increases, i.e. conditioning on X information increases. Comparing the goodness of fit measure ξ_α for different α allows to compare the information gain in different parts of the conditional distribution. For estimating this quantity in time series context $q_\alpha(X_i)$ is replaced by its kernel estimator $q_{n,\alpha}(X_i)$ and q_α by $q_{i-1,\alpha}$, which is the empirical quantile of the first $i-1$ data $Y_1, ..., Y_{i-1}$ and can be interpreted as a naive forecast of $q_\alpha(X_i)$.

2.2 Applications to Stock Returns

The applications of quantile smoothing to stock returns reveal a marked asymmetric behavior of volatility in the course of time, which is in contrast to the assumptions in many ARCH-, GARCH- and related models. Whereas upper quantiles proceed very smoothly in time, the corresponding lower ones show pronounced dips. Figure 2 shows the time series of daily DAX-returns with 1400 observations made between 2. January 1986 to 13. August 1993. The returns are defined by $Y_i = (\text{price}_i - \text{price}_{i-1})/\text{price}_{i-1}$.

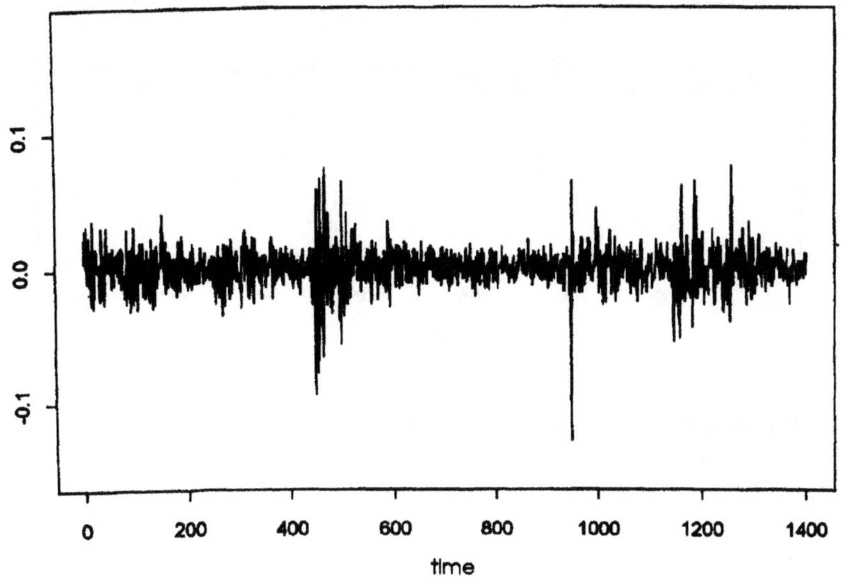

Fig. 2. Time series of daily DAX returns from 2. January 86 till 13. August 91 (n=1400)

For the kernel estimation of conditional quantiles, a bandwidth h_α depending on α must be chosen, therefore cross-validation is used. Also block-cross-validation is computed, but in the present application on daily DAX returns no remarkable influences of dependencies on the choice of bandwidth were observed. With the help of the maximum-likelihood method a GARCH-model for the DAX data is estimated with a Gaussian distribution as conditional distribution. The Gaussian distribution is completely determined by the mean and variance thats why the calculation of quantiles is possible. Looking at the estimated 0.25- and 0.75-quantiles in Fig. 3 or the estimated 0.10- and 0.90-quantiles in Fig. 4 shows that the asymmetrical distribution found by kernel estimation faces a symmetrical distribution by GARCH estimation with assumption of a conditional Gaussian distribution.

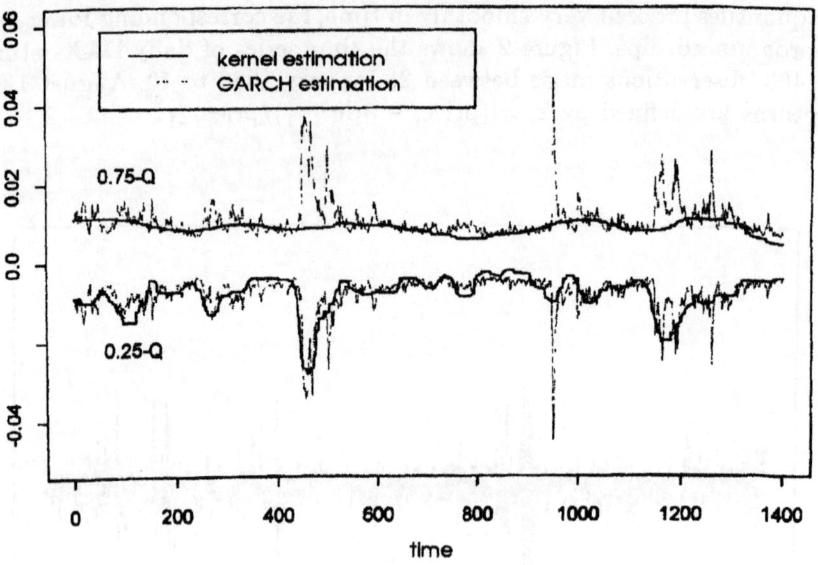

Fig. 3. Estimation of 0.25- and 0.75-quantiles of daily DAX returns

The nonparametric autoregression will be applied to the forecasting of stock return quantiles. A product kernel with identical bandwidths for each component of X will be used. Bisquare kernels of the form $K(u) = \frac{15}{16}(1 - u^2)^2 \mathbb{1}_{[-1,1]}(u)$ are chosen for the univariate case. The Daimler-Benz stock returns from 5. July 1987 until 30. April 1993 are presented in Fig. 5. One-step forecasts have been calculated for the last 450 returns, using past returns for predictors. In the simplest case, $X_i = Y_{i-1}$.

Adding further lags of Y as predictors can improve the forecasts. The dimension of X cannot, however, be inflated too much, since the kernel estimator otherwise does not find enough similar past courses in a neighbourhood of the actual course in the series. This problem is well known as "the curse of dimensionality".

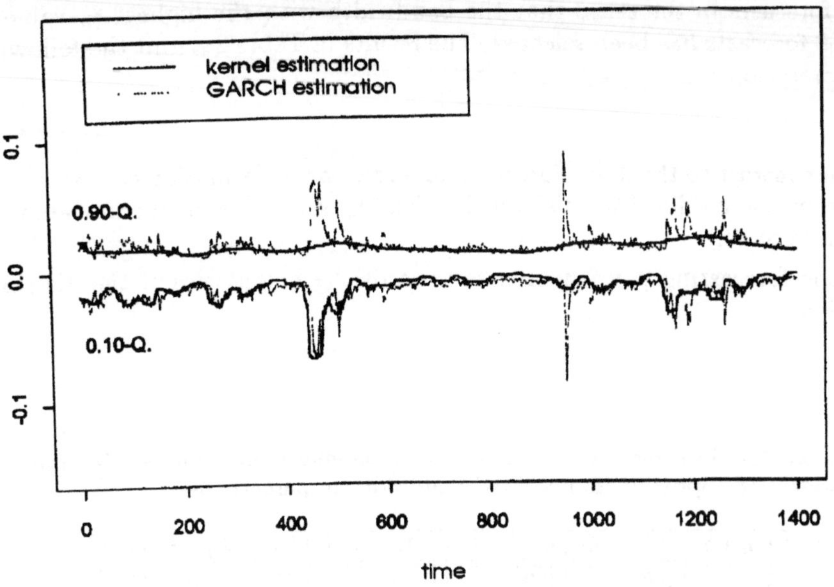

Fig. 4. Estimation of 0.10- and 0.90-quantiles of daily DAX returns

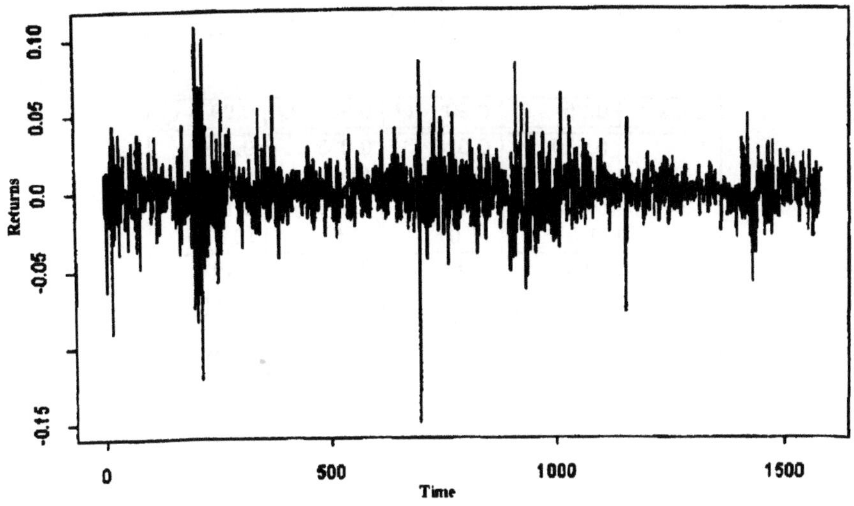

Fig. 5. Time series of Daimler-Benz returns from 5. July 1987 until 30. April 1993

Table 2 lists the ξ_α-values calculated from the 450 forecasts. They are ex-post forecasts in the sense that the bandwidth with the highest ξ_α-value of all 450 forecasts has been selected. The results in Table 2 admit the following interpretations:

1. In contrast to the naive forecast, no additional information can be gained from the middle of the distribution. The ξ_α-values for the median are very close to zero.
2. The forecasting power increases towards the boundaries of the distribution.

Table 2. ξ_α-values for the forecasts of 450 Daimler-Benz returns with optimal bandwidth (the optimal, standardized bandwidth in parentheses)

Lags Y	$0.01-Q$	$0.05-Q$	$0.10-Q$	$0.25-Q$	$0.50-Q$
1	0.044 (2.0)	0.036 (1.3)	0.036 (1.0)	0.01 (2.5)	0.0047 (2.3)
2	0.098 (1.5)	0.05 (2.0)	0.05 (1.5)	0.016 (2.5)	0.007 (2.5)
3	0.147 (1.2)	0.056 (1.8)	0.054 (1.5)	0.020 (2.5)	0.004 (3.0)
4	0.181 (1.3)	0.061 (2.5)	0.059 (2.5)	0.021 (3.0)	0.006 (3.5)

Lags Y	$0.75-Q$	$0.90-Q$	$0.95-Q$	$0.99-Q$
1	0.010 (1.8)	0.016 (4.0)	0.035 (3.0)	0.129 (1.3)
2	0.023 (1.0)	0.04 (1.0)	0.073 (1.0)	0.187 (0.8)
3	0.025 (2.0)	0.04 (1.8)	0.060 (1.8)	0.142 (3.5)
4	0.025 (2.8)	0.039 (2.8)	0.063 (1.8)	0.149 (4.0)

Figure 6 visualizes these interpretations. The maximal ξ_α-values are plotted against α, from what an u-shaped course results. Further examples of quantile smooting and quantile autoregression for various return series may be found in Abberger (1996).

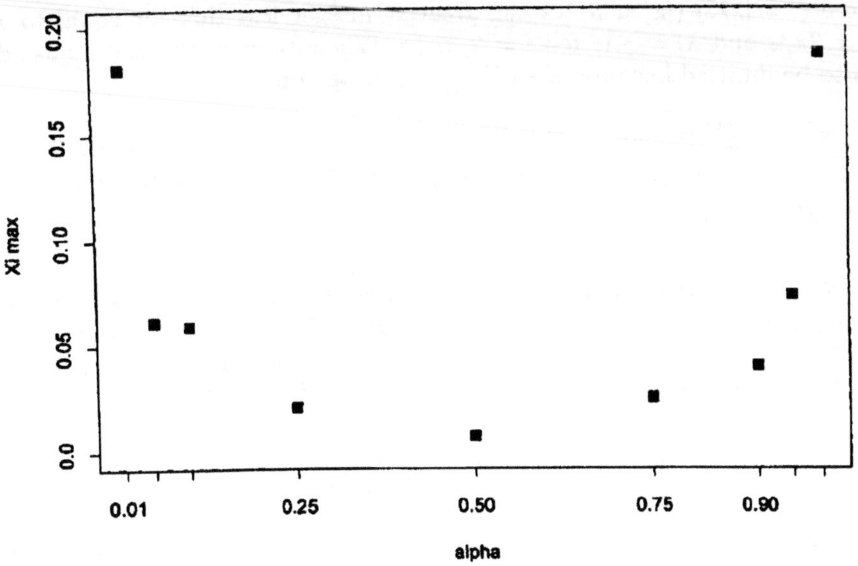

Fig. 6. Maximal ξ_α-values for one-step forecasts of Daimler-Benz returns

3 Time Series Decomposition with Locally Weighted Regression

We consider an equidistant time series Y_i, $i \in \mathbf{I}$, where $\mathbf{I} = \{1, 2, \ldots, n\}$ is the time index set. It is assumed that Y_i follows an additive components model

$$Y_i = G(i) + S(i) + \epsilon_i, \quad i = 1, 2, \ldots, n \ , \tag{7}$$

where the ϵ_i are assumed to be i.i.d. random variables with $E(\epsilon_i) = 0$ and $\text{var}(\epsilon_i) = \sigma^2$. G is the trend-cyclical component, S is the seasonal component with seasonal period s and $\mu := G + S$ is the nonrandom mean function. In the sequel we introduce a time series decomposition procedure using locally weighted regression based on the proposal of Heiler and Michels (1994).

The trend-cyclical component is assumed to be at least (p+1) times differentiable, so that it can be developed in a Taylor series around a point t, yielding a local polynomial representation of order p. In a similar way the seasonal component can be locally modelled by a Fourier series, i.e.

$$G(i) \doteq \sum_{j=0}^{p} \beta_{1j}(t)(i-t)^j \ , \tag{8}$$

$$S(i) \doteq \sum_{j=1}^{q} [\beta_{2j}(t) \cos \lambda_j (i-t) + \beta_{3j}(t) \sin \lambda_j (i-t)] \ , \tag{9}$$

where $q = [s/2]$ ($[x]$ denotes the greatest integer less than or equal to x), $\lambda_1 = 2\pi/s$ and $\lambda_j = j\lambda_1$ for $j = 2, \ldots, q$. When $\lambda_q = \pi$, the last sinusoidal has to be omitted because of $\sin(i\pi) = 0$, $i \in \mathbb{Z}$. Put

$$\begin{aligned}
\mathbf{x}_1(i) &= (1, (i-t), \ldots, (i-t)^p)' , \\
\mathbf{x}_2(i) &= (\cos\lambda_1(i-t), \sin\lambda_1(i-t), \ldots, \cos\lambda_q(i-t), \sin\lambda_q(i-t))' , \\
\mathbf{x}(i) &= (\mathbf{x}_1(i)', \mathbf{x}_2(i)')'
\end{aligned}$$

and $\mathbf{X} = (\mathbf{X_1} \vdots \mathbf{X_2})$ with the rows $x(i)'$, where $\mathbf{X}$ is the $n \times (p+s)$-regressor matrix. Let K be a symmetric density (a kernel of order 2) with compact support [-1, 1], $h \in \mathbb{N}$ be a bandwidth of the observation time, $\mathbf{y} = (Y_1, \ldots, Y_n)'$ denote the data vector and $\mathbf{K} = diag(k_i)$ be a weight matrix with

$$k_i = \begin{cases} K(\frac{i-t}{h+0.5}) , & i \in [t-h, t+h] , \\ 0 & , \text{ otherwise} . \end{cases}$$

Let $\mathbf{e}_j$ denote a $(p+1) \times 1$ vector with 1 in its jth entry and 0 elsewhere, $\boldsymbol{\Phi}_s$ denote an $(s-1) \times 1$ vector having 1 in its odd entries and 0 elsewhere, then the locally weighted least squares criterion

$$Q(t) = \sum_{i=1}^{n} [Y_i - \mathbf{x}_1(i)'\beta_1(t) - \mathbf{x}_2(i)'\beta_2(t)]^2 K(\frac{i-t}{h+0.5}) \Rightarrow \min$$

leads to the solutions

$$\hat{\mu}(t) = (\mathbf{e}_1', \boldsymbol{\Phi}_s')(\mathbf{X}'\mathbf{K}\mathbf{X})^{-1}\mathbf{X}'\mathbf{K}\mathbf{y} =: \mathbf{w}'\mathbf{y} , \quad (10)$$

$$\hat{G}(t) = (\mathbf{e}_1', \mathbf{0}')(\mathbf{X}'\mathbf{K}\mathbf{X})^{-1}\mathbf{X}'\mathbf{K}\mathbf{y} =: \mathbf{w}_1'\mathbf{y} \quad (11)$$

and

$$\hat{S}(t) = (\mathbf{0}', \boldsymbol{\Phi}_s')(\mathbf{X}'\mathbf{K}\mathbf{X})^{-1}\mathbf{X}'\mathbf{K}\mathbf{y} =: \mathbf{w}_2'\mathbf{y} , \quad (12)$$

where $\mathbf{0}$ denotes a vector of zeros of appropriate dimension. We see, $\hat{\mu}$, $\hat{G}$ and $\hat{S}$ are all linear smoothers and only observations with $i \in [t-h, t+h]$ obtain non-zero weights. In order to cope with boundary problems we distinguish a left bandwidth h_l and a right bandwidth h_r and put $h = \max(h_l, h_r)$. It is assumed that $h_l < t$, $h_r \le n-t$ and $h_T > p+s$, where $h_T = h_l + h_r + 1$ is called the total bandwidth. If h_l and h_r are fixed and $t \in [h_l+1, n-h_r]$, then all of $\hat{\mu}$, $\hat{G}$ and $\hat{S}$ work as moving averages with symmetric weights (the non-zero parts of $\mathbf{w}$, $\mathbf{w}_1$ and $\mathbf{w}_2$) for $h_l = h_r$ or asymmetric weights for $h_l \neq h_r$.

As in local polynomial fitting the νth derivative of the trend-cyclical component, $0 < \nu \le p$, can be estimated by

$$\hat{G}^{(\nu)}(t) = \nu!(\mathbf{e}_{\nu+1}', \mathbf{0}')(\mathbf{X}'\mathbf{K}\mathbf{X})^{-1}\mathbf{X}'\mathbf{K} =: \mathbf{w}^{\nu\prime}\mathbf{y} . \quad (13)$$

$\hat{G}^{(\nu)}$ works as a moving average, too.

4 Data-Driven Time Series Decomposition

In this section we will propose a data-driven procedure for time series decomposition. Estimation of variance and bias in a seasonal time series is discussed. A real example decomposed with the proposed procedure is given at the end of this section.

4.1 Estimation of the MASE

To select the parameters h_l, h_r and p we use the mean averaged squared error (MASE) as a distance between $\hat{\mu}(t)$ and $\mu(t)$ (error criterion),

$$M := n_1^{-1} \sum_{t \in \mathbf{I}^*} E[\hat{\mu}(t) - \mu(t)]^2 \ ,$$

where $\mathbf{I}^* \subseteq [h_l + 1, n - h_r]$ and n_1 denotes the number of observations in $\mathbf{I}^*$.

To estimate M we use the double smoothing (DS) criterion proposed by Müller (1985) (see also Härdle et al. 1992). The MASE splits up into a variance part and a bias part. The variance part of it is given by

$$V = n_1^{-1} \sum_{t \in \mathbf{I}^*} \mathrm{var}[\hat{\mu}(t)] = \sigma^2 n_1^{-1} \sum_{t \in \mathbf{I}^*} \sum_{i=1}^{n} w_i(t)^2$$

and can be estimated by

$$\hat{V} = \hat{\sigma}^2 n_1^{-1} \sum_{t \in \mathbf{I}^*} \sum_{i=1}^{n} w_i(t)^2 \ ,$$

where $\hat{\sigma}^2$ is an estimator of σ^2. The bias at each point is

$$b(t) = E[\hat{\mu}(t) - \mu(t)]$$

and the systematic bias part in MASE is $B = n_1^{-1} \sum_{t \in \mathbf{I}^*} b(t)^2$. Following the DS idea the bias $b(t)$ at a point $t \in \mathbf{I}^*$ is estimated by means of a pilot smoothing $\hat{\mu}_g$

$$\hat{b}_D(t) = \sum_{i=1}^{n} w_i(t) \hat{\mu}_g(i) - \hat{\mu}_g(t) \ .$$

The systematic bias part is estimated by

$$\hat{B}_D = n_1^{-1} \sum_{t \in \mathbf{I}^*} \hat{b}(t)^2 \ .$$

Hence the final DS estimation of M is

$$\hat{M}_D = \hat{V} + \hat{B}_D \ . \tag{14}$$

The optimal bandwidths on a given $\mathbf{I}^*$ following the DS criterion are defined as the pairs $\hat{H}_\mathrm{D} = \{\hat{h}_\mathrm{l}, \hat{h}_\mathrm{r}\}$ which minimize (14). Equation (14) involves a pilot estimate $\hat{\mu}_\mathrm{g}$, which will be evaluated with parameters selected by the R criterion of Rice (1984):

$$\hat{R} = n_1^{-1} \sum_{t \in \mathbf{I}^*} (\hat{\mu}(t) - y_t)^2 + (2w_{t^*}(t^*) - 1)\hat{\sigma}^2 \ , \qquad (15)$$

where t^* can be any observation time point in $\mathbf{I}^*$. If $\hat{R} < \hat{V}$ put $\hat{R} := \hat{V}$, since an error criterion should not be smaller than $\hat{V}$. The optimal bandwidths following this criterion are the pairs $\hat{H}_\mathrm{R} = \{\hat{h}_\mathrm{l}, \hat{h}_\mathrm{r}\}$ which minimize $\hat{R}$. Asymptotic properties of the DS bandwidth selector with the R criterion as pilot method can be found in Heiler and Feng (1997a, b).

4.2 Estimation of σ^2

To calculate $\hat{M}_\mathrm{D}$ or $\hat{R}$ one also needs a nonparametric estimator of the unkown variance σ^2. We will propose a bootstrap variance estimator $\hat{\sigma}^2_\mathrm{B}$ for seasonal time series, which is just an adaptation of the one proposed by Heiler and Feng (1997b). To calculate $\hat{\sigma}^2_\mathrm{B}$ an initial estimate of σ^2 is required. For this purpose a seasonal-difference-based variance estimator $\hat{\sigma}^2_\mathrm{G}$, an adaptation of the difference-based one of Gasser et al. (1986), is used. See Heiler and Feng (1996) for the definition of $\hat{\sigma}^2_\mathrm{G}$.

Following Heiler and Feng (1997b) the bootstrap variance estimator is defined by

$$\hat{\sigma}^2_\mathrm{B} = n^{-1} \sum_{i=1}^{n} r_i^2 \ ,$$

where r_i's are the residuals of a data-driven pilot smoothing. We see that this estimator is just equal to mean squared residuals of the pilot estimate. This variance estimator performs very well, if the pilot bandwidths are carefully selected (Heiler and Feng 1997b). To calculate $\hat{\sigma}^2_\mathrm{B}$ we need a constant based on the asymptotic analysis, called a bandwidth correction factor (CF). CF's for time series without seasonal component can be found in Table 1 of Heiler and Feng (1997b). For seasonal time series, assume that the bisquare kernel is used as weight function, we have CF=1.3188 for $s = 4$ and CF=1.3257 for $s = 12$, if the residuals r_i, $i = 1, \ldots, n$, are obtained from a local smoothing with $p = 3$. To calculate $\hat{\sigma}^2_\mathrm{B}$ an iterative procedure is proposed according to Heiler and Feng (1997b) using the k-NN (k-nearest neighbor) estimators. That is $h_\mathrm{T} =: k$ is fixed at each point t. h_l and h_r are taken as close to each other as possible. Starting with "$i = 0$", the algorithm proceeds as follows:

1. If $i = 0$, calculate $\hat{\sigma}^2_\mathrm{G}$ and put $\hat{\sigma}^2_{\mathrm{B},0} = \hat{\sigma}^2_\mathrm{G}$. Otherwise, let $\hat{k}_{v,i} = [\mathrm{CF}\hat{k}_i]$ and calculate $\hat{\sigma}^2_{\mathrm{B},i}$;
2. Select the bandwith for $p_\mathrm{P} = 5$ by the R criterion with variance $\hat{\sigma}^2_{\mathrm{B},i}$ and fit $\hat{\mu}_\mathrm{g}$;

3. Select the bandwidth $\hat{k}_i$ following the DS criterion with $p = 3$, $\hat{\mu}_g$ and $\hat{\sigma}^2_{B,i}$. If $i > 0$ and $\hat{k}_i = \hat{k}_{i-1}$, put $\hat{\sigma}^2_B = \hat{\sigma}^2_{B,i}$. Otherwise, put $i := i+1$ and go back to step 1.

Where $\hat{k}_i$ is the selected bandwidth for estimating μ with $p = 3$ and $\hat{k}_{v,i} = [CF\hat{k}_i]$ is the corresponding bandwidth used for calculating $\hat{\sigma}^2_{B,i}$ in the ith iteration, respectively.

4.3 Data-Driven Bandwidth Selection

Data-driven procedures to select the bandwidths are proposed by Heiler and Feng (1996). What follows is a brief description of their proposal. Here, only the procedure for selecting bandwidths in the central part with the R criterion is described. Procedures for selecting bandwidths at a boundary point or with the DS criterion are quite similar and hence are omitted. See Heiler and Feng (1996) for details. The highest polynomial order used in the main smoothing is p_m and it is given beforehand, e.g. $p_m = 3$. The polynomial order $p \leq p_m$ used in the central part or at a boundary point is chosen so that the error criterion is minimized. The polynomial order used in the pilot smoothing of the DS procedure, p_p, should be larger than p_m. In this paper $p_p = 5$ is used (as before). At the beginning we put $p = p_m$ and $h_{max0} = [n/4]$. If p_m is odd we put $p = p_m - 1$, because now $\hat{G}$, $\hat{S}$ and $\hat{\mu}$ are the same for $p = p_m$ and $p = p_m - 1$. However, at a boundary point the starting value of p is always p_m. Noting that now $h_l = h_r =: h_m$, the procedure reads as

1. Select an optimal bandwidth h_{m0} with given p and h_{max0} on the support $\mathbf{I}^*_0 = [h_{max0} + 1, n - h_{max0}]$;
2. Put $h_{max1} := h_{m0} + 1 + [\frac{1}{3}(h_{max0} - h_{m0})]$. Then select a new optimal bandwidth h_{m1} on the support $\mathbf{I}^*_1 = [h_{max1} + 1, n - h_{max1}]$. If h_{m1} is stable, then go to step 3, otherwise put $h_{max0} := h_{max1}$, $h_{m0} := h_{m1}$ and repeat this step;
3. Put $h_{max2} := h_{m1} + 1$ and select a new optimal bandwidth h_{m2} on the support $\mathbf{I}^*_2 = [h_{max2} + 1, n - h_{max2}]$. If h_{m2} is also stable, then go to step 4, otherwise put $h_{m1} := h_{m2}$ and repeat this step;
4. Put $h_{max3} := h_{m2}$ and select a new optimal bandwidth on the support $\mathbf{I}^*_3 = [h_{max3} + 1, n - h_{max3}]$. If h_{m3} is stable, too, put $h_m(p) := h_{m3}$ and go to step 5. If h_{m3} is not stable, put $h_{m1} := h_{m3}$ and go back to step 3;
5. If $p-2 < 0$, then go to step 6, otherwise put $h_{max0} := [1.2h_m(p)]$, $p := p-2$ and go back to step 1;
6. Select the bandwidth (called $\hat{h}_m$) corresponding to the smallest error criterion from all $h_m(p)$. $\hat{h}_m$, the respective $\hat{p}$ and $\hat{R}(\hat{h}_m, \hat{p})$ are the outcomes of this procedure in the central part.

The definition of h_{max} in steps 2 and 3 ensures that a larger optimal bandwidth can also occur in the next step. One can put different values for

h_{max} in steps 2, 3 or 5. It might cause a difference in the computing time, but it does not entail any change in the results. At a boundary point $p-1$ instead of $p-2$ should be used in step 5. For a given p the minimal bandwidth h_{min} remains the same in all steps. In all steps it must be $h_{max} << [n/2]$ such that there are enough observations in the support $\mathbf{I}^*$. Sometimes the $h_{m,j}$ do not converge but pendulate between two values. In this case we can take either of them as the optimal one.

To show how these procedures work we use the time series of the quarterly gross domestic product (GDP) in Germany from 1968 to 1994 (billion DM, at current prices) as an example. For this time series we have $\hat{\sigma}_G^2 = 10.998$ and $\hat{\sigma}_B^2 = 12.028$. Table 3 gives the selected left bandwidths, total bandwidths and orders of polynomials in the central part as well as at each right boundary point. The selected bandwidths in the central part with the R criterion and DS criterion are $h_m = 10$ and $h_m = 11$, respectively. The selected right bandwidth at each right boundary point $t > n - h_m$ is always $n - t$ and is omitted. With the parameters in Table 3 we can obtain $\hat{G}$, $\hat{S}$ and $\hat{\mu}$. The estimates at a left boundary point t, $t \leq h_m$, are obtained by symmetrically using the parameters for $t' = n - t + 1$. The decomposition results following the DS criterion are shown in Fig. 7. Figure 7 (a) displays $\hat{G}$ with the original data, while Fig. 7 (b) exhibits $\hat{S}$.

Note that the selected parameters are optimal in the sense of $\hat{\mu}$. They are not optimal for $\hat{G}$ or $\hat{S}$. Selection of parameters with respect to $\hat{G}$ or $\hat{S}$ will not be discussed in this paper.

Table 3. Selected bandwidths $\hat{h}_l$, $\hat{h}_T$ and orders of polynomials $\hat{p}$

Observation Time	Selected Parameters with R criterion			Selected Parameters with Double Smoothing		
(t)	$\hat{h}_l$	$\hat{h}_T$	$\hat{p}$	$\hat{h}_l$	$\hat{h}_T$	$\hat{p}$
n	8	9	1	8	9	1
$n-1$	13	15	2	15	17	2
$n-2$	12	15	3	15	18	3
$n-3$	12	16	3	15	19	3
$n-4$	12	17	3	15	20	3
$n-5$	12	18	3	15	21	3
$n-6$	10	17	3	16	23	3
$n-7$	9	17	2	17	25	3
$n-8$	11	20	2	18	27	3
$n-9$	10	20	2	11	21	2
$n-10$	10	21	2	11	22	2
$n-11$				11	23	2

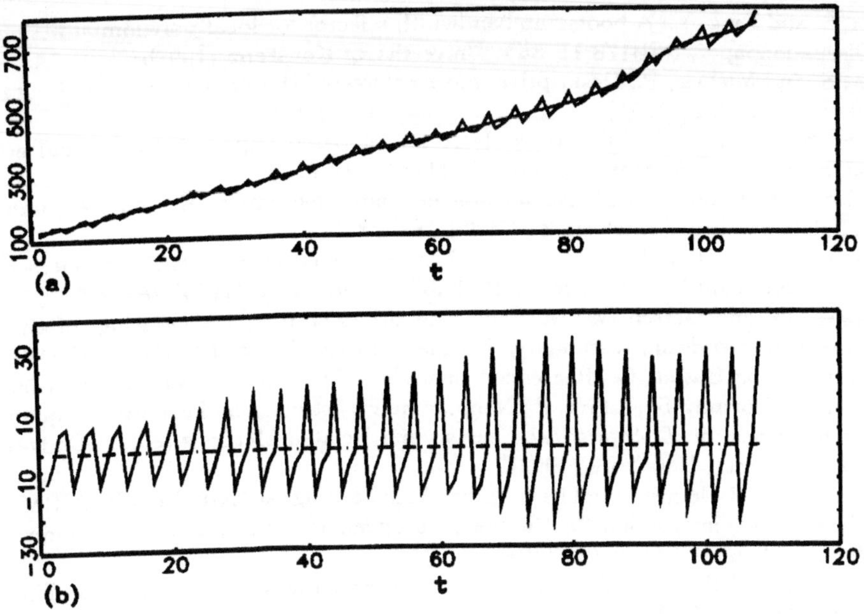

Fig. 7. Decomposition results of the time series GDP using the DS criterion. (a) The data and $\hat{G}$, and (b) $\hat{S}$.

References

Abberger, K.: Nichtparametrische Schätzung bedingter Quantile in Zeitreihen; Mit Anwendungen auf Finanzmarktdaten. Hartung-Gorre Verlag, Konstanz (1996).

Abberger, K.: Quantile Smoothing in Financial Time Series. Statistical Papers 38 (1997) 125-148.

Bollerslev, T., Chou, R.Y. and Kroner, K.F.: ARCH modelling in finance. Journal of Econometrics **52** (1992) 5–59.

Cheng, K.F.: Nonparametric estimators for percentile regression functions. Communications in Statistics, Theory and Methods **12** (1983) 681–692.

Gasser, T., Sroka, L. and Jennen-Steinmetz, C.: Residual variance and residual pattern in nonlinear regression. Biometrika **73** (1986) 625–633.

Härdle, W., Hall, P. and Marron, J.S.: Regression smoothing parameters that are not far from their optimum. J. Amer. Statist. Assoc. **87** (1992) 227–233.

Härdle, W., Lütkepohl, H. and Chen, R.: A review of nonparametric time series analysis. International Statistical Review **65** (1997) 49–72.

Harvey, A.C. and Todd, P.H.J.: Forecasting economic time series with structural and Box-Jenkins models – A case study. Journal of Business & Economic Statistics **1** (1983) 299–351.

Heiler, S. and Feng, Y.: Datengesteuerte Zerlegung saisonaler Zeitreihen. IFO-Studien **3/1996** (1996) 337–369.

Heiler, S. and Feng, Y.: A simple root n bandwidth selector for nonparametric regression. To appear in Journal of Nonparametric Statistics (1997a).

Heiler, S. and Feng, Y.: A bootstrap bandwidth selector for local polynomial fitting. Discussionpaper, SFB178 **II–344**, University of Konstanz (1997b).

Heiler, S. and Michels, P.: Deskriptive und Explorative Datenanalyse. Oldenbourg-Verlag, München (1994).

Hillmer, S.C. and Tiao, G.C.: An ARIMA–model–based approach to seasonal adjustment. J. Amer. Statist. Assoc. **77** (1982) 63–70.

Horváth, L. and Yandell, B.S.: Asymptotic of conditional empirical processes. Journal of Multivariate Analysis **26** (1988) 184–206.

Jones, M.C. and Hall, P.: Mean squared error properties of kernel estimates of regression quantiles. Statistics & Probability Letters **10** (1990) 283–289.

Kitagawa, G. and Gersch, W.: A smoothing priors state space modeling of time series with trend and seasonality. J. Amer. Statist. Assoc. **79** (1984) 378–389.

Koenker, R. and Basset, G.: Regression quantiles. Econometrica **46** (1978) 33–50.

Koenker, R., Portnoy, S. and Ng, P.: Nonparametric estimation of conditional quantile functions. In: L_1-Statistical Analysis and Related Methods (ed. Y. Dodge), North-Holland, New York (1992).

Maravall, A. and Goméz, V.: The program EASTS, 'Signal Extraction in ARIMA Time Series', instruction for the user, European University Institute, Florenz, Working Paper ECO **94/28** (1994).

Maravall, A. and Pierce, D.A.: A prototypical seasonal adjustment model. J. Time Series Analysis **8** (1987) 177–193.

Müller, H.-G. (1985): Empirical bandwidth choice for nonparametric kernel regression by means of pilot estimators. Statistics & Decisions, Supplement Issue No. **2** (1987) 193–206.

Rice, J.: Bandwidth choice for nonparametric regression. Annals of Statistics **12** (1984) 1215–1230.

Stier, W.: Zur Kanonischen Zerlegung von Zeitreihen – Bemerkungen zum Verfahren "SEATS". Allg. Statistisches Archiv **80** (1996) 313–331.

Stute, W.: Conditional empirical processes. Annals of Statistics **14** (1986) 638–647.

Tjøstheim, D.: Non-linear time series: A selective review. Scandinavian Journal of Statistics **21** (1994) 97–130.

Development of a Credit-Standing-Indicator for Companies Based on Financial Statements and Business Information with Backpropagation-Networks

Jörg Baetge[1] and Carsten Uthoff[2]

[1] Institut für Revisionswesen, Westfälische Wilhelms-Universität, Universitätsstr. 14-16, 48143 Münster
[2] BPV Baetge & Partner GmbH & Co. Verfahrensentwicklung KG, Mendelstr. 11, 48149 Münster

Abstract. The growing rate of company insolvency and growing losses of outstanding debt show the significance of a secure credit standing assessment of company clients for the creditor. Traditionally the balance sheet credit-standing analysis has been used by many creditors for debtor analysis, but analysis of non-financial indicators is gaining importance. Although the significance of analyzing non-financial indicators is seen in both practice and theory, the theoretic approach lacks a widely based empirical examination of a company's credit standing combining both financial and non-financial indicators. A study by the Institut für Revisionswesen (IRW) aims at solving this deficiency. The study's objective is to develop a credit-standing-indicator (CSI) using artificial neural networks (ANN) that might objectively and reliably identify threatened companies with a high degree of certainty. In cooperation with the Allgemeine Kreditversicherung AG (AKV) and the Verband der Vereine Creditreform e. V. (VVC), the IRW has created an extensive and representative database with approximately 16,000 financial statements and 13,200 pieces of business information concerning both solvent and insolvent companies. The following important aspects of debtor analysis are examined: (1) The efficiency of the Creditreform credit standing index should be analyzed and the index optimized. (2) In practice, financial statement and Creditreform information have highly differing importance in credit standing analysis. The importance of both sources for the credit-standing decision is determined. (3) The features from both datasources are analyzed together. The study examines whether debtor analysis is improved by using the combined features of the two different data sources.[1]

1 Introduction

Since 1991, the number of company insolvencies in Germany has continuously risen and reached a peak in 1996 with over 33,000 cases of insolvencies. In 1997 even more insolvencies are likely to occur.

[1] We kindly thank Win Neidlinger for his support during the translation of this article

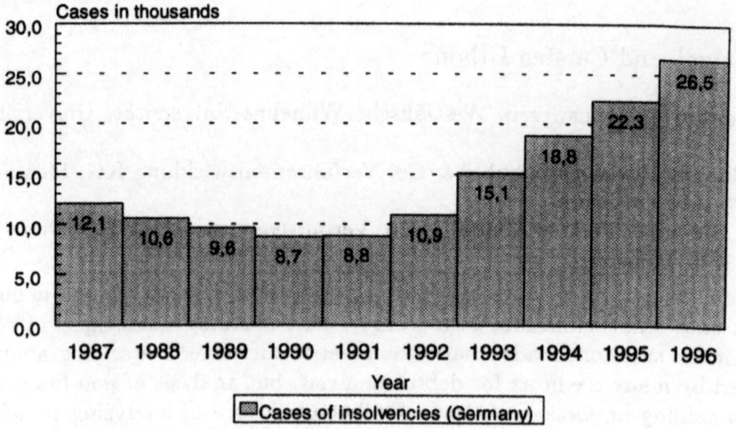

Source: Creditreform

Fig. 1-1: Cases of insolvencies in Germany 1987 - 1996

The creditor may experience considerable losses. The average losses for individual creditors increased from DM 610,000 per insolvency in 1990 to DM 1,280,000 in 1996[2]. The essential need for a secure credit-standing assessment for creditors becomes apparent not only in view of the growing cases of insolvency, but also in view of the distinct increase in **insolvency losses**. Only the creditor who can effectively assess the credit-standing risks well in advance and thus adopt precautionary measures will be successful on the credit market in the long run. Credit debt risks can be effectively identified by a CSI by means of which the creditor can assess any credit-standing risks objectively and reliably. The assessment of credit standing risks is part of the debtor analysis.[3]

Traditionally, the customer's financial statements form the basis of debtor analysis.[4] In order to determine the relevant ratios, statistical methods such as multivariate discriminant analysis or artificial neural network analysis (ANNA) can be used. Examples of very efficient discriminant functions, which are successfully used as a CSI, have been developed by the Institut für Revisionswesen (IRW) at the University of Münster in cooperation with the

---

[2] Cf. Creditreform, Unternehmensentwicklung 1996 alte und neue Bundesländer, p. 1-5.

[3] Cf. Büschgen, H. E., Bankgeschäfte und Bankmanagement, p. 803.

[4] Cf. Riebell, C., Die Praxis der Bilanzauswertung, p. 137; Deutsche Bundesbank, Die Untersuchung von Unternehmensinsolvenzen, p. 33.

Bayerische Vereinsbank, the Allgemeine Kreditversicherung and the Baden-Württembergischen Bank.[5] As we will show below, the classification results of the discriminant analysis were outperformed by the ANNA carried out by the IRW.[6]

Although the statistical methods that were developed using CSIs assess companies accurately, their efficiency is restricted by the nature of the financial statement. The financial statement is past-oriented, thus assessing the company's economic circumstance may be seen as incomplete.[7] Information such as management quality, product portfolio, industry outlook and future development cannot be derived from the financial statement. In order to overcome these financial statement deficiencies, extra information is needed.[8]

For the purpose of compiling a debtor analysis, the creditor can obtain additional information.[9] Sources for this kind of information are the companies themselves, banks, branch-services and other sources, such as the press, associations, business information agencies or research facilities. Because of their easy access and quick availability, business information centers are an important source and often the first data source used in debtor analysis.[10]

Within a cooperative project with Allgemeine Kreditversicherung AG (AKV), Verband der Vereine Creditreform (VVC) and the IRW, an extensive representative database was established, containing business information (VC-information) as well as financial statements of solvent and insolvent companies. The data supply consists of approximately 16,000 financial statements and over 13,200 pieces of VC-information of solvent and insolvent companies. In the cases of more than over 9,400 companies, both the financial statement and business information were available. From the VC-information a variety

---

[5] Cf. Niehaus, H.-J., Früherkennung von Unternehmenskrisen, p. 119-156; Feidicker, M., Kreditwürdigkeitsprüfung, p. 133-183; Hüls, D., Früherkennung insolvenzgefährdeter Unternehmen, p. 184-252.

[6] Cf. Krause, C., Kreditwürdigkeitsprüfung mit Neuronalen Netzen, p. 211; Baetge, J./Schmedt, U./et. al., Bonitätsbeurteilung von Jahresabschlüssen nach neuem Recht, p. 340 f.; Baetge, J./Hüls D./Uthoff, C., Früherkennung der Unternehmenskrise, p. 26 f.; Baetge, J./Kruse, A./Uthoff, C., Bonitätsklassifikationen von Unternehmen mit Neuronalen Netzen, p. 275 f.

[7] Cf. Baetge, J., Früherkennung negativer Entwicklungen, p. 652 f.

[8] Cf. Burgard, H., Bilanzanalyse aus Sicht der Kreditinstitute (Teil II), p. 325.

[9] Cf. Bühler, W., Informationsressourcen für die Bonitätsanalyse, p. 92; Schmoll, A., Theorie und Praxis der Kreditprüfung unter besonderer Berücksichtigung der Klein- und Mittelbetriebe (I), p. 102-104; Beuter, H. B./Reiss, I./Rust, H. J., Erfahrungen mit formalisierten Verfahren bei der Kreditwürdigkeitsprüfung, p. 58.

[10] Cf. Fischer, J. H., Computergestützte Analyse der Kreditwürdigkeit auf Basis der Mustererkennung, p. 177; Schmoll, A., Theorie und Praxis der Kreditprüfung unter besonderer Berücksichtigung der Klein- und Mittelbetriebe (II), p. 174 f.; Beuter, H. B./Reiss, I./Rust, H. J., Erfahrungen mit formalisierten Verfahren bei der Kreditwürdigkeitsprüfung, p. 63-65.

of quality indicators (VC-features) were extracted and then analyzed together with financial statement ratios.

When debtor analysis is used, financial statement and VC-information differ greatly in their significance. In order to determine the value of information for credit standing assessment, the VC-features and financial statement ratios must be analyzed separately. The separate analysis examines the following objectives:

- What performance can be achieved by a classifier based on information by the VVC?
- What performance can be achieved by a classifier based on financial statement data provided by the AKV?

Considering the features of both data sources, it becomes apparent that a company's credit standing can be assessed using various sources of information rather than using information from just one source. We will examine whether multiple information can improve the classification of a company. Depending on the aggregation level, the combined analysis applies first to the decision level and second to the features level. The combined analysis examines the following objectives:

- Can performance be improved by connecting the decision components of both classificators?
- Can performance be further improved by connecting the features directly?

2 Dataprocessing

2.1 Data

The dataset of the present study contains the financial statements and VC-information of a total of 11,248 companies. The dataset consists of 10,162 solvent and 1,086 insolvent companies. An insolvent company complies with the following facts: protest of a bill, moratorium, composition, bankruptcy. Only data relating to the time before insolvency was used for the analysis. Due to the fact that financial statements and VC-information was not available for all companies, the companies are dissimilarly divided throughout the dataset. It is the existing information about a company rather than the company itself that makes up the reference quantity.

The sample consists of 27,350 records of solvent companies and 2,054 records of insolvent companies. Out of the total dataset, 22,035 patterns are allotted to the financial statement dataset and 16,842 to the dataset on VC-information. 9,473 patterns out of the VC-information/financial statement data contain ratios as well as chronologically suitable information. The proportion of data concerning insolvent companies varies between 6.6% and 7.4%.

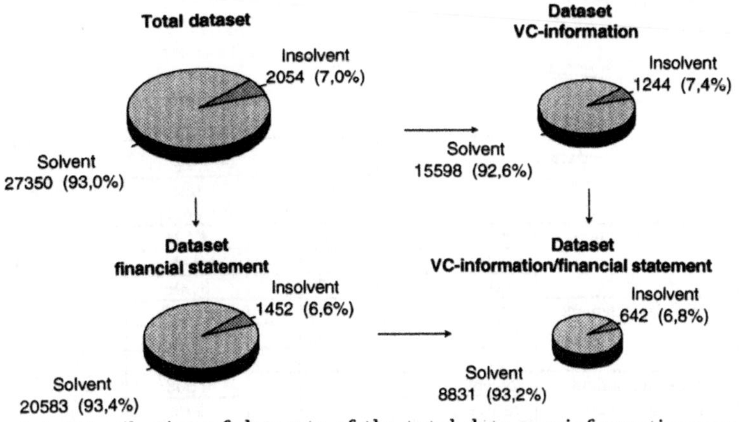

Fig 2-1: Distribution of datasets of the total data per informationsource

The pattern outlined in the three datasets shape the foundation of the empirical analysis. First a classifier is developed based on the VC-features (dataset "VC-information") and then a classifier based on ratios (dataset "financial statement"). For the combined analysis (the development of a classifier based on business information and financial statement ratios) the dataset "VC-information/financial statement" is used.

For neural network development purposes the dataset was split into three random samples. For the sample setup the companies in each dataset were leveled according to the credit-standing criteria "solvent" or "insolvent". At the "insolvent" level, patterns are drawn randomly and then distributed evenly over the three samples. During the learning process the classifier needs to learn the typical structures of solvent as well as insolvent companies, and therefore solvent companies are drawn at random until their number reaches the number of sets of insolvent companies. The remaining pattern are distributed randomly into the test and the validation set. Data from the test set are used to examine the performance of the network during development. Only when the network is completely developed, its performance is tested on unseen data (validation set).[11] Because the test and the validation set are used to test or validate the classifier, the sets of credit standing data are distributed in non-equal parts. The patterns of the two other datasets are divided into the training set and the test sample in the same manner.

[11] Cf. separation of the dataset into a learn-, test-, and validation set: Hecht-Nielsen, R. Neurocomputing, p. 115 f.

The distribution of solvent and insolvent companies is shown in the following table:[12]

dataset		sample		
		learning sample	test sample	validation sample
VC-information	solvent	413	7.591	4.310
	insolvent	413	415	214
	total	826	8.006	4.524
financial statement	solvent	485	10.047	4.310
	insolvent	485	485	214
	total	970	10.532	4.524
VC-information/ financial statement	solvent	213	4.308	4.310
	insolvent	213	215	214
	total	426	4.523	4.524

Table 2-1: Distribution of patterns per dataset over the sample

2.2 Features of VC-information and financial statement ratios

In cooperation with representatives of Creditreform and Allgemeine Kreditversicherungs AG, a catalogue containing 91 features was established based on the existing business information. The catalogue is divided into four parts. The "*institutional-based general features*" contains 25 features such as company size, industry, company age and legal form. The subject-area "*personnel-based features*" contains 51 features such as shareholder structure, supervisory board and management. Five features such as bank credit, guarantees and mode of payment make up the "*financial features*". The "office-area-based features" comprise ten features representing the distribution sector and the personnel sector. Based on the defined features catalogue, the feature-values were extracted from the datasources.[13]

The companies' financial statements were registered according to the AKV pattern. This forms the basis of the ratio-definition. Ratio-definitions published in specialized literature or in other studies were matched as closely as possible with the AKV-pattern and then added to the ratio catalogue.

The ratio-catalogue is devided into three parts: success situation, financial stability and growth. The success situation is described by 72 ratios

[12] Cf. Uthoff, C., Erfolgsoptimale Kreditwürdigkeitsprüfung, p. 199.

[13] Cf. feature-catalogue: Uthoff, C., Erfolgsoptimale Kreditwürdigkeitsprüfung, p. 124-131.

showing revenue structure, expenditure structure, profitability structure and value creation; 143 ratios describe capital structure, asset structure, margin structure and financial strength. Ten ratios such as investment attitude and financing potential comprise information about a company's growth.[14]

2.3 Missing values and encoding

In the case of incomplete records we must consider how the record is to be treated in the analysis. On the one hand, if incomplete features are excluded from the analysis, part of the information potential is lost. On the other hand, incomplete records or records containing estimated values do not represent the actual situation of the company, thus creating distortions. Therefore, the degree of distortion must be weighed against the potential information gain. As a general rule in this study, information is distorted when a feature contains missing values by more than half of a record. This applied to 48 features, thus leaving only 43 features for analysis. When a value was missing in the remaining 43 features, it was replaced by the mean value. Because group-specific mean values as valuers represent ex-post knowledge, missing test data and validation data have to be estimated with the feature-specific mean value. Missing values of the training set can be estimated with either the group-specific mean value or the feature-specific mean value, the former being the more plausible alternative.

Missing values in the training set were replaced by the groupspecific mean, missing values in the test- and validation set were replaced by a feature specific mean.

For the empirical analysis the VC-information and financial statement information are encoded in two steps. Each feature is first scaled and then transformed according to the scale chosen. Specific transformation rules are applied to each scale. Transformation implies giving each feature an adequate numeric value in order that it can be processed mathematically. The companies are classified according to the features chosen. These features are prepared for neural network processing by means of data encoding. The number of inputs is determined by the transformation rule.

With categorical data, we proceed in a different way. Variants of 'one of n encoding' were used for encoding of categorical features. For the one of n encoding the principle of local representation, meaning that one input neuron presents one expression is applied. The number of expressions determines the number of input neurons.[15] Depending on the scale different encoding-functions can be used. In addition to the treatment of a probable existing order, the encoding-functions differ in how exactly their expressions can be

---

[14] Cf. ratios-catalogue: Uthoff, C., Erfolgsoptimale Kreditwürdigkeitsprüfung, p. 139-145

[15] Cf. Schumann, M./Lohrbach, T./Bährs, Versuche zur Kreditwürdigkeitsprognose mit Künstlichen Neuronalen Netzen, p. 6.

deliminated. If the feature-expressions cannot be clearly defined or are subject to subjective judgments on the investigators side, it makes sense to code the feature in a fuzzy way. the following table shows the scale and rule of transformation used.[16]

scale	transformation rule
nominal	one of n encoding
ordinal	fuzzy one of n encoding gradient one of n encoding thermometer encoding
metrical	directly

Table 2-2: Scale and transformation rule

3 Data analysis

3.1 Determining the objective function

For the development of the credit-standing index we used multilayer perceptrons trained with backpropagation (BPN). Modelling a BPN for classification problems can be described as a multi-stage process. Each stage consists of several different procedures. After each procedure the classifier has to be evaluated using a representative testset. In our study we used the mean-square-error (MSE) to optimize neural nets. As the MSE is not the appropriate function to evaluate the performance we introduced a different function. After each stage the classifiers are to be assessed with respect to quantifiable objectives which are defined below and the best neural net is to be selected.[17] The form of the objective function of the classifier depends on parametervalues (which might not be given) for a specific credit allocation process. If the classifier cannot be developed for a concrete credit allocationprocess with known parameter values, the classifier has to be optimized independently of a credit allocation process.[18]

Usually the values of the parameter are not obtainable for a commercial information service such as an information agency or rating-agency. The

[16] Cf. concerning the encoding: Neural Ware Inc., Using NeuralWorks, p. 137-140.

[17] Cf. Pytlik, M., Diskriminanzanalyse und künstliche neuronale Netze zur Klassifizierung von Jahresabschlüssen, p. 267-279; Cf. the minimizing the beta-error for a fixed alpha-error: Krause, C., Kreditwürdigkeitsprüfung mit Neuronalen Netzen, p. 131-191.

[18] Cf. Uthoff, C., Erfolgsoptimale Kreditwürdigkeitsprüfung, p. 27-30.

information service sells the information, the credit-standing judgment, to many creditors.[19] An information service will not only sell the information exclusively to one creditor, but to as many clients as possible. Thus, an information service develops a classifier, which results can be sold to many clients.

Because the objective function "minimizing the total error" or "minimizing the beta-error (possibility of classifying a solvent company as insolvent) for a fixed alpha-error (possibility of classifying an insolvent company as solvent)" requires information about the credit allocation process, the classifiers are evaluated with respect to process-independent calculation, the error-area. By varying the cut-off-point, different combinations of alpha-error and beta-error can are obtained from the outputs for solvent and insolvent companies. In an alpha-error/beta-error-diagram the determined $\alpha\beta$-combinations for a certain classifier can be represented as a so called $\alpha\beta$-function. The error-area is described as the area created from both axis (alpha-error and beta-error) and the $\alpha\beta$-function of the classifier. The smaller the error-area is the better the classifier.[20]

The error-area calculation is defined for the BPN used. A BPN is according to the error-area calculation to be evaluated meaningfully, if the BPN produces just one outputvalue spreading almost constantly across a defined interval.

The error-area of a classifier is the value (measured in per cent) of the standard square. The standard square is the $\alpha\beta$-diagram across a value-interval from 0% to 100%. A classifier like C1 whose beta-error is 100% divides the data with an error-area of 100% regardless of the alpha-error. The classifier C2 yields an error-area of 50%. The empirically developed classifier C3 shows the typical convex shape of the $\alpha\beta$-function. Only if the function is almost convex to the origin and enough $\alpha\beta$-tupel are determined, the error-area of an empirical produced classifier is sensibly determined. The classifier C3 depicted above divides companydata with an error-area of 15.5%.[21]

3.2 Testarrangement

At the beginning of the testseries the initial settings of the different components (network architecture, standardization and scaling of inputdata, data-

---

19 Cf. Schmoll, A., Theorie und Praxis der Kreditprüfung unter besonderer Berücksichtigung der Klein- und Mittelbetriebe (I), p. 102-104; Bühler, W., Informationsressourcen für die Bonitätsanalyse, p. 92; Bulling, H., Entscheidungshilfen der Diskriminanzanalyse bei der Gewährung von Konsumentenkrediten und gewerblichen Krediten, p. 147; cf. to different ways of outsourcing of the debtor analysis: Kuhlmann, J., Die Problematik einer Ausgliederung der Kreditwürdigkeitsprüfung im Firmenkreditgeschäft der Banken, p. 44-53.

20 Cf. Uthoff, C., Erfolgsoptimale Kreditwürdigkeitsprüfung, p. 222 f.

21 Cf. Uthoff, C., Erfolgsoptimale Kreditwürdigkeitsprüfung, p. 222 f.

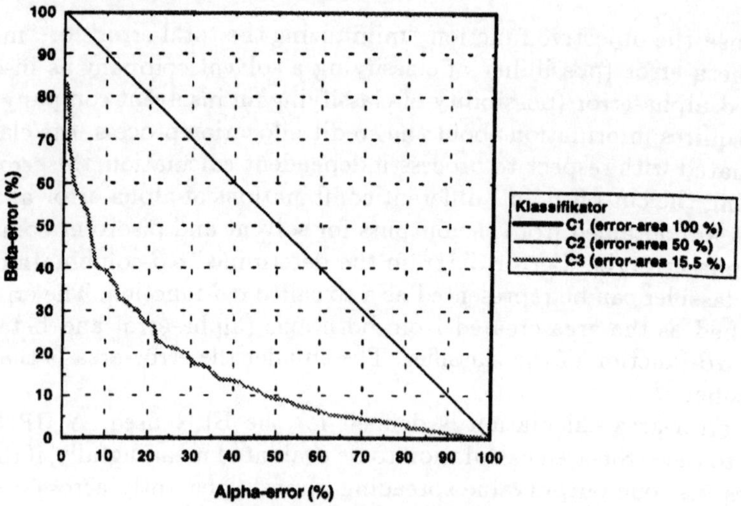

Fig 3-1: Standard square

processing on the hidden layer and the output layer, learning rule) of the BPN have to be set.[22] The network with the initial settings represents the basic networktype. Starting with the initial setting we used two pruning procedures, which led to the identification of the credit-standing relevant features.

Approaches to reduction (pruning) a BPN differ in two respects: firstly, they apply to different components of the BPN and secondly, they are of different complexity.[23] In this study weight pruning and relevance pruning have been used. When weight pruning is used the connections with the smallest absolute weights are identified and a certain number of the connections with the smallest weights are cut off.[24] If all outgoing connections from an input neuron are eliminated, it is withdrawn from further information processing. In contrast to weight pruning the importance of a neuron is directly checked with relevance pruning. With the relevance pruning those neurons are identified, which either increase the performance of the BPN; those neurons which reduce the performance of a BPN are to be eliminated[25] The pruning algo-

[22] Cf. Krause, C., Kreditwürdigkeitsprüfung mit Neuronalen Netzen, p. 149-174.

[23] Cf. Zell, A., Simulation Neuronaler Netze, p. 319-334.

[24] Cf. Zell, A., Simulation Neuronaler Netze, p. 320; Zimmermann, H.-G., Neuronale Netze als Entscheidungskalkül, p. 66 f.

[25] Cf. Mozer, M. C./Smolensky, P., Using Relevance to Reduce Network Size Automatically, p. 3.

rithms reduce successively the architecture of a BPN until the error-area is decreased to its lowest level. The result is a BPN which outperforms the primary network (which processes all features) on the test set using considerably less features.

3.3 Separate analysis

Analysis of the credit reporting information data

Based on plausibility-considerations and on a careful comparison of feature combinations of solvent and insolvent companies, the VVC identified credit-standing relevant features and weighted these according to their relevance.[26]As a result, the Creditreform-credit standing index (VC-index), consists of 15 features. The features are either part of the information or are obtained by branch-comparison values out of the VVC-database.[27]

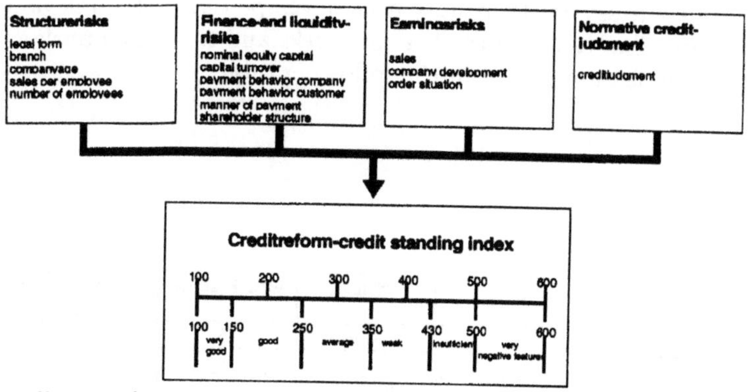

Fig 3-2: Structure of the Creditreform-credit-standing-index

The VC-index is calculated by evaluating the individual value of the feature with their riskfactor and is then multiplied with the accompanying weightfactor. The weight of each feature varies according to legal form and marketsituation.[28] The credit-standing index of a company is the weighted sum of the 15 feature expressions evaluated at their risk factor. The lower the value of the VC-index is the higher the credit-standing. The VC-index with

[26] Cf. Creditinform, Creditreform Informationssystem, p. 1.

[27] Cf. Creditinform, Das Creditreform Informationssystem, p. 4.

[28] Cf. Creditinform, Creditreform Informationssystem, p. 3; Creditinform, Das Creditreform Informationssystem, p. 4.

its six classes is comparable to schoolgrades (German schoolgrade system), reaching from very good (= A+, US-high school grade) to inadequate (= F-).[29] In the following analysis of the VC-information, the VC-index is used as a benchmark. The classification using the benchmark produced an error area of 32.075%.

The analysis of the VC-features consists of three stages. In stage 1 the features of the VC-index were examined. The VC-index consists of a total of 15 different features. In this study only 9 out of 15 features have been drawn from the VC-information. In stage 1 each of the 9 features are classified, either as credit-standing relevant or as nonsignificant. If a feature appears to be of little significance it might deteriorate the performance of the classifier. During stage 2 additional features (next to the features which have been selected during stage 1) are considered which are derived (computer-aided) from the VC-information. Predominantly "personnel-based features" are added to the list of features. The list of stage 2 consists of 19 features. In stage 3 this list is extended by 15 features (per company of the dataset) collected carefully by a work-group of the IRW. These so called manually encoded features again concern "personnel-based features". The new features focus on the professional qualification of the shareholders and management.

For each stage several BPN have been developed using the relevance pruning algorithm. The results of the best networks and the benchmark evaluated with the VC-index are shown in the following table:

Testseries	Name	#VC-Features	Error-area (%) test sample
Benchmark	VC-Index	15	32,075
Basictype	FA01-9	9	26,007
Stage 1	FA12-7	7	25,949
Stage 2	FA21-10	10	24,868
Stage 3	FA31-16	16	24,518

Table 3-1: Per stage selected features

As depicted in table 3-1 the increased performance compared to the benchmark can be explained by adjusted weights, features which are either eliminated or added. The network basictypes outperform the VC-index due to a different weighting and/or the omission of features. The performance can be further increased by adding additional features from stage 2 and stage 3 into the network and eliminating disturbing features. Starting from the basic

[29] Cf. Creditinform, Creditreform Informationssystem, p. 3.

network, the error-area decreased from 26.007% to a level of 24.518%. The features chosen in stage 2 led to the biggest decrease of the error-area of the optimized networks.

In spite of the convincing performance of the generated BPN, the results, however, have to be judged cautiously, because the results reported in table 3-1 were obtained on the test set. However, the test set is part of the data on which the BPN was optimized (which is not case for the VC-index). Whether the developed BPN delivers similar results using unseen data, has to be tested on the validation set.

Analysis of the financial statement

The BP-13 is a backpropagation-network which used 13 financial statement ratios as inputs. The BP-13 is chosen as the benchmark for the BPN, which are developed on the basis of the financial statements received from the AKV. The BP-13 was developed by the IRW with company data, which also originated from the AKV and was collected by Feidicker[30], just like the data used during the following analysis. However, the data of the BP-13 differs from the data used in this study: The BP-13 was developed using financial statements which were valid according to the old German Stock Corporation Law 1965. The analyzed companies in this project state the balance sheet according to the new Commercial Code 1985. Furthermore, very small businesses were excluded before at a turnover limit of DM 500,000, in this study the limit is set at DM 1,000,000.

The financial ratios of the BP-13 describe information such as earnings structure, profitability structure, capital structure, asset structure and financial strength. The outputvalue of the BP-13 is to be interpreted as an indicator for the financial statement standing. The BP-13 estimates the testdata with an error-area of 19.868%. This result is the benchmark for the financial statement data analysis.

In addition to the 13 ratios mentioned above the list consists of 211 ratios. Out of the total of 224 ratios the relevant combinations have to be evaluated. 224 ratios can be presented to the network, thus there are 224-1 different ratiocombinations[31]. In order to limit the combinations that have to be tested, pruning algorithms are used, which successively determine the relevant ratios from a ratio list. First the ratios are preselected with the weight pruning algorithm, because the weight pruning algorithm is less costly to calculate than the relevance pruning. Second the leftover ratios are analyzed using the prunesort algorithm.

The number of relevant ratios was reduced from 224 to 150 ratios by the weight pruning. A BPN with only 15 ratios was determined with the relevance pruning. Compared to the benchmark of the BP-13 with an error-area value of 19.868%, the new combination of ratios reduces the error-area value. The

---

[30] Cf. Feidicker, M., Kreditwürdigkeitsprüfung, p. 35-54; cf. Krause, C., Kreditwürdigkeitsprüfung mit Neuronalen Netzen, p. 104-106.

[31] It makes no sense to process no ratios, thus the quantity is reduced by one.

lowest error-area value of 17.551% was reached by a network using 15 ratios (FB21-15).

3.4 Combined analysis

Benchmark During the separate analysis of the company data a benchmark resulted from the VC-index for analyzing the VC-data and the BP-13 for analyzing financial statement data. Because no external existing benchmark was suitable for the combined analysis, a benchmark was constructed for the combined analysis from the VC-index and the BP-13. Combining both benchmarks equally weighted results in the overall index VCIBP13. This index processes (as a benchmark) data from both informationsources[32] On the test sample a benchmark is an error-area of 20.161% for the constructed benchmark VCIBP13 using company data.

Combining assessments based on partial information sets In regard to the structure of the benchmarkfunction VCIBP13 the assessments based on partial information sets often are aggregated to a total assessment.

The network FA31-16 performed best on the test sample using VC-features only. On the other hand, the network FB21-15 classifies the balance sheet data with the lowest error-area value.

The question arises whether the performance can be improved if the partial judgments described above are combined. To answer this question there are two possibilities: first, we simply combine the partial judgments equally weighted to a total judgment. Second, the weights are adjusted using backpropagation. The data set (training- and test set) used for analysis consists of companies from which VC-information as well as financial statements are available for the same year.

The following table shows the results of the networks using the respective information source (1 - 2), the equally-weighted combined networks (4) and the networks weighted by ANNA (5) compared to the benchmark (3).

The results of the combined networks show, that with the equally-weighted combined network VCIB13 can be improved, if the partial judgments of the respective informationsource developed classifiers are connected to a total judgment by the ANNA. This result is yet not surprising, because classifier FB21-15 which was developed using balance sheet data outperforms VCIBP13 on the test set. Overall the combined network where the partial judgments weights are adjusted by the ANNA appears to be the superior model on the test set.

Connection on the featurelevel So far just the assessments based on subsets are combined, thus connections between a VC-feature and a ratio

[32] Cf. Uthoff, C., Erfolgsoptimale Kreditwürdigkeitsprüfung, p. 252 f.

Testseries	Name	#VC-Features-/Ratios	Error-area (%) test sample
1. Partial assessment VC-inform.	FA31-16	16	24,518
2. Partial assessment financial statement	FB21-15	15	17,551
3. Combined benchmark	VCIBP13	28	20,161
4. Equally-weighted assessments	FK01-31	31	17,679
5. Weighted assessment by ANNA.	FK02-31	31	17,247

Table 3-2: Selected features per stage and testresults

couldn't be represented by a BPN. During the separate analysis of the sources 'financial statement' and 'business information' just data of one source was examined, leaving it impossible for the network to determine interactions between VC-features and ratios. In the second analysis the feature list is extended over all features, in order to examine connections between all VC-features and ratios.

The results of the networks, which connect the partial assessments to a total assessment, show that the performance on the test set can be improved by combining partial assessments. Because the integration of partial assessments doesn't represent the interactions between the features and the ratios, those networks are examined which connect, in a hidden layer, the features and the ratios, in order to represent possible credit-standing relevant relations between the non-financial and financial information. By connecting the VC-features and the ratios the performance improves, compared to a connection on assessment level.

The results of the generated networks show, that the VCIBP13 benchmark is outperformed, if the weights of the partial assessment are adjusted using ANNA to the training and test data. By combining the features directly, the testresults can be further improved. As in the preceding section, networks which represent on the featurelevel the structures between the data from both informationsources, are superior to networks connecting just partial assessments per source.

4 Benchmarktest

4.1 Improvement of the VC-index

As the VC-index processes only a few features of an information, it was examined, whether the features of the VC-index are actually sufficient or inferior compared to a new combination of features of the VC-information.

Testseries	Name	#VC - features /ratios	Error-area (%) test sample
Partial assessment VC-information	FA31-16	16	24,518
Partial assessment financial statement	FB21-15	15	17,551
Combined benchmark	VCIBP13	28	20,161
Equally-weighted assessments	FK01-31	31	17,679
Objectively weighted assessments	FK02-31	31	17,247
Preselected features	FK12-20	20	17,008
Selection of all features	FK21-36	36	16,990

Table 3-3: Selected features per level

The results of the networks which were developed using VC-data show, that the benchmark can be out-performed, by processing more features from the VC-information. The benchmarktest of the developed networks at various stages depicted in figure 4-1 confirms the results.

The VC-index separates the data on the validation set with an error-area of 31,458%. The BPN FA12-7 outperforms the VC-index by six percent using only seven features. Compared to the testresults, the BPN FA21-10 is the superior classifier on the validation set. The error-area-value of the BPN FA21-10 is decreased by seven percent compared to the benchmark. The BPN FA31-16, which processes manually collected data in addition to the features of the FA21-10, separates the data with an error-area of 24.385% on the validation set, slightly poorer than the FA21-10.[33]

The VC-information consists of credit-standing relevant features which so far have not been processed by the VC-index. The computer-aided derived features as well as the manually encoded features improve the classifiers performance, if optimized according to its error-area.

4.2 Business information vs. financial statement

The BPNs, which are developed in the separate analysis, process the information either from the VC-information or from the financial statement. In this section we determine, whether an informationsource is of great importance for the credit-standing judgment of companies. 43 features from the VC-information are analyzed. 224 ratios from different information-areas are determined from the financial statements.

The most efficient BPN FA31-16 considering the VC-testdata processes 16 out of 43 features. The most efficient BPN FB21-15 considering balance sheet

[33] Cf. Uthoff, C., Erfolgsoptimale Kreditwürdigkeitsprüfung, p. 266 f.

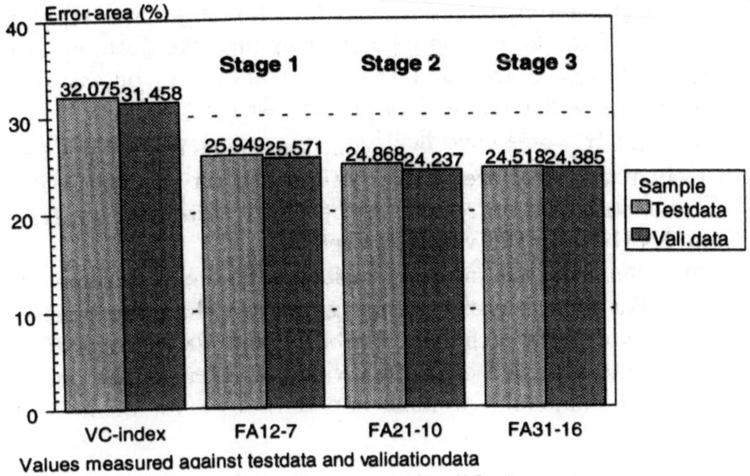

Fig 4-1: Results on the test and validation set per stage

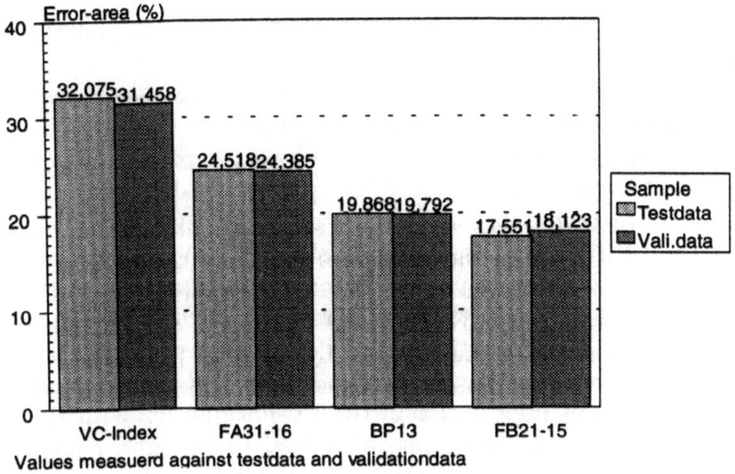

Fig 4-2: Results on the test and validation set of classifiers developed on different information sources

data consists of 15 ratios. The results of both classifiers are only comparable relative to each other. In order to compare the performance of both classifiers, their error-area is measured on the validation set and compared with the benchmark.

The classification of the data on the test set using the BPN FB21-15 gives an error-area of 18.123%. The network classifies the data less accurate on the validation set (error area of 17.551%). Thus it can be assumed that the networkfunction overfitted the data on the test set.

Despite the slightly worse classification, the error-area of the FB21-15 is still smaller than the error-area-value determined on the validation set for the BP-13. The classification results of the FB21-15 appear superior to the results of the FA31-15 on the validation set.[34]

The information from the informationsource 'financial statement' seem to be more important for the credit-standing analysis than the business information. If decision criteria such as accessibility, actuality, processing or costs are ignored and the achieved error-area value is taken as the only decision criteria, the source financial statement is preferred to the VC-information.

4.3 Combination of assessments based on subsets vs. combination of features

Networks using data from both sources as well as networks using data from one source only were developed during the combined analysis. Because of the aggregation levels partial judgment level and feature level, it is to be examined whether the interactions between the information have to be connected directly or on the judgment level to model an efficient BPN. The following figure 4-3 shows which approach classifies the companies with higher accuracy on the validation set.

The VCIBP13 sets a benchmark of 18.386% on the validation set. The benchmark improves clearly the BPN FK02-31 (which connects the assessments based on subsets to a total judgment) by 3.5 percent to an error-area of 15.331%. Although the BPN FK12-20 (which only directly processes 20 features of the FK02-31) separates the testdata slightly better, it performs worse than the FK02-31 on the validation set. Only the FK21-36 separates the data on the test set as well as the data on the validation set with a lower error-area. In contrast to the FK12-20 the features of the FK21-36 were not preselected by the separate analysis, but by the total list of the features.

The results on the validation set show, that a direct connecting of the credit-standing relevant information of the VC-information with the credit-standing relevant financial statement ratios lead to an improvement of the performance. The networks which only combine the partial assessments and connect them to a total judgment, classify the data with less accuracy on the validation set than networks which directly process the features to a total

[34] Cf. Uthoff, C., Erfolgsoptimale Kreditwürdigkeitsprüfung, p. 268 f.

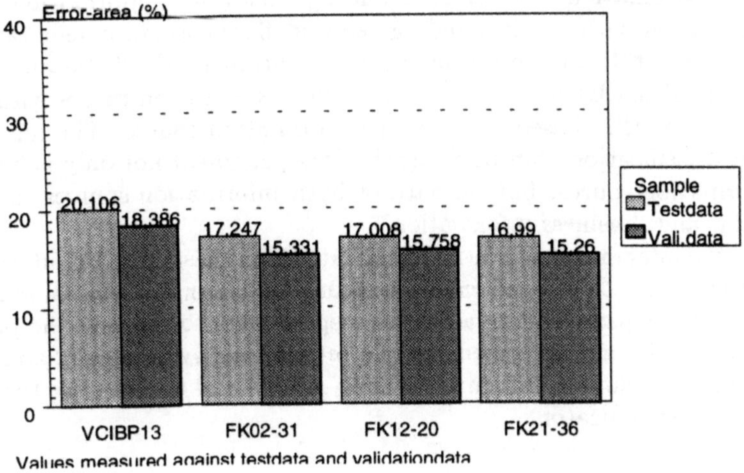

Fig 4-3: Test- and validation results of the combined developed classifiers

judgment. The results also show that networks which directly process the features, lead to an overfitting of the test set.

5 Conclusion

The VC-index is a suitable instrument for the classification of solvent and insolvent companies. For the data used, the average time period between the date of insolvency and the date of the information obtained from the VVC was 22 months. By optimizing the weights of the already used features, the performance of the VC-index can be improved; whereas a separation of VC-index features cannot improve the performance significantly. The VC-information includes features which previously were not part of the so called VC-index. By elimination of some features of the VC-index and adding further VC-information features, it is possible to increase significantly the performance based on VC-information compared to the VC-index. Especially VC-information such as industry type, company age, legal form, age distribution of the managers, qualification of the managers and manner of payment seem to be credit-standing relevant.

An efficient BPN for credit-standing judgment of companies was developed on the basis of the financial statement data. The BPN based on financial statement data assesses companies with a higher performance than a BPN based on VC-information. Thereby, the relatively high importance of the financial statement as an informationsource for the credit-worthiness analysis (in practice) is confirmed.

The combination of credit-standing relevant information of a VC-information with credit-standing relevant financial statement ratios increases significantly the performance of the optimized BPN. Because of the combined analysis, BPN were developed which assessed companies with higher accuracy than BPN based on only one information source. The number of incorrect classifications can be reduced, if the judgment not only is based on one information source, but on data of both information sources 'financial statement' and 'business information'.

With the integration of non-financial data from the source VC-information in an empirically developed credit-standing indicator for the standardized judgment of company clients, a further step is taken to improve the practice of assessment. Future improvement can be achieved by analyzing additional credit-standing relevant information using ANNA and to integrate them in a credit-standing-indicator.

Acknowledgement

We kindly thank the Institut für Statistik und Mathematische Wirtschaftstheorie especially Tae Horn Hann for many useful discussions and his support concerning the translation of this article.

References

Baetge, Jörg, Früherkennung negativer Entwicklungen der zu prüfenden Unternehmung mit Hilfe von Kennzahlen, Die Wirtschaftsprüfung, No. 22/23 1980, pp 651-665.

Baetge, Jörg/Hüls, Dagmar/Krause, Clemens/Uthoff, Carsten, Rationalisierung des Firmenkundenkreditgeschäfts auf der Basis empirisch-statistisch gewonnener Bilanzbonität, VBO-Informationen, No. 1 1994, pp 4-21.

Baetge, Jörg/Hüls, Dagmar/Uthoff, Carsten, Früherkennung der Unternehmenskrise, Forschungsjournal der Westfälischen Wilhelms-Universität, No. 2 1995, pp 21-29.

Baetge, Jörg/Kruse, Ariane/Uthoff, Carsten, Bonitätsklassifikationen von Unternehmen mit Neuronalen Netzen, Wirtschaftsinformatik, No. 3 1995, pp 273-281.

Baetge, Jörg/Krause, Clemens, The Classification of Companies by Means of Neural Networks, Journal of Information Science and Technology, Vol. 3, No. 1, 1993, pp 96-112.

Baetge, Jörg/Schmedt, Ulrich/Hüls, Dagmar/Krause, Clemens/Uthoff, Carsten, Bonitätsbeurteilung von Jahresabschlüssen nach neuem Recht (HGB 1985) mit Künstlichen Neuronalen Netzen auf der Basis von Clusteranalysen, Der Betrieb, No. 7 1994, pp 337-343.

Beuter, Hubert B./Reiss, Ingo/Rust, Hans Joachim, Erfahrungen mit formalisierten Verfahren bei der Kreditwürdigkeitsprüfung, Interaktion. Modellierung, Kommunikation und Lenkung in komplexen Organisationen, (ed.) Bernd Schiemenz, Berlin 1994, p. 55-73.

Bühler, Wilhelm, Informationsressourcen für die Bonitätsanalyse, Kreditmanagement. Rechtliche und betriebswirtschaftliche Instrumente zur Risikoreduzierung in der Kreditpraxis, (ed.) Bühler, Wilhelm/Schmoll, Anton, Wien 1987, pp 69-116.

Burgard, Horst, Bilanzanalyse aus Sicht der Kreditinstitute (Part II), Möglichkeiten und Grenzen der Bilanzanalyse im Kreditgeschäft der Banken, Deutsches Steuerrecht, No. 10 1991, pp 324-327.

Büschgen, Hans E., Bankgeschäfte und Bankmanagement, 4th. Edition, Wiesbaden 1993.

Creditreform, Unternehmensentwicklung 1996 alte und neue Bundesländer. Eine Creditreform-Untersuchung über Unternehmensneueintragungen, Unternehmenslöschungen, Insolvenzen, Verband der Vereine Creditreform, Neuss 1997.

Creditreform, Das Creditreform Informationssystem. Ein entscheidungsunterstützendes System zur Bonitätsbeurteilung von Unternehmen, Verband der Vereine Creditreform, Neuss 1995.

Creditreform, Creditreform Informationssystem, Creditreform-Bonitätsindex zur Risikobestimmung für Kreditmanagement und Marketing. Ein System zur Bonitätsbeurteilung von Unternehmen. Entwickelt von Creditreform in Zusammenarbeit mit Wissenschaft und Kreditmanagement, Verband der Vereine Creditreform, Neuss 1986.

Kreditwürdigkeitsprüfung durch die Deutsche Bundesbank, Monatsberichte der Deutschen Bundesbank, No. 1 1992, pp 30-36.

Feidicker, Markus, Kreditwürdigkeitsprüfung. Entwicklung eines Bonitätsindikators, dargestellt am Beispiel von Kreditversicherungsunternehmen, Schriftenreihe des Instituts für Revisionswesen der Westfälischen-Wilhelms Universität Münster, (ed.) Jörg Baetge, Düsseldorf 1992.

Fischer, Jürgen H., Computergestützte Analyse der Kreditwürdigkeit auf Basis der Mustererkennung, Betriebswirtschaftliche Schriften zur Unternehmensführung, Vol. 23, Kreditwesen, Düsseldorf 1981.

Hecht-Nielsen, Robert, Neurocomputing, Reading et. al. 1990.

Hüls, Dagmar, Früherkennung insolvenzgefährdeter Unternehmen, Schriftenreihe des Instituts für Revisionswesen der Westfälischen-Wilhelms Universität Münster, (ed.) Jörg Baetge, Düsseldorf 1995.

Krause, Clemens, Kreditwürdigkeitsprüfung mit Neuronalen Netzen, Schriftenreihe des Instituts für Revisionswesen der Westfälischen-Wilhelms Universität Münster, (ed.) Jörg Baetge, Düsseldorf 1993.

Kuhlmann, Jürgen, Die Problematik einer Ausgliederung der Kreditwürdigkeitsprüfung im Firmenkreditgeschäft der Banken, Europäische Hochschulschriften, Vol. 5, Volks- und Betriebswirtschaft, No. 1253, Frankfurt a. M. 1992.

Mozer, Micheal C./Smolensky, Paul, Using Relevance to Reduce Network Size Automatically, Connection Science, Vol. 1, No. 1, Abingdon 1989.

Neural Ware Inc. (Hrsg.)., Using NeuralWorks, Pittsburgh 1993.

Niehaus, Hans-Jürgen, Früherkennung von Unternehmenskrisen. Die statistische Jahresabschlußanalyse als Instrument zur Jahresabschlußprüfung, Schriftenreihe des Instituts für Revisionswesen der Westfälischen-Wilhelms Universität Münster, (ed.) Jörg Baetge, Düsseldorf 1987.

Pytlik, Martin, Diskriminanzanalyse und künstliche neuronale Netze zur Klassifizierung von Jahresabschlüssen. Ein empirischer Vergleich, Europäische Hochschulschriften, Vol. 5, Volks- und Betriebswirtschaft, No. 1688, Frankfurt a. M. et. al. 1995.

Riebell, Claus, Die Praxis der Bilanzauswertung, 5th. Edition, Stuttgart 1992.

Schmidt-von Rhein, Andreas/Rehkugler, Heinz, KNN zur Kreditwürdigkeitsprüfung bei Privatkundenkrediten, Neuronale Netze in der Ökonomie, Grundlagen und finanzwirtschaftliche Anwendungen, (ed.) Heinz Rehkugler/Hans Georg Zimmermann, München 1994, pp 491-545.

Schmoll, Anton, Theorie und Praxis der Kreditprüfung unter besonderer Berücksichtigung der Klein- und Mittelbetriebe (I), ÖBA, No. 3 1983, pp 87-106.

Schmoll, Anton, Theorie und Praxis der Kreditprüfung unter besonderer Berücksichtigung der Klein- und Mittelbetriebe (II), ÖBA, No. 5 1983, pp 165-191.

Schumann, Matthias/Lohrbach, Thomas/Bährs, Peter, Versuche zur Kreditwürdigkeitsprognose mit Künstlichen Neuronalen Netzen, paper No. 2 of the department Wirtschaftsinformatik II, Georg-August-Universität-Göttingen, (ed.) Matthias Schumann, Göttingen 1992.

Uthoff, Carsten, Erfolgsoptimale Kreditwürdigkeitsprüfung auf der Basis von Jahresabschlüssen und Wirtschaftsauskünften mit Künstlichen Neuronalen Netzen, Stuttgart 1997.

Zell, Andreas, Simulation Neuronaler Netze, Bonn/Paris 1994.

Zimmermann, Hans-Georg, Neuronale Netze als Entscheidungskalkül, Grundlagen und ihre ökonometrische Realisierung, Neuronale Netze in der Ökonomie, Grundlagen und finanzwirtschaftliche Anwendungen, (ed.) Heinz Rehkugler/Hans Georg, Zimmermann, München 1994, pp 1-87.

Data Warehousing and OLAP: Delivering Just-In-Time Information for Decision Support

Christoph A. Breitner

Universität Karlsruhe, Postfach 6980, D-76128 Karlsruhe

Abstract. Global competition and rapidly changing markets require up-to-date, high-quality and complete information for decision support. A data warehouse offers an integrated view on all business-relevant data, and thus, provides the data foundation upon which an enterprise-wide information system can be based. OLAP is an analytical processing technology, which creates new business information from existing data, through a rich set of business transformations and numerical calculations.

1 Introduction

"*Computer promised a fountain of wisdom but delivered a flood of data!*"

1.1 The need for an effective management of information

Flexibility to react on rapidly changing general conditions of the environment has become a key factor for economic success of any company. As the speed of technology progress has accelerated and consumer trends and product-life-cycles get significantly shorter, operational as well as strategic decisions have to be taken in decreasing intervals. The competitiveness of an enterprise is therefore to a large extent determined by the ability of its executives to deliver exact and just-in-time decisions. These decisions must be based on high-quality, up-to-date and complete information.

Today, in almost all lines of business - be it trade and industry, financial services, or public administration - data-base management systems (DBMS) provide a robust and persistent storage *of* as well as an efficient access to *huge* and quickly increasing volumes of data. Although this data resource represents a valuable asset of the enterprise, it is often only used for the day-to-day operation and control purposes, but not for analysis and strategic decision making. The reason is that - despite of the myriad of data - DBMS are not prepared to provide means for retrieving *essential* information, at least not in a form which is necessary for creative analysis or decision support.

1.2 Characteristics of information systems

What distinguishes systems that are able to deliver decision relevant data from systems which support the so-called online transaction processing (OLTP)?

In OLTP-applications many user concurrently access data collections held in a central repository or distributed over only a few database systems at most. The processing can be characterized by a - more or less - fixed number of pre-defined and very simple read/write-transactions (debit/credit-transactions), which usually touch only a comparably small amount of records (one-record-at-a-time). The size of the data sources is typically restricted to gigabytes of today's current data.

In contrast to OLTP-systems decision support systems (DSS) are usually used by only a small number of data-analysts or executives and far less frequent than OLTP-systems. Data access is confined to read-only-transactions which in turn are very complex, i.e. have to consider a very large amount of data including historical information (many-records-at-a-time) spread over a multitude of data sources. Consequently, the total volume of data rises into magnitudes of giga- up to terabytes of data. In addition, these 'transactions' are not pre-defined but rather generated highly dynamic and flexible.

Figure 1 summarizes the differences between transactions for supporting the day-to-day operations of an enterprise (OLTP) and decision support.

Characteristics	**OLTP**	**Decision Support**
Number of User/Frequency	high	low
Complexity of Transactions	simple	complex
Kind of Transactions	stactic, predifined	dynamic, flexible
Number of Sources	few	many
Size of Sources	Gigabyte	Gigabyte-Terabyte
Actuality	current-valued	current-valued and historical

Fig. 1. Characteristics of OLTP- and Decision Support Systems

The remainder of the paper is organized in two major parts which illustrate new concepts and solutions of information technology to meet the demands of modern decision support systems. The first part introduces an architecture for information systems, called data warehouse, which serves as a platform for analytical processing techniques. One of them, OLAP (Online Analytical Processing) is examined in more detail in part two. A more comprehensive description of the role of information management, as well as advanced methods for information processing, can be found in Breitner C. et al (1997).

2 Data warehousing

Due to the different requirements it seems sensible to physically and logically separate operational data of OLTP-systems from 'intelligent' information processing. The connection between operational data stores and decision support systems is then established by an intermediate database, the data warehouse, which acts as a common, enterprise-wide information base.

An exact definition of what really makes a data warehouse is hard to obtain. According to W.H. Inmon, considered the father of data warehouses, a data warehouse is characterized by the following requirements [Inmon W. H. (1996)]:

- *Subject-orientation:* Whereas classical database design emphasizes the function of the transactions, a subject-oriented representation of data emphasizes the structuring of data according to the core-business areas and the strategic goals of a company, e.g., products, customer, turnover, ...
- *Integration:* The change from function- to subject-orientation leads to a modification of database organizations across several database systems. The result is an integrated store of data which includes all relevant aspects.
- *Time-variance:* The term 'time-variance' has two different but related meanings: From a more static point of view, the data contains time aspects, e.g., in form of time-stamps, time related attributes etc. Dynamically speaking time-variance relates to the strategy of ensuring ongoing refreshment of data.
- *Non-volatility:* Once a data record is brought into the data warehouse it remains unchanged. Refreshing data does always translate to adding new data, i.e. all data accesses are read-only transactions.

Though rather short and intuitive, this definition hides a lot of organizational as well as technical tasks and concepts. Organizational challenges include the gathering and integration of the high number of supplementing, overlapping, redundant and even inconsistent data sets which might be collected over decades, and the design of tools and methodologies for their exploitation. Technical challenges mainly concern the management during operation.

Building a data warehouse is a very complex task which requires a variety of different skills, e.g., business user, database-, hardware- and network-specialists, application developer etc. In order to break down the complexity of the design and development in manageable pieces and to be able to coordinate the work done by the different groups, a framework in form of an systems architecture has come up, which can serve as a common ground for coordination and demonstration of the tasks to be done. In the following we present the architecture of such a data warehouse environment.

2.1 Architecture of a data warehouse enviroment

The term data warehouse environment subsumes the data warehouse as well as additional components needed for its building and exploitation. A data warehouse environment consists of four layers (see figure 2).

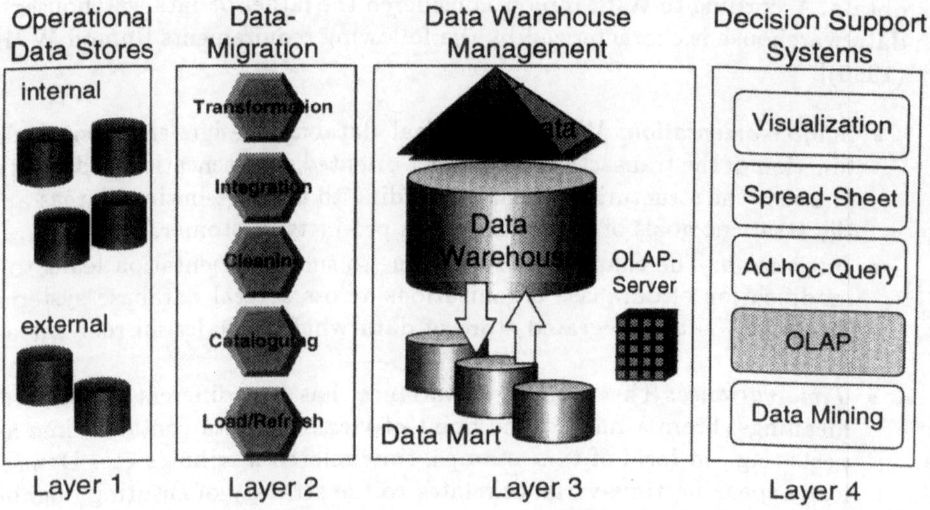

Fig. 2. Architecture of a Data Warehouse Enviroment

2.2 Layer 1: Operational data stores

The operational data stores represent the sources which feed the data warehouse. Most often the data reside in heterogeneous systems ranging from simple file systems, hierarchical and relational up to object-oriented database systems. Although it might be reasonable in some cases to access those OLTP-systems directly, there are more arguments which suggest otherwise:

1. Today, OLTP-systems are usually 'right-sized' to meet the demands of the day-to-day transactions and, thus, cannot cope with additional work.
2. Existing systems do not offer special support for the different access-profile and functional requirements of decision support applications.
3. OLTP-system only hold present-day information, i.e. data of the last 60-90 days. Historical information is not available on-line.
4. Design and implementation of direct access to heterogeneous data sources is yet under research. Only few commercial solutions with very restricted functionality are available.

2.3 Layer 2: Data migration

Layer 2 is responsible for migrating the data from the operational data stores into the integrated data warehouse. The problems encountered can be roughly categorized into five classes: transformation, integration, cleaning, cataloging and loading/refreshing data.

Transformation: The first step on the way to an integrated collection of data is to bridge the gap between the different data models of the operational data stores. Since the the data warehouse will in most cases be managed by a relational database system (RDBMS), the relational model can be used as a common basis for integration. The task is then to transform the schema of each data source to a relational schema (or more general: a schema following the common data model). Techniques for transforming schemas onto data models can be found in [Lang S., Lockemann P. C. (1995)].

Integration: After transformation the schemas have to be consolidated into a single global schema [Lang S., Lockemann P. C. (1995)]. Consolidation covers a wide range of activities including the identification and resolution of attribute-equivalence and attribute-overlapping, and handling differences in coding, scales of measurement and definition of key-fields:

- Coding: For instance, the line of business can be encoded in various ways: finance/automobile/chemistry/trade, f/a/c/t, 1-2-3-4, etc.
- Scales of measurement: The turnover of a company can be specified in different currencies (dollar, pound, francs) or units (thousands, millions).
- Key-Fields: Data sources may define their own rules for identifying data records, e.g. use a unique customer-identifier, a name together with address, or even the relative position of a record within a file.

Following the schema integration, procedures for migrating the actual data into the new schema are to be designed and implemented.

Cleaning: Since one cannot expect the data stores to be error-free, data cleaning is an essential phase which might have huge payoffs afterwards as reliable data is one of the prerequisites for user acceptance. Data cleaning comprises measures to detect errors by specifying integrity rules, and complement missing data by filling in default-values or values derived by cross-relating information from other sources. Sometimes, erroneous data may make up to 30% of the whole data set, which may cause the complete data set to be dismissed. If all compensation steps like identification of alternative data sets or even the definition of procedures for automatic or manual data collection fail, the schema may have to be truncated afterwards.

Cataloguing meta-data: Meta-data provide information about the origin, actuality and quality, and mechanisms for the extraction and transport of the data. Meta-data becomes particularly important, if the data model has changed over time, but for certain analyses older data has to be accessed. Meta-data must be self-explanatory in order to enable the user to ask the 'right' questions and formulate those questions according to their needs. Meta-data, therefore, represent the 'shop window' or 'order catalogue' of the data 'products' of the data warehouse.

Further administrative and operational meta-information is transformation- and cleaning-procedures, integrity rules, business related data like glossaries, ownership of the data, authorization, and statistical usage information.

Loading and refreshing: In addition, strategies for loading and refreshing the data must be designed and implemented. Because of the huge volume of data traditional techniques for sequential loading of databases quickly reach their limit, e.g., loading 1 terabyte of data in sequential batch-modes would last several weeks! As a consequence, most database vendors offer facilities for parallel and incremental loading. During loading many other procedures can be carried out, e.g., creation of indexes, sorting, aggregating and partitioning of the data as well as additional integrity checks.

Data refreshing ensures that ongoing changes in operational data stores are reflected by the data warehouse. Since opposed to the large volume of data there is only a small time-window - usually at night - where the database can be taken off-line, refreshing the complete data warehouse in a single step is far beyond what is technically practicable. Consequently, data refreshing must consider two aspects, namely in which time intervals and to what extent the data has to be updated. In database research this field corresponds to the well-known view-update problem and the maintenance of materialized views [Zhuge Y. et al et.al. (1995), Gupta A. et al et.al. (1995)].

2.4 Layer 3: Data warehouse management

After integration, consolidation and cleaning of the data collections the data warehouse covers all business relevant data. In order to easier manage and disseminate information on the department level, meaningful subsets of data are identified and shipped to the departments in form of small data warehouses, so-called data marts, which e.g. contain only information related to a certain customer or product.

In analogy to the classical methodology of view consolidation data marts can also represent the starting point for designing an enterprise-wide data warehouse: At first, departments individually build data marts which are then combined into an enterprise-wide data warehouse. In most cases building a data mart instead of a data warehouse can be accomplished in a significant shorter time and is far less expensive. Furthermore, data marts can serve as a prototype for demonstration and evaluation of user requirements.

The management of the data warehouse and the accompanying meta-data is usually realized by a high-performance database server on a SMP- (Symmetric Multi-Processing) or even MPP-basis (Massive Parallel Processing).

Other reasons for using parallel processing facilities are the possibility to scale the database according to increasing data volumes and number of users, as well as the availability of the data warehouse which will often evolve to a mission critical application.

The data warehouse management layer is complemented by additional systems which support the functionality of succeeding applications, e.g. special OLAP-Server (see next section).

2.5 Layer 4: Decision support systems

Exploitation of the data collected in the data warehouse is done by decision support applications, among them desktop-based tools for visualization, spread-sheet applications, statistics packages, query- and report-generators, as well as OLAP and data mining systems.

The applications are located at the client-site and access the data via diverse middleware platforms, which hide different networks and protocols. A quickly emerging trend is the use of internet/intranet-technology.

Decision support systems can be roughly classified along two dimensions: The first dimension is the kind of processing a tool supports, which is either analytical or discovering. Analytical processing means that the user generates hypotheses about the underlying data and uses the tool to verify (or falsify) them, while discovering tools are more autonomous, i.e. are able to formulate and evaluate hypotheses in order to detect detecting patterns, peculiarities and relationships in the data without the user guiding each single step. A second dimension is whether the tool can be applied for pre-defined tasks or can be used in an ad-hoc-manner for flexible and dynamic specified tasks.

For instance, management-/executive information systems (MIS/EIS), as well as report- and query generators fall into the class of analytical tools for pre-defined tasks, whereas OLAP is suited for ad-hoc analyses. Their discovering counterpart is formed by tools, which employ techniques from artificial intelligence to solve special tasks (neural nets for credit scoring), and data mining systems, which can be used for a variety of purposes (for a comprehensive overview of data mining techniques see [Nakheizadeh G., Reinartz T. & Wirth R. (1997), Nakheizadeh G. (1997)]).

Figure 3 shows a classification of tools, where the bottom row contains tools for operational data processing. Note: The boundaries between the classes are not entirely fixed.

Discovering	(Customized) AI-Solution	Data Mining
Analytical (Verification)	MIS, EIS, Reprot-Query-Generators	OLAP
Operational	OLAP	Dynamic SQL
	Pre-defined	**Ad-hoc**

Fig. 3. Classification of decision support applications

3 OLAP

One of the most basic requirements in analytical processing is the ability to view data from different angles and on different aggregation levels. Take, for instance a data analyst who has to look for peculiarities in a table containing weekly sales figures for 20 products, which are sold in 100 regions (or countries) over a period of three years (150 weeks), and thus, consisting of 300.000 tuples (20x100x150). A very intuitive method for getting acquainted with the data would be to look at aggregated figures, e.g. summed up by region and product, and then to dig into detail on outliers. Whereas in classical RDBMS this approach would result in a multitude of highly complex aggregate-queries which can be prohibitively expensive to answer, online analytical processing systems (OLAP) [1] support the navigational and layered process of examining multidimensional data.

3.1 Multidimensional representation and operators

Closely related to the term OLAP is the term of multidimensionality of data. The idea is to logical separate independent attributes (dimensions) from dependent attributes (facts). Typical dimensions are customer, departments, regions, products as well as time, typical facts can be sales figures, turnover, fault probabilities, etc. Each independent attribute forms a dimension of ist own (in graphical notation: an axis), which contains the values of the attribute as axis labels. The independent attributes therefore form a data space, wherein the values of the dependent attributes are positioned according to the values of the independent attributes in the data record. In the case of three-dimensional data the data space forms a *data cube*. In the meanwhile, the term data cube is often used as a synonym for multidimensional data. Figure 3 illustrates the transformation from relational to two- and three-dimensional representation.

Dimensions are further characterized by aggregation hierarchies, i.e. their values can be stepwise refined or generalized. For instance, hierarchies for time and geographic attributes are day-month-quarter-year, or city-region-country-continent, etc. The basic idea of OLAP is the *automatic* aggregation (summation, counting,...) of information along several dimensions and hierarchies.

In order to utilize the data cube in a flexible and intuitive fashion special operators are provided. The most popular operations are slice/dice, rotate, drill-down/roll-up:

- Slice/Dice: The focus of the user is shifted to a certain layer or sub-cube by slicing or dicing the data-cube. For instance, the turnover figures

[1] See also http://www.olapcouncil.org

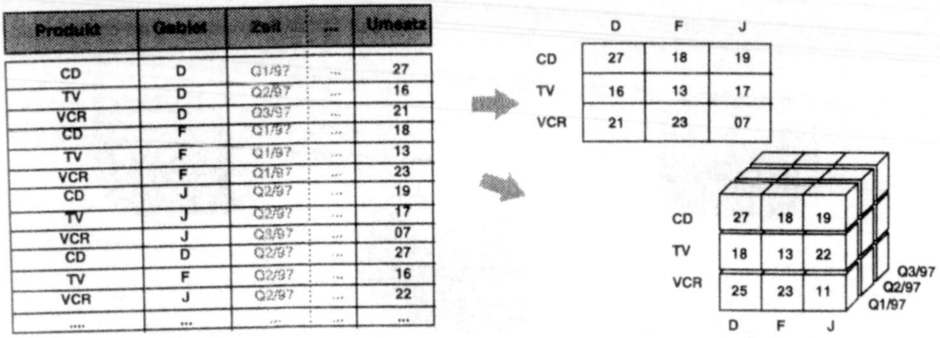

Fig. 4. Changing from relational to multidimensional view

for the second quarter of 1997 for products and regions are obtained by taking the second slice of the cube in figure 3.

- Rotate: This operation changes the orientation of the cube. Usually rotation is followed by a slice-operation, i.e. to analyze the data for a certain product (VCR) with respect to regions and time, the cube is rotated to get regions and time to the front and the slice 'product = VCR' is chosen.
- Drill-Down/Roll-Up: The drill-down/roll-up-operation utilizes attribute hierarchies and enables a stepwise refined/aggregated view on the data. Drill-down is often preceded by a dice-operation, e.g. to look further into the low turnover figures for VCRs in the first quarter of 1997 in Japan, the dice (cell) 'VCR-Japan-Q1/97' is selected using a dice-operation and then refined by drilling down along the time and region axis.

Since these operations are very data-intensive and expensive to compute, they are often supported by special hard- and software products. Relational OLAP (ROLAP)-Server extend existing relational DBMS by mapping the multidimensional operators to standard operators of an RDBMS. The computational requirements are met by implementing special mapping strategies, which e.g. combine several group-by-statements into one construct, intelligently pre-compute queries (materialization), and make use of parallel hardware (SMP/MPP-systems).[Gray J. et al (1996)].

Multidimensional OLAP (MOLAP)-Server represent autonomous database systems which directly implement multidimensional data structures and operators, i.e. do without expensive mapping strategies. MOLAP-server use optimized n-dimensional arrays for storing data and implement special strategies and techniques for data access. Currently, the main limitation of MOLAP-server is that data processing is done exclusively in main-memory, which restricts the application to comparable small data volumes.

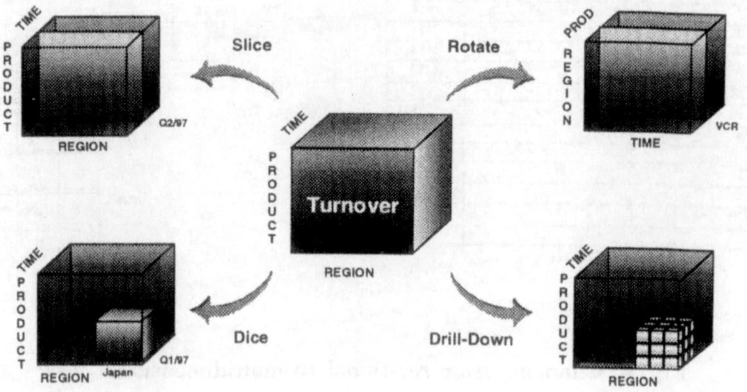

Fig. 5. Multidimensional Operators

Selected techniques for database support: Current support for OLAP-functionality includes techniques for parallel query processing, special implementations of operators (intelligent scan, star-join), access paths (bitmap-, join-index) as well as techniques for automaticallyp re-computing query results. We will take a closer look at two techniques which in the meanwhile can be found in several database servers: bitmap-index and materialization of aggregate queries.

Bitmap-Index Bitmap-indexes are designed for data sets which contain attributes with only a small number of different values, which is e.g. the case for the dimension attributes (see previous section). For every value of an attribute, which is to be indexed, a separate list - a bitlist or bitmap - is maintained, which contains an entry for each row of the table. Suppose, the bitlist j corresponds to the value a_j of an attribute a. Then the ith entry of bitlist j is true (1), if the value of the attribute a of the ith row of the table is a_j, and false (0) otherwise.

Bitmap-Indexes are particularly suited for answering queries consisting of AND- or OR-connected expressions which use indexed attributes. Evaluating such a query simply means to use the appropriate bitlist and compute the expression by applying the boolean AND-, OR- and NOT-operators.

Example: For the attributes product and region of the turnover table in figure 3 a bitmap-index is to be created. Since the table only contains 3 products and 3 regions, the ollowing 6 (3+3) bitlists are generated:

Region:	F:	(0, 0, 0, 1, 1, 1, 0, 0, 0, 0, 1, 0, ...)
	D:	(1, 1, 1, 0, 0, 0, 0, 0, 0, 1, 0, 0, ...)
	J:	(0, 0, 0, 0, 0, 0, 1, 1, 1, 0, 0, 1, ...)
Product:	CD:	(1, 0, 0, 1, 0, 0, 1, 0, 0, 1, 0, 0, ...)
	TV:	(0, 1, 0, 0, 1, 0, 0, 1, 0, 0, 1, 0, ...)
	VCR:	(0, 0, 1, 0, 0, 1, 0, 0, 1, 0, 0, 1, ...)

A query which determines the "turnover of TV-sets in France" can be answered by combining the bitlists for the value 'F' and 'TV' of the attributes region and product as in:

F&TV: (0, 0, 0, 0, 1, 0, 0, 0, 0, 0, 1, 0, ...).

For the computation of the overall turnover for TV-sets in France only those tuples have to be accessed which fulfill both conditions, i.e. the 5th, 11th, ...

Materialization of aggregate queries Computation and manipulation of the data cube generally involves a great number of tuples being accessed and aggregated. For instance, the data cube shown in figure 3 can be generated via the following SQL-query:

```
select product, region, time, sum(turnover)
from turnover_tab
group-by product, region, time;
```

These aggregate queries are often repeated in exactly the same or at least very similar way and are very expensive to compute, leading to unacceptable response times.

A popular technique for avoiding the overhead of frequent re-computation is to store results in form of materialized views for later reuse. Since materialization of all queries is not practicable, only those views have to be selected which are most promising with respect to the profit/storage-ratio. Another question is to decide, which view can be used for answering a given query at the lowest costs.

An approach to solve this problem is given in [Harinarayan, V.n et al (1996)], which introduces a query-dependency-graph as depicted in figure 6. The graph can be further augmented to account for selections (slice-operations).

Figure 6 shows the dependency graph for the attributes region and time. A solid arch from A to B is used to indicate that an aggregation B can be computed by using A, e.g. (region) from (region,time), a dotted arch describes additional dependencies resulting from aggregation hierarchies. For instance, the materialization of the query:

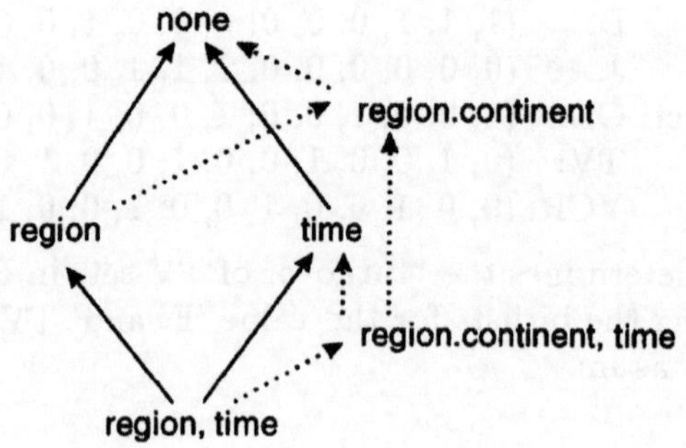

Fig. 6. Dependency graph for aggregate queries

```
select product, region, time, sum(turnover)
from turnover_tab
group-by product, time;
```

can be used to answer Q1 but not for Q2:

```
Q1:= select region.continent,      Q2:= select region,
     sum(turnover)                      sum(turnover)
     from turnover_tab                  from turnover_tab
     where time='Q1/97'                 where time='Q1/97'
     group-by region.continent;         group-by region;
```

4 Conclusion

Despite the initial costs for building a data warehouse and providing facilities for interactive data analysis, the key factor for initiating a data warehouse project is the huge potential for gaining a competitive edge, be it by cutting down costs, shortening time-to-market for new products and services, improving products and quality of service, or innovating market strategies. In particular, data warehouse solutions are essential for all service- and consumer-oriented companies, e.g. from financial service (banking, insurance), retail, and consumer-industry.

The benefit stems from the fact that information can be uniformly accessed from a single, consistent, integrated and up-to-date source of high-valued information. A data warehouse thus lays the foundation for a decision support environment. In addition, OLAP-facilities are an emerging technique for exploiting data on-line by offering comfortable means for manipulating data and computing new information tailored to individual needs. Therefore, OLAP is a promising way for extracting just-in-time information in order to be able to make just-in-time decisions.

References

Barquin R. C. & Edelstein H. A. (1997): Planning and Designing the Data Warehouse, Prentice Hall.

Breitner C., Lockemann P. C. & Schlösser J. (1997). Die Rolle der Informationsverwaltung im KDD-Prozeß. To appear in Nakheizadeh (Hrsg.): Data Mining, theoretische Aspekte und Anwendungen. Physica.

Datamation (1995), May 1996.

Gray J., Bosworth A., Layman A., & Pirahesh H. (1996). Data Cube: A Relational Aggregation Operator Generalizing Group-By, Cross-Tab, and Sub-Totals. Proceedings of the Intl. Conference on Data Engineering (ICDE), pp. 152-159.

Gupta A., Mumick I. S., & Ross K. A. (1995). Adapting Materialized Views after Redefinitions. Proceedings of the ACM SIGMOD Intl. Conference on Management of Data, pp. 211-222

Harinarayan, V., Rajaraman, A., & Ullman, J. D. (1996). Implementing Data Cubes Efficiently. Technical Report, University of Standford, Computer Science Department.

Inmon, W. H. (1996). Building the Data Warehouse. John Wiley.

Kimball R. (1996). The Data Warehouse Toolkit, John Wiley.

Lang, S., Lockemann, P. C. (1995). Datenbankeinsatz. Springer.

Nakhaeizadeh G. (1997). Einführung in Data Mining. Proceeding of the 6th Oeconometric Workshop. Springer.

Nakhaeizadeh G., Reinartz T. & Wirth R. (1997). Wissensentdeckung in Datenbanken und Data Mining. Ein Überblick. To appear in Nakheizadeh (Hrsg.): Data Mining, theoretische Aspekte und Anwendungen. Physica.

Zhuge Y., Garcia-Molina H., Hammer J., & Widom J. (1995). View Maintenance in a Warehousing Environment. Proceedings of the ACM SIGMOD Intl. Conference on Management of Data, pp. 316-327

A Supplier of data warehouse components

The following (incomplete) list of vendors of data warehouse components is partly taken from [Datamation (1995)] and [Barquin R. C. & Edelstein H. A. (1997)]. The layers correspond to the architecture of the data warehouse environment shown in figure 2.

- **Layer 1: Connection of operational data stores:** Apertus, CA-Ingres Gateway, Information Builders EDA/SQL, IBM Data Joiner, Informix Enterprise Gateway, Microsoft ODBC, Oracle Open Connect, Platinum InfoHub, SAS Connect, Software AG Entire, Sybase Enterprise Connect, Trinzic InfoHub

- **Layer 2: Data migration:** CA-Ingres Replicator, Carleton Passport, Evolutionary Tech Inc. ETI-Extract, Harte-Hanks Trillium, IBM Data Joiner/Data Propagator, Oracle 7, Platinum InfoRefiner/InfoPump, Praxis OmniReplicator, Prism Warehouse Manager, Redbrick TMU, SAS Access, Software AG Sourcepoint, Sybase Replication Server, Trinzic InfoPump
  - *Metadata Management:* HP Intelligent Warehouse, IBM DataGuide, Platinum Repository, Prism Directory Manager
- **Layer 3: Data warehouse management:**
  - *Data Warehouse Servers:* CA-Ingres, IBM DB2, Information Builders Focus, Informix,Oracle, Praxis Model 204, Redbrick, Software AG ADABAS, Sybase MPP, Tandem, Teradata
  - *ROLAP Servers:* HP Intelligent Warehouse, Information Advantage Asxys, Informix Metacube, MicrosStrategy DSS Server
  - *MOLAP Servers:* Arbor Essbase, Comshare Commander OLAP, Oracle IRI Express, SAS System
- **Layer 4: Decision support systems:**
  - *Query/Reporting Environments:* Brio/Query, Business Objects, Cognos Impromptu, CA Visual Express, IBM DataGuide, Information Builders Focus Six, Informix ViewPoint, Platinum Forest & Trees, SAS Access, Software AG Esperant
  - *Multidimensional Analysis:* Andyne Pablo, Arbor Essbase Analysis Server, Business Objects, Cognos PowerPlay, Dimensional Insight CrossTarget, Holistic Systems HOLOS, Information Advantage Decision Suite, IQ Software IQ/Vision, Kenan Systems Acumate, Lotus 123, Microsoft Excel, Microstrategy DSS, Pilot Lightship, Platinum Forest & Trees, Prodea Beacon, SAS OLAP++, Stanford Technology Group Metacube
  - *Data Mining:* Angoss KnowledgeSeeker, Cross/Z Query Objects, Epsilon Data Management EQL, IBM Intelligent Miner, Information Discovery IDIS3, ISL Clementine, NCR Knowledge Discovery Workbench, SAS Institute NNA.

Financial Calculations on the Net

Wolfgang Härdle and Stefan Sperlich

Institut für Statistik und Ökonometrie, Wirtschaftswissenschaftliche Fakultät, Humboldt-Universität zu Berlin, D-10178 Berlin, Germany

1 Introduction to XploRe

XploRe is an interactive statistical computing environment. As the name indicates, support for exploratory data analysis, for simulation and intensive calculation scenarios is given by a variety of interactive tools. Like most of the statistic software packages it is based on vector- and matrix calculation, but unlikely many such packages, in XploRe matrices can be of up to nine dimensions. Further, XploRe offers extraordinary opportunities in dynamic and interactive graphics. For our purpose, that is calculating in finance, the interactivity of the user interface and the techniques of visualization are of special importance; particulary since this interactivity is net based and easy to implement for programmers who want to modify or extend existing methods.

A noteworthy quality in XploRe are the capabilities of network facilities, which in software development have become more and more important. From the very beginning, XploRe was geared towards inter- or intranet compatability. Thus, for example, the help system is available in HTML. Furthermore, the newly developed Java interface endows the user with the possibility to work with XploRe via inter- or intranet and thus without the necessity of installing it locally. Thus, a complete session can be opened just with a freely available WWW browser like Netscape (try the http address: *http://www.xplore_stat_de/*).

2 The `finance` library of XploRe

There is growing interest in quantifying and simulating economic processes, particulary in the statistical analysis of the behaviour of financial markets. The library **`finance`** is designed for this purpose. It offers macros to predict and estimate (time series) processes, for example stock returns, to determine option prices and to evaluate different scenarios (e. g. for portfolio strategies). To give a brief overview of this library we will present and illustrate some macros implemented in **`finance`**. It is called by **`library("finance")`** on the free Java interface of XploRe. We will pay special attention to the simulation of processes of stock returns.

The elements of `finance`

The basic method to determine option prices of the European kind is to use the analytic solution of Black and Scholes. Their formula allows to calculate the option price under strong assumptions on the model. The analogue for American options is the approximation by McMillan. Distributions of dividends for assets like stocks usually lead to changes in the value of the corresponding derivative. Unfortunately this often can not be taken into account in the analytic solutions and thus has to be done numerically for European as well as for American options. Therefore binomial trees are used. These thechniques can be performed interactively via the macro **`optstart()`**, as presented in figure 1 for illustration. It is also of interest to investigate and

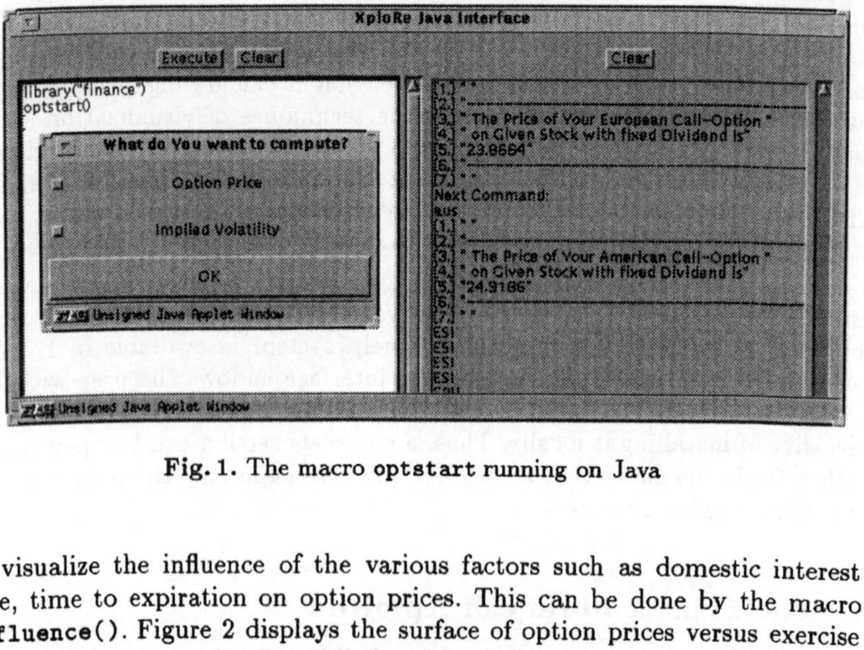

Fig. 1. The macro **`optstart`** running on Java

to visualize the influence of the various factors such as domestic interest rate, time to expiration on option prices. This can be done by the macro **`influence()`**. Figure 2 displays the surface of option prices versus exercise price and time to expiration while keeping all the other parameters fixed.

Obviously, scenarios for portfolios strategies like the construction of spreads or arbitrage possibilities can easily be obtained by table calculation. It has long be recognized that stock returns do not obey a simple Wiener process. Various alternatives have been introduced, including among others, jump processes or a mixture of Wiener and jump processes, see for example Streller (1995). Macros for simulating such processes and for estimating the parameters in those models are also available, e. g. **`stocksim()`, `stockest()`**.

A different approach to model stock prices is using typical time series models such as ARCH, GARCH, EGARCH, T-ARCH etc. , see Gourieroux (1997).

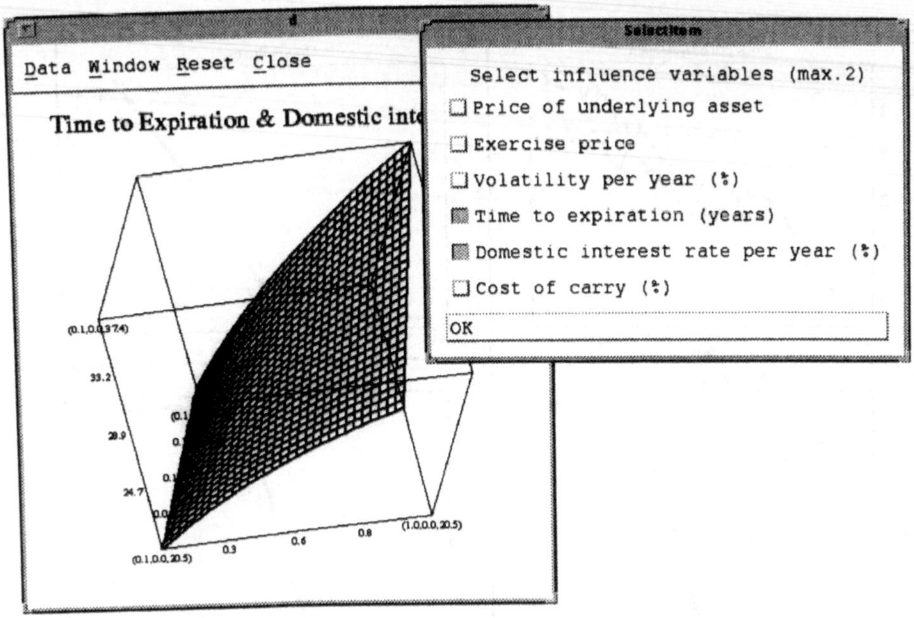

Fig. 2. The macro `influence` in action

The class of ARCH models, built upon the concept of stochastic volatility, is much more flexible than classic models in fitting financial data. In particular, the time discrete model of Duan (1995) using the GARCH(1,1) process is appealing for the theory of option pricing.

A crucial question is the correct specification of the stochastic volatility. Therefore it is necessary to develop programs to model the volatility more flexible, e. g. , for asymmetry of shocks. Examples of such a development are the Threshold ARCH model of Rabemananjara and Zakoian (1993) and the extension of it to T-GARCH by Härdle and Hafner (1997). The program to analyse and compare these models is implemented in **tgarsim()**.

In Figure 3 we present results of a comparison of GARCH and T-GARCH to the Black and Scholes option pricing model. Displayed are the generated processes, option prices, absolute and relative differences to Black and Scholes versus moneyness (S/K with S=stock price, K= strike price). The greyish curve indicates the T-GARCH model.

All programs are available on the net.

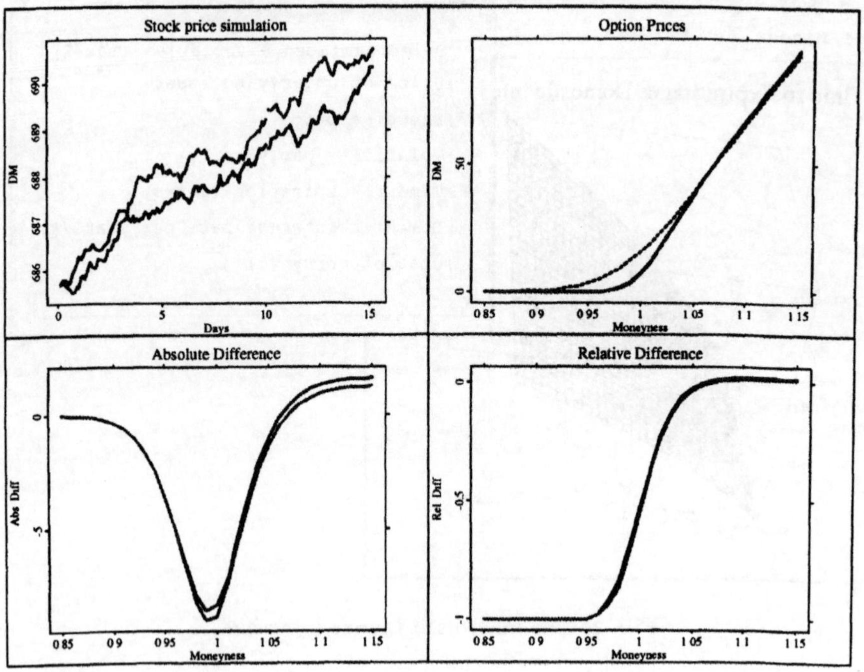

Fig. 3. The result of the macro **tgarsim**

References

Duan, J. C.: The GARCH option pricing model, Mathematical Finance. **5** (1995) 13–32

Gourieroux, C.: ARCH Models and Financial Applications, Springer Verlag, Heidelberg & Berlin. (1997)

Härdle, W. , Hafner, C.: Discrete Time Option Pricing with Flexible Volatility Estimation, forthcoming. (1997)

Rabemananjara, R. , Zakoian, J. M.: Threshold ARCH Models and Asymmetries in Volatility, Journal of Applied Econometrics. **8** (1993) 31–49

Streller, A.: Modellierung von Aktienkursen durch Diffusionen mit Sprüngen – Modelldiskussion und ein Weg zur Schätzung der Parameter, Discussion Paper 56., sfb 373. (1995)

The Durbin-Watson Test for Neural Regression Models

William Holt, Paul Refenes

Department of Decision Sciences, London Business School,
Sussex Place, Regent's Park, London NW1 4SA, England

Abstract. The need to subject neural regression models to residual diagnostic tests is justified. A generalisation of the Durbin-Watson statistic is introduced for neural network regression models. The distribution of the statistic is estimated using approximation and exact methods (Imhof algorithm).

1 Introduction

Neural network models are increasingly used in econometric applications where the parametric relationship between variables is poorly understood. However, the use of such models in critical situations may be limited due to the lack of diagnostic tests available to verify that the final model is a true representation of the underlying functional relationship between variables. A neural network regression model may be described by

$$\boldsymbol{y} = f(\boldsymbol{X}, \boldsymbol{w}) + \boldsymbol{\varepsilon} \tag{1}$$

where $\boldsymbol{y}$ is the vector of observations on the dependent variable, $\boldsymbol{X}$ is the non-stochastic matrix of independent variables, $\boldsymbol{w}$ is a vector of model parameters (weights), f is the non-linear function and $\boldsymbol{\varepsilon}$ is the vector of residual errors. The lack of diagnostic checks is particularly important for neural networks due to the universal approximation properties of the model and estimation(training) inefficiencies such as multiple minima.[1] Both of these properties mean that a neural model may easily estimate a function that is not a valid approximation to the data. Traditional model checking using out- of-sample (test set) validation may be misleading when performed on a small sample and may not be possible if we do not have enough data. In contrast, linear regression models have a wide variety of statistical tests for variable and parameter significance to enable parsimonious models to be produced. Statistics are available to test every model assumption; the F-test may be used to assess the significance of the whole model, t-tests examine the significance of individual model parameters and residual diagnostics (e.g. Durbin-Watson, Breusch- Pagan) test the assumptions of the residual errors. These tests are particularly important in situations where model selection is based on small data sets, possibly without an out-of-sample data set.

[1] These features mean that a neural network may easily overfit the data or mis-estimate the data generating function.

2 The error term in regression models

The error term, ε, is included in a regression model for a number of reasons including :

1. One cannot know all the factors influencing the dependent variable so we must recognise that some independent variables have been omitted from the model; many variables will have only a slight intermittent influence on the system.
2. The data used to estimate the model may not be completely representative of the population and may contain measurement errors, missing observations[2] or outliers.
3. The error term captures approximation error that arises because the functional form of the regression will only approximate the true data generating process; the magnitude of the error term is unknown. Econometric models must always be viewed as a more or less crude approximation to the (usually highly complex) process by which the data is generated.

Thus, the disturbance term may be seen as the summation of many small independent events. These factors are grouped into the residual error term which is generally assumed to be random and small in comparison to the variation in the data.

Residual testing aims to check that the regression errors are indeed random. One way of testing the randomness of residuals is to attempt to model them using a regression model and to check the significance of the model parameters. If these parameters are statistically significant, there is a deterministic component in the residuals that should be incorporated in the model. Any deterministic pattern in residuals shows a relationship in the data that could be captured by a different model and is therefore strong evidence of model mis- specification.

It is important to note that the magnitude of the error term is unknown, making exact calculation of the model parameters impossible. Model parameters are therefore estimated in the presence of noise. The specification of the model must include specific assumptions about the residuals which usually include:

1. $E[\varepsilon_i] = 0$ The error term has an expected value of zero.
2. $E[\varepsilon_i^2] = \sigma^2$ The error term has constant variance.
3. $Cov[\varepsilon_i, \varepsilon_j] = E[\varepsilon_i \varepsilon_j] = 0, \ \forall i, j \ ; \ i \neq j$ The error terms at different points in time are uncorrelated.

These assumptions define a weak stationary variable. In linear regression it is vital that these assumptions are not violated for the parameter estimates

[2] Which may be dealt with by interpolation or complete removal from the model.

to be optimal (i.e. unbiased and efficient). The distribution of the error process is assumed normal as the error is generated by small random shocks and thus by the Law of Large Numbers may be considered normal.

These properties of the errors for a well specified model are usefully shown in the variance-covariance matrix given by

$$E\left[\boldsymbol{\varepsilon\varepsilon}'\right] = \begin{bmatrix} E[\varepsilon_1\varepsilon_1] & E[\varepsilon_1\varepsilon_2] & \dots & E[\varepsilon_1\varepsilon_n] \\ E[\varepsilon_2\varepsilon_1] & E[\varepsilon_2\varepsilon_2] & \dots & E[\varepsilon_2\varepsilon_n] \\ \dots & \dots & \dots & \dots \\ E[\varepsilon_n\varepsilon_1] & E[\varepsilon_n\varepsilon_2] & \dots & E[\varepsilon_n\varepsilon_n] \end{bmatrix} = \begin{bmatrix} \sigma^2 & 0 & \dots & 0 \\ 0 & \sigma^2 & \dots & 0 \\ \dots & \dots & \dots & \dots \\ 0 & 0 & \dots & \sigma^2 \end{bmatrix} \tag{2}$$

The diagonal elements are the variances of the individual disturbances and the off-diagonal terms are the covariances between them. For the errors to be homoscedastic, the diagonal elements should be constant and for no autocorrelation the off-diagonals should be zero.

2.1 Autocorrelation

Probably the most important assumption concerning the errors is that of no autocorrelation. Autocorrelated errors lead to serious underestimation of the standard errors of the regression coefficients (inefficiency) and prediction intervals that are excessively wide. At its most basic, a significant autocorrelation in the residuals represents a deterministic error component that may be included in the model. Violating the assumption of no autocorrelation for residuals may invalidate other model specification tests, for example, the significance of independent variables or model parameters. Kennedy (1992) observes the main causes of autocorrelated disturbances:

Spatial Autocorrelation - Autocorrelation may occur in cross- sectional regression. Consider a cross sectional study of regional economic data. An event in one region may affect adjacent regions due to economic ties, resulting in the error terms between these regions being autocorrelated.

Prolonged Influence of Shocks - In time series data, a disturbance may persist over a number of time periods. For example, a natural disaster will affect economic performance over a long time period.

Inertia - Inertial or psychological factors mean that past actions or phenomenon have a strong effect on current decisions. Examples include market crashes.

Data Manipulation - Much data is smoothed or interpolated which may average the true distrubances over several time periods.

Mis-specification There are two sources of mis-specification that lead to autocorrelated errors

- *Omission of a relevent variable.* A mis-specified model in the form of incorrect functional form or missing independent variables is likely to produce autocorrelated errors.

- *Incorrect functional form.* Although neural models are capable of universal approximation, this problem can arise if the network has insufficient complexity.

It is usual to assume that the first four of these factors are small or have been satisfactorally accounted for, so that autocorrelation may be interpreted only as model mis-specification in the form of missing variables or incorrect functional form.

The Durbin-Watson statistic is the most commonly used test for autocorrelation in regression models and has a solid body of theory. The test considers autocorrelation in the first lag of the residuals for two reasons. Firstly, if autocorrelation is present it is often strongest at the first lag. Secondly, it makes tractable the problem of estimating the large number of off-diagonal elements of the error variance-covariance matrix.

3 The Durbin-Watson statistic

For a single equation linear regression model

$$y_t = \beta_0 + \beta_1 x_{1t} + \beta_2 x_{2t} + \ldots + \beta_k x_{kt} + \varepsilon_t \tag{3}$$

we may test the residual errors for first order autocorrelation. We assume a stationary first order error process given by

$$\varepsilon_t = \rho\varepsilon_{t-1} + \nu_t \ , \quad |\rho| < 1 \tag{4}$$

where ν_t is *i.i.d.* $N(0, \sigma_\nu^2)$ and test the hypothesis $H_0 : \rho = 0$, i.e. the residuals are white noise, against a 1 or 2 tailed alternative hypothesis of autocorrelation[3]. The Durbin-Watson(1950, 1951, 1970) statistic is given by the following equation

$$d = \frac{\sum_{t=2}^{n}(\hat{\varepsilon}_t - \hat{\varepsilon}_{t-1})^2}{\sum_{t=1}^{n}(\hat{\varepsilon}_t)^2} \tag{5}$$

As $\hat{\rho}$ takes values in the range $[-1, 1]$ the DW statistic may take the following values

$d \simeq 4$: $\hat{\rho} = -1$, negative autocorrelation
$d \simeq 2$: $\hat{\rho} = 0$, no autocorrelation
$d \simeq 0$: $\hat{\rho} = 1$, positive autocorrelation.

Values of d close to 2 constitute evidence of an absence of autocorrelation. Values away from 2 suggest first order linear autocorrelation. With knowledge of the distribution of d a statistical test may be constructed to detect autocorrelation.

[3] It may appear that a test statistic is not required as we may simply test ρ for significance from the error model regression. The variance-covariance matrix of the residuals is given by $E[\hat{\varepsilon}\hat{\varepsilon}'] = \sigma_\varepsilon^2 M$ (Johnston, 1984). Thus, even when the null hypothesis is true, i.e. $E[\varepsilon\varepsilon'] = \sigma_\varepsilon^2 I$, the off-diagonal terms in M will not vanish and the residual errors will display autocorrelation. It is therefore impossible to derive an exact finite sample test on $\hat{\varepsilon}$ that would be valid for every X.

3.1 Deriving the Durbin-Watson statistic

We write the single equation linear regression model as

$$\boldsymbol{y} = \boldsymbol{X\beta} + \boldsymbol{\varepsilon} \tag{6}$$

where $\boldsymbol{y}$ is the $n \times 1$ vector of observations on the dependent variable, $\boldsymbol{X}$ is the $n \times k$ matrix on the k explanatory variables $\boldsymbol{X}_i$ of rank k,
$\boldsymbol{\beta}$ is the $k \times 1$ vector coefficients on the k explanatory variables,
$\boldsymbol{\varepsilon}$ is the $n \times 1$ vector of white noise errors, assumed to have the characteristics $E[\boldsymbol{\varepsilon}] = \boldsymbol{0}$, $V[\boldsymbol{\varepsilon}] = E[\boldsymbol{\varepsilon}^{\boldsymbol{i}}\boldsymbol{me\varepsilon}] = \sigma^2\boldsymbol{\Omega}$, where the diagonal elements of $\boldsymbol{\Omega}$ are constant and the off-diagonals are non-zero.

The least squares estimates of $\boldsymbol{\beta}$ are given by (e.g. Greene, 1993)

$$\hat{\boldsymbol{\beta}} = (\boldsymbol{X}'\boldsymbol{X})^{-1}\boldsymbol{X}'\boldsymbol{y} \tag{7}$$

and the estimated residual errors by

$$\hat{\boldsymbol{\varepsilon}} = \boldsymbol{y} - \boldsymbol{X}\hat{\boldsymbol{\beta}} \tag{8}$$

Substituting 7 into 8 gives the least squares estimator of $\boldsymbol{\varepsilon}$

$$\begin{aligned} \hat{\boldsymbol{\varepsilon}} &= \boldsymbol{y} - \boldsymbol{X}(\boldsymbol{X}'\boldsymbol{X})^{-1}\boldsymbol{X}'\boldsymbol{y} \\ &= [\boldsymbol{I} - \boldsymbol{G}]\boldsymbol{y} \qquad (9) \\ &= \boldsymbol{My} \\ &= \boldsymbol{M}(\boldsymbol{X\beta} + \boldsymbol{\varepsilon}) \\ \hat{\boldsymbol{\varepsilon}} &= \boldsymbol{M\varepsilon} \qquad (10) \end{aligned}$$

where $\boldsymbol{I}$ is an identity matrix of order n and $\boldsymbol{M}$ is a symmetric idempotent matrix with the property $\boldsymbol{MX} = \boldsymbol{0}$. We call $\boldsymbol{G}$ the influence matrix and $\boldsymbol{M}$ the projection matrix. Following Greene(1993) pp10-11, we may write

$$\sum_{t=1}^{n} \hat{\varepsilon}_t^2 = \hat{\boldsymbol{\varepsilon}}'\hat{\boldsymbol{\varepsilon}} \tag{11}$$

$$\sum_{t=2}^{n} (\hat{\varepsilon}_t - \hat{\varepsilon}_{t-1})^2 = \hat{\boldsymbol{\varepsilon}}'\boldsymbol{A}\hat{\boldsymbol{\varepsilon}} \tag{12}$$

where $\boldsymbol{A}$ is the real symmetric $n \times n$ matrix

$$\boldsymbol{A} = \begin{bmatrix} 1 & -1 & 0 & . & . & 0 \\ -1 & 2 & -1 & 0 & . & 0 \\ . & & & & & \\ . & & & & & \\ 0 & . & 0 & -1 & 2 & -1 \\ 0 & . & . & 0 & -1 & 1 \end{bmatrix} \tag{13}$$

which controls the lag at which the test is performed. We may therefore write the Durbin-Watson statistic as

$$d = \frac{\hat{\varepsilon}' A \hat{\varepsilon}}{\hat{\varepsilon}' \hat{\varepsilon}} \tag{14}$$

This is the value of the Durbin-Watson statistic which we calculate from the model residuals. Using the result $\hat{\varepsilon} = M\varepsilon$ we may write Equation 14 as

$$d = \frac{\varepsilon' M' A M \varepsilon}{\varepsilon' M' M \varepsilon'} = \frac{\varepsilon' M A M \varepsilon}{\varepsilon' M \varepsilon} \tag{15}$$

This expression gives the exact value of the Durbin-Watson statistic and implies that the test is data specific, i.e. the distribution of d needs to be evaluated for each data set X. A single critical value of the test cannot be used since the values of ε and therefore d depend on the values taken by the explanatory variables in the model equation. The expression obtained is not enough to calculate the exact value of the Durbin-Watson statistic as the true errors ε are unknown. We therefore must derive a computable form of d to obtain the true distribution of d, providing the significance points of the test.

3.2 The distribution of the Durbin-Watson statistic

In this section we present the main procedures to calculate the distribution of d. The exact distribution of the statistic may be computed using an algorithm such as the Imhof(1961) method. Alternatively, a beta distribution approximation may be used. Both of these methods (especially Imhof) require reasonably complicated computational procedures to be calculated for each regression matrix X to be tested. For these reason, Durbin and Watson derived tables, based on the fact that the true value of d can be shown to lie between two beta distributions independently of the particular set of regressors.

Exact methods We first examine the exact distribution calculated by the Imhof(1961) method. A new error term ζ is defined as

$$\zeta = P' \varepsilon \tag{16}$$

which are independent and distributed with $N(0, \sigma_\varepsilon^2)$ as P is an $n \times n$ orthonormal matrix. The matrix P diagonalises both quadratic forms MAM and M. The statistic can therefore be written as

$$d = \frac{\varepsilon' M A M \varepsilon}{\varepsilon' M \varepsilon} = \frac{\zeta' P' M A M P \zeta}{\zeta' P' M P \zeta} = \frac{\sum_{j=1}^{n-k} \lambda_j \zeta_j^2}{\sum_{j=1}^{n-k} \zeta_j^2} \tag{17}$$

where λ_j are the $n-k$ non-zero eigenvalues of $\boldsymbol{MA}$[4]. Using the fact that $P[d > d_0] = P[Q > 0]$ where

$$Q = \sum_{j=1}^{n-k} \lambda_j \zeta_j^2 - d_0 \sum_{j=1}^{n-k} \zeta_j^2 = \sum_{j=1}^{n-k} (\lambda_j - d_0)\zeta_j^2 \tag{18}$$

and d_0 is a particular calculated value of d, L'Esperance *et al*(1976) showed that the distribution of Q may be represented in the following way. The characteristic function of Q is given by

$$\Phi(t) = \frac{1}{\prod_{j=1}^{n} (1 - 2ia_j t)^{\frac{1}{2}}} \tag{19}$$

where $a_j = \lambda_j - d_0$ for $d_0 < 2$ and $a_j = \lambda_j - (4 - d_0)$ otherwise and $i = \sqrt{-1}$. The cumulative distribution function is given by

$$F(0) = P[d > d_0] = P[Q > 0] = \frac{1}{2} + \frac{1}{2\pi} \int_0^\infty \frac{\Phi(-t) - \Phi(t)}{it} dt \tag{20}$$

Thus a p-value may be obtained for a particular regression matrix. Durbin and Watson(1971) present another method based on the Fourier inversion integral. The p-value may be represented as

$$P[Q > 0] = \frac{1}{2} - \frac{1}{\pi} \int_0^\infty \frac{\sin\{\theta(t)\}}{t\rho(t)} dt \tag{21}$$

where

$$\theta(t) = \frac{1}{2} \sum_{j=1}^{m} \tan^{-1}\{(\lambda_j - d_0)t\} \tag{22}$$

and

$$\rho(t) = \prod_{j=1}^{m} \{1 + (\lambda_j - d_0)^2 t^2\}^{\frac{1}{4}} \tag{23}$$

With both of these procedures, problems include the choice of step size t and the truncation of the range of the integral. They are also potentially computationally expensive, especially for large n and when the step size of integration is small. For a discussion of numerical integration and associated problems, see Press et al. (1992).

[4] For the linear regression model, using the fact that the roots of the product of two matrices are independent of the order of multiplication and M is idempotent, $M^2A = MA$. Note that in the neural networks case, λ_j are the n eigenvalues of MAM.

Approximation using a beta distribution Alternatively, Durbin and Watson (1951) approximate the density function of d on the basis of the first two moments by a beta distribution:

$$\frac{1}{B(p,q)}\chi^{p-1}(1-\chi)^{q-1} \quad , 0 \le \chi \le 1 \; ; \; 0 \;\; otherwise \tag{24}$$

As the DW test gives values in the range $[0,4]$ we approximate the density function $\frac{1}{4}d$ by

$$\frac{1}{B(p,q)}\left(\frac{1}{4}d\right)^{p-1}\left(1-\frac{1}{4}d\right)^{q-1} \tag{25}$$

which has mean and variance

$$E[d] = \frac{4p}{(p+q)} \tag{26}$$

$$V[d] = \frac{16pq}{(p+q)^2(p+q+1)} \tag{27}$$

The moments of d may be calculated using

$$E[d^s] = \frac{E(\sum_{i=1}^{n-k} v_i\zeta_i^2)^s}{E(\sum_{i=1}^{n-k}\zeta_i^2)^s} \tag{28}$$

for any positive integer s, as Durbin and Watson (1950, p419) showed that the numerator and denominator are independent (the moments of the ratio are the ratios of the moments). So, the first and second moments (mean and variance) are given by

$$E[d] = \frac{1}{n-k}\sum_{i=1}^{n-k} v_i \tag{29}$$

and

$$V[d] = \frac{2\sum_{i=1}^{n-k}(v_i E[d])^2}{(n-k)(n-k+2)} \tag{30}$$

where

$$\sum_{i=1}^{n-k} v_i^s = tr(\boldsymbol{MA})^s \tag{31}$$

By equating these expressions with those for the beta distribution we obtain p and q. We have enough information to calculate the distribution of the Durbin-Watson statistic. However, this is specific to a particular $\boldsymbol{X}$ and would have to be recalculated for each data set and model which is a computationally expensive procedure. We would like to obtain a distribution which is independent of $\boldsymbol{X}$. Durbin and Watson's (1950,1951) solution was to calculate upper and lower bounds on d that do not depend on $\boldsymbol{X}$. Thus the true value of the Durbin-Watson statistic for any $\boldsymbol{X}$ lies between two beta distributions

$$d_L \leq d \leq d_U \tag{32}$$

where

$$d_L = \frac{\sum_{i=1}^{n-k} \lambda_i \zeta_i^2}{\sum_{i=1}^{n-k} \zeta_i^2} \tag{33}$$

and

$$d_U = \frac{\sum_{i=1}^{n-k} \lambda_{i+k-1} \zeta_i^2}{\sum_{i=1}^{n-k} \zeta_i^2} \tag{34}$$

where λ_i are the eigenvalues of $\boldsymbol{A}$. It is important to note that these bounds were obtained under the assumption that $\boldsymbol{X}$ contains a vector of ones (i.e. a constant) and that no other column of $\boldsymbol{X}$ is an eigenvector of $\boldsymbol{A}$.

There is now a region where the test gives an inconclusive result, where we are incapable of accepting or rejecting the null hypothesis. In this situation, three options are available: i) Compute the exact distribution using the Imhof method, ii) Use the the beta approximation to provide the significance points & iii) Use an alternative test (e.g. the runs test).

4 The Durbin-Watson test for non-linear models

The Durbin-Watson statistic was first generalised to non-linear models by White (1992). The A.N.D. (Approximate Nonlinear Durbin-Watson) statistic generalises the influence matrix $\boldsymbol{M}$ by computing the matrix of derivatives $\boldsymbol{Z}(\boldsymbol{\beta}) = \partial f(\boldsymbol{X}, \boldsymbol{\beta})/\partial \boldsymbol{\beta}'$ to produce a non-linear estimate of $\boldsymbol{M} = \boldsymbol{I} - \boldsymbol{Z}(\hat{\boldsymbol{\beta}}) \left[\boldsymbol{Z}(\hat{\boldsymbol{\beta}})'\boldsymbol{Z}(\hat{\boldsymbol{\beta}})\right]^{-1} \boldsymbol{Z}(\hat{\boldsymbol{\beta}})'$. The eigenvalues of this expression can be used to estimate the exact distribution of d. White's approach is not generally applicable to neural models for three reasons; i) he only considers parametric models, ii) the model cannot be biased and iii) no regularisation term is included. In this paper, a more general influence matrix is derived and the Durbin-Watson statistic modified to account for the limitations of the A.N.D. test. The proposed statistic in this paper may thus be viewed as a more general version of the A.N.D. test.

5 Generalising the projection matrix for neural models

In the linear regression model all parameters contribute towards the degrees of freedom of the model. In contrast, neural network models are overparameterised, i.e. a certain number of weights are not controlled by the data and their value will be predominantly controlled by the initial weight matrix. In the absence of any data, the model parameters should not differ from zero. Information in the data drives so called *well defined parameters* away from the small random initial values chosen for the minimisation algorithm. It is clear then that the number of model parameters is a poor estimate of the number of well defined parameters. It is therefore important to estimate the number of weights that are controlled by the data. In this section we state an expression for the projection matrix $\boldsymbol{M}$ and the influence matrix $\boldsymbol{G}$ presented by Moody (1992) that is valid for neural networks with regularisation terms. We thus obtain an estimate of the degrees of freedom in the model and may generalise the Durbin-Watson test to neural models.

The parameters of the model are estimated such that the error function U is minimised

$$U(\boldsymbol{X}, \hat{\boldsymbol{\beta}}, \boldsymbol{y}; \lambda) = T(\hat{\boldsymbol{\beta}}) + \lambda R(\hat{\boldsymbol{\beta}}) \tag{35}$$

where $T(\hat{\boldsymbol{\beta}}) = \frac{1}{2}\sum_{i=1}^{n}(y_i - f(\boldsymbol{X}'_{[i]}, \hat{\boldsymbol{\beta}}))^2$, R is the regularisation function such as weight decay and λ is a constant which controls the trade-off between residual error and function complexity. Moody (1992) shows that a generalised expression for the projection matrix $\tilde{\boldsymbol{M}}$ is given by

$$\tilde{\boldsymbol{M}} = \boldsymbol{I} - \tilde{\boldsymbol{G}} = \boldsymbol{I} - \boldsymbol{T}_{y\hat{\beta}}(\boldsymbol{U}_{\hat{\beta}\hat{\beta}})^{-1}\boldsymbol{T}_{\hat{\beta}y} \tag{36}$$

When the second order derivatives are continuous for all $\boldsymbol{y}, \hat{\boldsymbol{\beta}}$, we can replace $\boldsymbol{T}_{\hat{\beta}y}$ by $\boldsymbol{T}'_{y\hat{\beta}}$ in $\tilde{\boldsymbol{M}}$

$$\tilde{\boldsymbol{M}} = \boldsymbol{I} - \boldsymbol{T}_{y\hat{\beta}}(\boldsymbol{U}_{\hat{\beta}\hat{\beta}})^{-1}\boldsymbol{T}'_{y\hat{\beta}} \tag{37}$$

and both $\tilde{\boldsymbol{M}}$ and $\tilde{\boldsymbol{G}}$ will be symmetric. Therefore, the degrees of freedom of the neural model are given by

$$\begin{aligned} tr(\tilde{\boldsymbol{M}}) &= tr(\boldsymbol{I}_n)tr\left(\boldsymbol{T}'_{y\hat{\beta}}\boldsymbol{T}_{y\hat{\beta}}(\boldsymbol{U}_{\hat{\beta}\hat{\beta}})^{-1}\right) \\ &= n - p_e \end{aligned} \tag{38}$$

where p_e is the effective number of well defined parameters.

6 The modified Durbin-Watson statistic

A modification to the Durbin-Watson statistic may now be written as

$$d = \frac{\boldsymbol{\varepsilon}'\tilde{\boldsymbol{M}}'\boldsymbol{A}\tilde{\boldsymbol{M}}\boldsymbol{\varepsilon}}{\boldsymbol{\varepsilon}'\tilde{\boldsymbol{M}}'\tilde{\boldsymbol{M}}\boldsymbol{\varepsilon}'} \tag{39}$$

where

$$\tilde{\boldsymbol{M}} = \boldsymbol{I} - \boldsymbol{T}_{y\hat{\beta}}(\boldsymbol{U}_{\hat{\beta}\hat{\beta}})^{-1}\boldsymbol{T}'_{y\hat{\beta}} \tag{40}$$

We now examine by the assumptions concerning the original Durbin-Watson statistic.

The model equation includes an intercept term For neural networks, this assumption is not violated due to their universal approximation characteristics (White, 1990), i.e. if there is a constant in the true function, it will be in the correctly specified neural model.

The error process is 1st order autoregressive, i.e. AR(1) No violation as we assume this in the non-linear case.

The model equation does not include lagged dependent variables e.g. y_{t-1}. This is potentially most serious violation due to the widespread use of lagged independent variables as network inputs.

None of the explanatory variables are stochastic No violation as this is also an assumption of neural networks. However, we must check that no irrelevent variables have been included in the model by checking that the Hessian is positive definite.

There are more than 15 observations This assumption is only relevent if the Durbin-Watson tables are used and will not affect results when the exact distribution is calculated. However, it is inadvisable to use few observations with neural models due to model overparameterisation and estimation inefficiencies.

The residual error terms are weak stationary No violation as this is also an assumption of neural networks.

Calculation of the probability density of the statistic may be carried out in a similar way as the linear case, i.e the Imhof procedure or Durbin and Watson's method of approximation to a beta distribution. The statistic is written

$$d = \frac{\boldsymbol{\zeta' P' \tilde{M}' A \tilde{M} P \zeta}}{\boldsymbol{\zeta' P' \tilde{M}' \tilde{M} P \zeta}} = \frac{\sum_{j=1}^{n} \lambda_j \zeta_j^2}{\sum_{j=1}^{n} \zeta_j^2} \tag{41}$$

where λ_j are the eigenvalues of $\boldsymbol{\tilde{M}' A \tilde{M}}$. Note that all the eigenvalues of the expression must be calculated. The distribution can then be computed using the method in Equation 21. Note that the calculated distribution is exact and will only be incorrect due to mis-estimation of $\boldsymbol{M}$.

7 Conclusion

We have introduced a generalisation of the Durbin-Watson test statistic which may be used with feedforward neural networks to detect for the presence of first order autocorrelation. The exact distribution of the test may be calculated with moderate computational cost.

It is hoped that the use of this statistic as part of a diagnostic testing procedure will lead to both greater care in the specification of neural models and aid their acceptance in critical applications. This paper is part of ongoing work at London Business School to produce a complete set of residual diagnostics for neural regression models.

Acknowledgments

Financial support from BT Laboratories and the U.K. Economic and Social Research Council (E.S.R.C.) is gratefully acknowledged.

References

Amari S., 1995, Learning and statistical inference,
Handbook of Brain Theory and Neural Networks, MIT Press.

Durbin J. and G. Watson, 1950, Testing for serial correlation in least squares regression, I,
Biometrika 37, 409-428.

Durbin J. and G. Watson, 1951, Testing for serial correlation in least squares regression, II,
Biometrika 38, 159-178.

Durbin J. and G. Watson, 1971, Testing for serial correlation in least squares regression, III,
Biometrika 58, 1-19.

Greene W., 1989, Econometric Analysis,
Prentice Hall International.

Johnston J., 1984, Econometric Methods,
McGraw-Hill Book Company.

L'Esperance W., D. Chall and D. Taylor, 1976, An algorithm for determining the distribution function of the Durbin-Watson test statistic,
Econometrica 44, 1325-1326.

Moody J., 1992, The effective number of parameters: An analysis of generalization and regularization in nonlinear learning systems,
in: Moody J., S. Hanson and R. Lippmann (eds.), Advances in Neural Network Processing Systems 4, 847-854. San Mateo,CA: Morgan Kaufmann.

White H., 1990, Connectionist nonparametric regression: Multilayer feedforward networks can learn arbitrary mappings,
Neural Networks 3, 535-549.

White K., 1992, The Durbin-Watson test for autocorrelation in nonlinear models,
Review of Economics and Statistics 74, 362-365.

Neuro-Fuzzy Methods in Finance Applied to the German Stock Index DAX

Rudolf Kruse[1], Stefan Siekmann[2], Ralph Neuneier[2], and Hans Georg Zimmermann[2]

[1] Otto-von-Guericke University Magdeburg
Faculty of Computer Science
D-39106 Magdeburg, Germany
phone: +49 391-18 706

[2] Siemens AG, Corporate Technology
D-81730 Munich, Germany
Stefan.Siekmann@mchp.siemens.de
phone: +49 89-636 44495

Abstract. We present an extended neuro-fuzzy system, combined with a semantic-preserving backpropagation-based learning algorithm, which makes effective use of prior knowledge and of historical data. After the network is initialized with a set of rules, the learning algorithm optimize the rule base without destroying the initial semantic. Due to a sparse initial network structure the effective number of parameters is small which prevents the network from overfitting.

Our extended architecture enables us not only to change the shape and the position of the membership functions but also to apply typical pruning algorithms to delete or insert single premises. With insertion of premises, the rule base is improved by creation of new rules. Deletion of premises leads to a further reduction of the system complexity. Both strategies have a positive effect on the generalization performance. These methods are implemented in the software enviroment for neural networks SENN of Siemens Nixdorf Advanced Technologies. We have tested the neuro-fuzzy approach on two financial series. Here we present the encouraging results on the task to predict daily returns of the German Stock Index DAX.

Keywords. neuro-fuzzy, daily prediction of DAX, semantic-preserving learning algorithm, pruning of rules and premises, rule generation

1 Application of Neuro-Fuzzy Methods in Finance

In finance, there are many traders who are able to express their experience and theories in form of fuzzy rules. Such expert knowledge is often ignored when applying classical statistic methods as well as neural networks to time series prediction. An interesting approach is to combine fuzzy methods with neural network learning algorithms (Nauck D. et al (1997)) and to study so called neuro-fuzzy architectures. The main problem with straight forward combination is that typical neuro-fuzzy systems destroy their initial rule base during learning or have to use penalties which restrict the optimization (Tresp V. et al (1993)). Our architecture is combined with a semantic-preserving learning algorithm which ensures that the rules always allow a meaningful interpretation after learning. Our experience is that traders highly appreciate these useful insights in the dynamics of the system because the approach relates their theories to the real data of the financial markets. Furthermore, it turns out that the integration of a priori knowledge into a neural network can be useful to reduce the effect of overfitting. Overfitting and bad generalization performance are typical in the field of financial analysis because the data is very noisy and often not available to the necessary extent (Finnoff W. et al (1992)). Since rules given by an human expert are generally low dimensional and each rule takes only a small number of input variables into account, the resulting architecture of our neuro-fuzzy system is very sparse. The complexity of the network is further reduced by the mentioned semantic-preserving learning algorithm which forces the parameters to meet some constraints. Furthermore, the applicability of pruning techniques to delete non relevant premises and rules results in a even smaller number of parameters.

2 Fuzzy-Rules and Neural Networks

We indicate how a rule base is transformed into network parameters. Assume that an expert has formulated a set of rules of the form

Rule R_i:		
IF	Δdax_t	$=$ *increasing*
AND	Δusd_t	$=$ *increasing*
THEN	Δdax_{t+1}	$=$ *increasing*

where Δdax_t presents the change of the DAX and Δusd_t the change of the US-\$/DM-exchange rate at time step t. Like Tresp V. et al (1993), Takagi T., Sugeno M. (1985) and Wang L. X., Mendel J. M. (1991) we transform a rule base consisting of a finite number of fuzzy rules $R_1,..,R_i,...$ into a neural network of the form of eq.(1). Each rule R_i corresponds to a basis function b_i (i.e. neuron) in the hidden layer. The output y for a given input vector $\underline{x} = (x_1, ..., x_j, ...)$ is the weighted average of the conclusions $w_i(\underline{x})$ where the weighting factor depends on the activity level of the rules or basis functions b_i due to the input $\underline{x}$.

$$y(\underline{x}) = \frac{\sum_i w_i(\underline{x}) b_i(\underline{x})}{\sum_i b_i(\underline{x})} \quad \text{with} \quad b_i(\underline{x}) = \kappa_i \Pi_j m_{ij}(x_j) \tag{1}$$

The weight κ_i, which express the measure of "importance" of the rule i (Kruse R. et al (1991)), are positive real numbers with their sum being equal to some a priori fixed constant. In this way unimportant or inconsistent rules are pruned automatically as the learning algorithm drives κ_i of such rules to zero (Tresp V. et al (1993), Neuneier R., Tresp V. (1994)). The linguistic expressions or fuzzy sets, e. g. *increasing* and *decreasing* are described by one-dimensional membership functions, e. g. Gaussian or logistic functions.

$$m_{ij} = \exp\left(-0.5\frac{(x_j-\mu_{ij})^2}{\sigma_{ij}^2}\right) \text{ (Gaussian)}$$
$$m_{ij} = \frac{1}{1+\exp\left(-4\frac{(x_j-\mu_{ij})}{\sigma_{ij}}\right)} \text{ (logistic)} \qquad (2)$$

The location of a membership function is determined by the parameter μ_{ij}, which represents the mean of the Gaussian resp. the turning point of the logistic function. The scope of a membership function depends on $\frac{1}{\sigma_{ij}}$, which is the inverse of the deviation of a Gaussian membership function, resp. is proportional to the slope at the turning point of the logistic function. Within this very general formulation we are free in specifying the form of a parameterized membership function membership function. If we define the conclusions to be independent of the input $\underline{x}$, $w_i(\underline{x}) = w_i$, and the membership functions to be Gaussians (Fig. 1, b) we obtain a typical normalized radial basis function network like eq. 3 (Neuneier R., Tresp V. (1994)).

$$b_i(\underline{x}) = \kappa_i \Pi_j m_{ij}(x_j) = \kappa_i \exp\left(-0.5\sum_j\left(\frac{x_j-\mu_{ij}}{\sigma_{ij}}\right)^2\right) \qquad (3)$$

Especially if describing economical relationships, monotonous fuzzy statements expressed by a logistic membership function may be more appropriate (Fig. 1, a).

3 The Extended Neuro-Fuzzy Architecture

In order to eliminate single premises of rules, we extend the neuro-fuzzy approach such that the resulting neural net has an additional layer which represents the type of membership functions available for each input signal. This "membership function layer" is connected with the "rule layer" (Fig. 2. If rule b_i take the fuzzy set m_{kj} into account then the weight p_ji is set to 1, otherwise $p_ji = 0$, which means that there is no connection. k represents the index of the fuzzy sets for each input variable x_j. This network works like a normal feed forward net with one input cluster, one output cluster and two hidden layers.

4 The Semantic-Preserving Learning Algorithm

After initializing the network with expert knowledge, training on actual data can improve the network performance. In the following we assume that

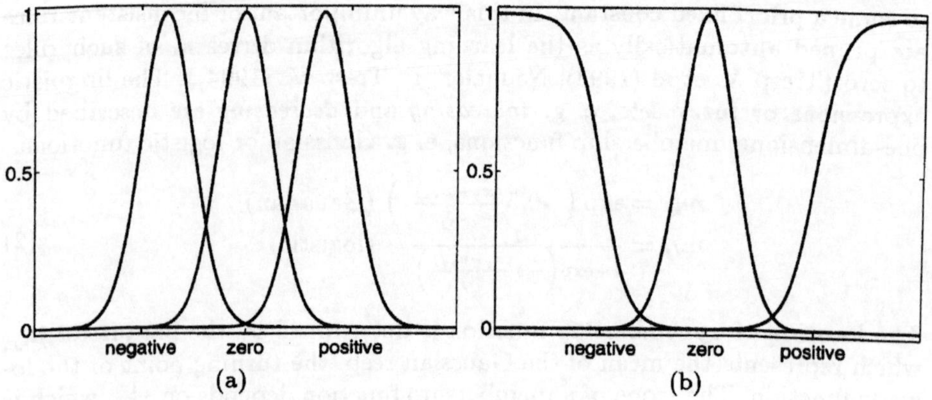

Fig. 1. Gaussian membership function (a) and in combination with logistic membership function (b)

the learning algorithm is based on a gradient descent technique (i.e. backpropagation, conjugate gradient and others, see Zimmermann H. G. (1994)). Normally, the learning process will destroy the initial rules. In Tresp V. et al (1993), several ways are described to ensure that the initial knowledge is preserved during training. These methods use penalties, stop-training methods or dynamical allocation of additional neurons in order to avoid that the parameters of the memerbship functions move too far from their original values. We propose a different approach. In addition to the rules, the user can express two types of relations between the parameters of the membership functions. First, one may demand that the parameters of different membership functions to have the same value, i. e. belong to one equivalence class. For example, a rule describing a trend behavior looks like

IF	Δdax_{today}	$= increasing$
AND	$\Delta dax_{yesterday}$	$= increasing$
THEN	$\Delta dax_{tomorrow}$	$= increase$
WEIGHTING 0.8,		

where Δdax_{today} measures the relative difference between the last two days of the DAX. Since the fuzzy set *increasing* concern similar time series, the user may link all or some of the parameters of these membership function, i.e. the parameter μ:

$$\mu^{increasing}_{\Delta dax_{today}} = \mu^{increasing}_{\Delta dax_{yesterday}} \tag{4}$$

Note, that the affected fuzzy sets can be part of different rules.

Second, with order constraints we mean that the position parameters (i. e. the center μ_i for a Gaussian membership function) shall be kept in their original semantic order or within some bounds, i.e.:

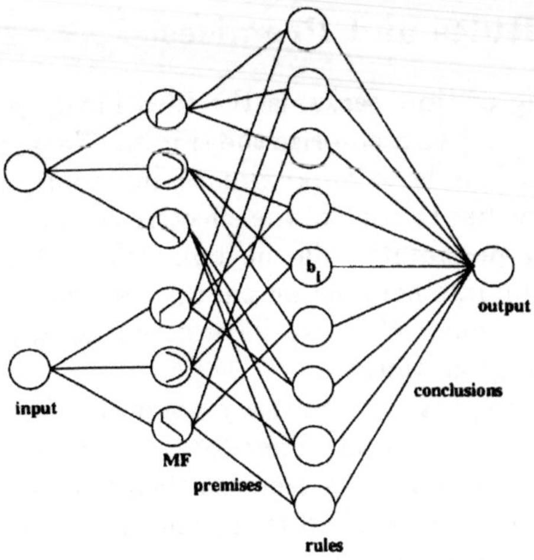

Fig. 2. The neuro-fuzzy architecture

$$-5 < \mu_{\Delta dax}^{decrease} < \mu_{\Delta dax}^{zero} < \mu_{\Delta dax}^{increase} < 5 \tag{5}$$

After one training cycle the parameters are adapted in a direction computed by the learning algorithm. Before the new training cycle starts we have to insure that the user given constraints are fulfilled. Instead of using general constraint-optimization methods which would heavily enlarge the computing time, we solve that problem by ranking the constraints and applying a variation of a active set technique (Gill P. E. et al (1981)). If two parameters have reversed the original order, they must be corrected. The length of the correction step of a parameter depends on how many members exist in its equivalence class. As the correction can lead to a situation where other constraints are not met, the process might have to be iterated. If all constraints can not be solved within a given number of iteration the user is given a warning that the rule base is not consistent with the data. Then one can change the rules or release some constraints. In our experiments the algorithm needs rarely more than one iteration and has always converged.

We want to stress that in contrast to the methods in Tresp V. et al (1993), our technique preserves the initial semantic given by the relations among the fuzzy expressions in the rule base. The semantic-preserving learning algorithm ensures that reinterpretation after training is always possible and can give useful insights for an improved system understanding. In addition, the constraints reduce the effective number of parameters which can avoid overfitting.

5 Pruning of Rules and Premises

The task of modeling of time series is the identification of the underlying stochastic process and not to memorize the training data. Since the proposed neuro-fuzzy approach is able to approximate any continuous function with arbitrary accuracy, one has to control the effective complexity of the model to avoid overfitting. One possibility to achieve models with high generalization performance is to eliminate neurons or parameters which are only approximating the noise in the data. Now, we show how typical pruning algorithms apply to the neuro-fuzzy network.

The positive rule weights κ_i are always unconstrained except that their sum has to be equal to a constant. Therefore, if the learning algorithm increases a rule weight, others have to decrease their values. Ideally, this might lead to automatical elimination of unimportant or inconsistent rules. Alternatively, the user can prune rules with low κ.

Due to the extended neuro-fuzzy architecture it is possible to use typical (inexpensive) pruning algorithms (Finnoff W. et al (1992), Zimmermann H. G. (1994), Tresp V. et al (1996)) which compute a test value for each premise. Low values indicate that these premises are not relevant or inconsistent with the data. Based on these statistics the user can delete some premises, say 10%, reducing the complexity of the neuro-fuzzy system. Thereafter, one trains the system again to convergence or until the error on a cross-validation set increases again (Finnoff W. et al (1992)). These two steps might be iterated until one achieves a minimal network with good performance. Note, that the semantic-preserving learning algorithm still applies and keeps the optimized rule base interpretable.

It may happen that after pruning of premises rules have been derived which are identical. These rules are merged by adding their rule weights κ. It is also possible that the resulting rules differ only in their conclusions. The rule with lower κ could be deleted or the user can leave these rules unchanged because inconsistent knowledge can appear in financial analyses. The merging of inconsistent rules is only possible by creating new conclusions and therefore changing the initial semantic of the rule base. Typically, an initial rule base for financial analysis has many rules. We also used the pruning algorithm to find some misleading initial rules which might be the result of an artifact (i.e. a typing error, missing data, etc.).

Another (expensive) possibility to prune premises is to compute test values for all active premises (weight p_{ij}=1) as following :

$$t_{ij} = E(\mathrm{NN}(x)|p_{ij} = 0) - E(\mathrm{NN}(x)|p_{ij} = 1) \tag{6}$$

where $E(.)$ is the error function or another cost function. If $t_{ij} <= 0$ then the weight p_{ij} should be set to zero, because the error of the system will be reduced. It is also possible to insert premises (weight $p_{ij} = 0$) into the rule base:

$$t_{ij} = E(\mathrm{NN}(x)|p_{ij} = 1) - E(\mathrm{NN}(x)|p_{ij} = 0). \tag{7}$$

If $t_{ij} <= 0$ then the weight p_{ij} should be inserted. An iteration of deletion and insertion of those premises, which have an positive effect on the error function, will improve the initial rule base. After pruning the network has to be trained on historical data again. In future we use this pruning algorithm for systematic rule generation.

6 Experiments

We tested our neuro-fuzzy approach intensively on two financial time series. The experiments with the daily returns of the US-$/DM exchange rates are described in Neuneier R. and Zimmermann H. G. (1995). Here we present the results of the experiments to the German stock index (see also Dichtl H. (1996), Zimmermann H. G. et al (1996), Siekmann S. et al (1997)). For the experiments, we used the neural network simulator SENN of Siemens Nixdorf Advanced Technologies, SNAT, because it integrates all mentioned methods. See webpage http://www.snat.de/pns.htm for more information.

6.1 The German Stock Index DAX

The recently growing dynamics of the international stock markets has made portfolios consisting of such assets more attractive than bonds or other investments. In this article, we focus on the German stock index DAX because the 30 companies forming the index are responsible for about 70% of the turnover at the stock market in Frankfurt. Since the DAX is a weighted mixture of stocks it behaves like a real portfolio.

The task is to predict the daily returns (relative differences) of the DAX given 12 inputs. These input variables describe a so called fundamental model using the DAX itself and other external time series (table 1). We have chosen

Table 1. Inputs for predicting the daily DAX returns.

DAX	composite DAX
German 3 month interest rates	return Germany
Morgan Stanley index Germany	Dow Jones industrial index
US-$/DM	US treasury bonds
gold price	Nikkei index Japan
Morgan Stanley index Europe	price earning ratio

these time series because the data is available on a daily basis and the number of national and international variables are balanced. More details about the selection of relevant input information can be found in Dichtl H. (1996). These inputs are preprocessed by computing the relative differences, i. e.:

$$\Delta dax_t = \left(\frac{\text{DAX}_t}{\text{DAX}_{t-1}} - 1 \right) * 100. \tag{8}$$

DAX_t represents the German Stock Index at time step t. The data has been transformed as described in eq. 8. Due to the time difference the information of the US market is lagged by one day.

6.2 The Initial Rule Base

For each input variable of table 1 three fuzzy sets are defined: *decreasing*, *stable* and *increasing*. Only the fuzzy set *stable* is realized by a Gaussian membership function, the others by logistic membership functions. The parameters of the membership functions are defined by analyzing the distributions of the variables. The center of the fuzzy set *stable* is set to zero and could not be adapted. By relating the input Δdax_t to each of the 11 other input variables using all combination of the fuzzy sets we construct $99 = 9 * 11$ rules. Note, that only pairwise correlations are used. The rules are constructed according assumed structural relationships (for details, see Dichtl H. (1996)). Due to the fact that each rule takes only two inputs into account, the neural network has only 36 fuzzy sets and 99 rules.

6.3 Partitioning of Data

The partitioning of the data in training, cross-validation, and test set is shown in table 2.1. The cross-validation set is not used for training of the system, but gives a hint when to stop training (Finnoff W. et al (1992)). If the error on this portion of the data starts to increase we usually stop training and try to reduce the complexity of the network by pruning. Afterwards we train again until the next increase of the error. To evaluate the resulting neuro-fuzzy

Table 2. Partitioning of the data.

Set	Start	End	Number of patterns
Training	1986/1/1	1993/2/28	1888
Cross-Validation (CV)	1993/3/1	1994/2/28	261
Test	1994/3/1	1997/3/25	800

models we construct a trading strategy going long if the network prediction is positive and selling short if the prediction is negative.

6.4 Benchmarks

The benchmarks which are used to compare the performance are the following trading systems respectively models:

1. **Buy&Hold:** buy the DAX at the beginning of the test set and sell it at the end.This strategy assumes an efficient capital market which does not allow excess return because the conditional expectation of the returns are zero. The Buy&Hold strategy gains only by exploiting the market trend.
2. **naive prediction:** buy or hold the DAX if the last difference is positive and sell otherwise. The naive prediction assumes that the market behaves like a random walk.
3. **linear model:** A linear model with 36 inputs.
4. **MLP:** We construct a (state of the art) neural network with 36 inputs, 40 hidden neurons using the interaction layer topology of Zimmermann H. G., Weigend A. (1997) and the adaptive noise training scheme of Zimmermann H. G. et al (1997) to achieve an optimized model with 12 inputs and 27 weights.

6.5 Measurements

The performance of the strategies are compared by computing the return on invest (*roi*) on the test set. This curve measures the return of a trading system by

$$roi = \sum_{t=1}^{T}(\frac{\mathrm{DAX}_{t+1}}{\mathrm{DAX}_t} - 1) * \ \mathrm{sign}(\mathrm{NN}(\underline{x_t}) \tag{9}$$

where $\mathrm{NN}(\underline{x_t})$ is the output of the network for time step t and $\mathrm{sign}(\mathrm{NN}(\underline{x_t})$ is the predicted direction of the DAX. *roi* increases if the predicted direction is correct, otherwise it decreases. The annualized return on invest is computed by

$$roi_{annu} = \frac{1}{T} * roi * 260 \tag{10}$$

Alternatively one can compute return on invest in DAX-points by

$$rt = \sum_{t=1}^{T}(\mathrm{DAX}_{t+1} - \mathrm{DAX}_t) * \ \mathrm{sign}(\mathrm{NN}(\underline{x_t}) \tag{11}$$

Additionally, the hit rate (*hr*) is given, which measures, how often the sign of the change of the DAX (*increase* or *decrease*)is correctly predicted:

$$hr = \frac{1}{T}\sum_{t=1}^{T} \ \mathrm{sign}(\mathrm{DAX}_{t+1} - \mathrm{DAX}_t) * \ \mathrm{sign}(\mathrm{NN}(\underline{x_t}) \tag{12}$$

The Sharpe ratio (SR), a popular measure of performance for trading systems, has a higher value for high returns with low volatility (Refenes A. P. (1994) and is computed by

$$SR = \frac{\overline{roi}}{\frac{1}{T}\sqrt{\sum_{i=1}^{T}(roi_i - \overline{roi})^2}} \tag{13}$$

where

$$roi_t = (\frac{\mathrm{DAX}_{t+1}}{\mathrm{DAX}_t} - 1) * \mathrm{sign}(\mathrm{NN}(x_t)) \quad \text{and} \quad \overline{roi} = \frac{1}{T}\sum_{t=1}^{T} roi_t \tag{14}$$

T represents the number of predicted patterns on test set.

6.6 Performance of the Initial Rule Base

The performance of the fuzzy-system on the training generalization set is displayed in Fig. 3.

The performance of the fuzzy system on training set is satisfying. This does not hold for the validation and generalization set.

6.7 Optimization

We use the following steps to optimize the initial fuzzy system.

1. Training until the minimal MSE on validation set is reached
2. Pruning of rules with small rule weight (e. g. $\kappa_i < 0.1$)
3. Use pruning methods (e. g. EBD Tresp V. et al (1996)) for deletion or insertion of premises
4. Iteration of training and pruning until the model can not be further improved

The quality of the model during optimization is measured by the values described in section 6.5. Our experience is that the model with the best performance does not corresponds with the minimal error model.

7 Results

We present two neuro-fuzzy-models. The first has been optimized with the EBD-Pruning method to delete premises. The optimized neuro-fuzzy model consists of only 31 rules and 22 fuzzy sets, the other has been pruned. It is possible to further delete all rules with a *stable*-conclusion and with minimal rule weights achieving a model with 17 rules and 15 fuzzy sets without a significant decrease of the performance.

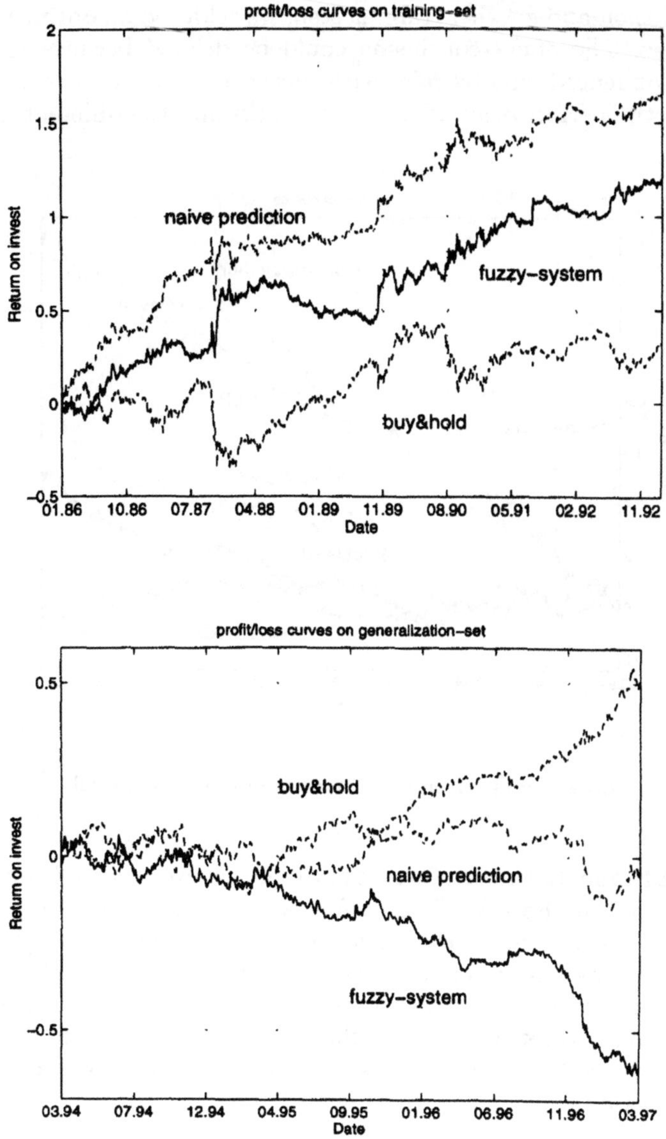

Fig. 3. Performance of the Fuzzy-System on training set(a) and generalization set (b).

The second model was optimized with the expensive pruning method described above. We iterate deletion and insertion of premises This neuro-fuzzy model consists of 52 remaining rules. After deletion of rules with a *stable*-conclusion and another training step, we achieved an optimized system with 36 rules. The *stable*-conclusion could be deleted because the trading strategy is influenced only by rules with *increase*- and *decrease*-conclusions. Fig. 4 show the performance of the benchmarks and the different models.

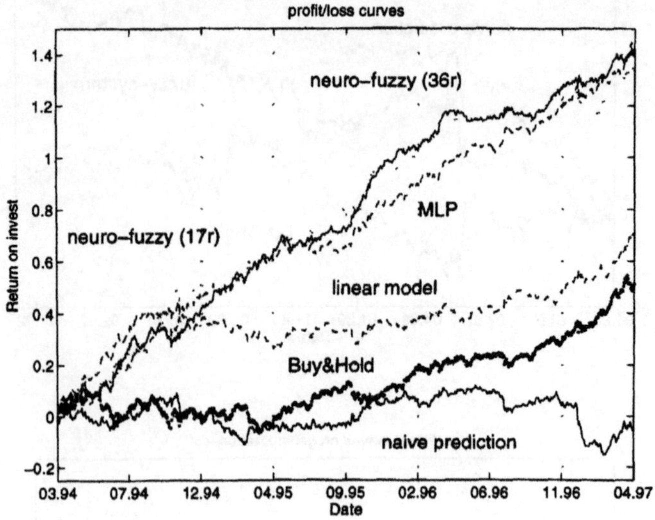

Fig. 4. Comparison of *roi*-curves of different models.

The MLP and the neuro-fuzzy approach achieve a significantly higher profit than the benchmarks. The performance of the neuro-fuzzy approach is comparable to a neural net which has been trained with highly sophisticated algorithms Zimmermann H. G., Weigend A. (1997). Table 3 and table 4 shows different measurements (see section 6.5) of our models and the benchmarks on test set. Comparing the trading signals for the first week of the test-set shows that both models give the same signal which is also the correct prediction.

8 Conclusions

We have shown that the proposed neuro-fuzzy approach in combination with pruning methods is able to construct superior trading systems. The low number of parameters of the resulting system might be responsible for the high generalization performance. The few rules are still interpretable due to the

Table 3. Results.

Model	roi	rt
naive prediction	−0.016	−62.62
Buy & Hold	+0.537	1351.07
Linear	+0.730	1852.64
Neuro-Fuzzy(17r)	+1.331	3088.12
Neuro-Fuzzy(36r)	+1.429	3281.16
MLP	+1.430	3388.47

Table 4. Additional measurements

Model	roi_{annu}	SR	hr
naive prediction	−0.0052	-	-
Buy & Hold	+0.1745	-	-
Linear	+0.2373	0.1008	0.498
Neuro-Fuzzy(17r)	+0.4325	0.1921	0.529
Neuro-Fuzzy(36r)	+0.4644	0.2039	0.559
MLP	+0.4647	0.2072	0.572

semantic-preserving learning algorithm and have to be further analyzed. The application of pruning algorithms for rule generation seemed to be another interesting idea.

References

Dichtl H., Zur Prognose des Deutschen Aktienindex DAX mit Hilfe von Neuro-Fuzzy-Systemen, Institut für Kapitalmarktforschung, J. W. Goethe-Universität, Frankfurt, Germany, 1996.

Finnoff W., Hergert F., Zimmermann H. G., Improving Generalization by Nonconvergent Model Selection Methods. Neural Networks 6, 1992.

Gill P. E., Murray W., Wright M. H., Practical Optimization. Academic Press, 1981.

Kruse R., Schwecke E., Heinsohn J., Uncertainty and vageness in knowlegde based systems, Springer, Heidelberg, 1991.

Nauck D., Klawonn F., Kruse R., Foundations of Neuro-Fuzzy Systems, Wiley, 1997 (to appear)

Neuneier R., Tresp V., Radiale Basisfunktionen, Dichteschätzungen und Neuro-Fuzzy. In H. Rehkugler, H. G. Zimmermann (editors), Neuronale Netze in der Ökonomie, Vahlen Verlag, München, 1994.

Neuneier R. and Zimmermann H. G., A semantic-preserving learning algorithm for neuro-fuzzy systems with applications to time series prediction, ICANN-Workshop, 1995.

Refenes A. P., Neural Networks in the Capital Market, Wiley & Sons, 1994.

Siekmann S., Kruse R., Neuneier R., Zimmermann H. G., Advanced Neuro-Fuzzy Techniques in Finance, Proc. of EFDAN'97, Dortmund, 1997, (to appear)

Takagi T., Sugeno M., Fuzzy identification of systems and its applications to modelling and control. IEEE Trans. on Syst., Man and Cybernetics, 1, Vol. 15, 1985.

Tresp V., Hollatz J., Ahmad S., Network Structuring and Training Using Rule-Based Knowledge. In S. J. Hanson, J. D. Cowan, C. L. Giles (editors), Advances in Neural Information Processing Systems 5, 1993.

Tresp V., Zimmermann H. G., Neuneier R., Early Brain Damage, Advances in Neural Information, Proc. NIPS, 1996.

Wang L. X., Mendel J. M., Generating Fuzzy Rules from Numerical Data, with Applications. Univ. of Southern Cal., tech. report USC-SIPI REPORT 169, 1991.

Zimmermann H. G., Neuronale Netze als Entscheidungskalkül. In H. Rehkugler, H. G. Zimmermann (editors), Neuronale Netze in der Ökonomie, Vahlen, 1994, München

Zimmermann H. G., Neuneier R., Siekmann S., Dichtl H., Modeling the German stock index DAX with Neuro-Fuzzy, Proc. of EUFIT'96, 1996.

Zimmermann H. G., Weigend A.: How to represent dynamical systems in feed-forward networks: a six layer architecture, Decisisons Technologies for Financial Engineering: Proc. of NNCM'96, World Scientific, 1997

Zimmermann H. G., Mayer P., Neuneier R., Adaptive Noise and Clearning, forthcoming, 1997.

Statistical Process Control and its Application in Finance

Thomas Severin and Wolfgang Schmid

Europe-University Viadrina, Department of Statistics, D-15207 Frankfurt (Oder), Germany

Abstract. Analyzing a financial time series it is of great interest to monitor changes in its structure. In this paper it is shown how control charts can be used to detect such positions. Several control schemes for GARCH processes are introduced and compared with each other. Furthermore these methods are applied to stock market returns.

Keywords. Financial time series, GARCH process, statistical process control, control charts, stock market returns.

1 Introduction

Statistical Process Control (SPC) is one of the most important statistical tools in quality management. It is applied to control a production process. The aim is to detect deviations from a supposed target process as soon as possible. However, such problems arise in many other fields like e.g. in environmental sciences (changes in the ozone content) or in finance (early hints in the structure of stock market returns). Especially in economics the variables of the target process are in most cases no longer independent. In recent years several control schemes for time series have been proposed. In principle it is distinguished between residual charts and modified control schemes. Residual charts (e.g. Alwan/Roberts (1988), Harris/Ross (1991), Montgomery/Mastrangelo (1991)) base on a transformation of the data such that the transformed data are independent. Thus all well-known control procedures like the Shewhart, EWMA and CUSUM schemes can be applied on the residuals. Modified control schemes are direct extensions of the classical charts (e.g. Yashchin (1993), Schmid (1995, 1997a, 1997b)). A comparison of several control schemes is given in Schmid (1997b) assuming that the target process is an autoregressive process of order 1. Furthermore recommendations on the optimal choice of the smoothing parameters are given.

Modeling stock market returns and exchange rates GARCH (generalized autoregressive conditional heteroscedasticity) models (Engle (1982), Bollerslev (1986)) have obtained an enormous importance and popularity. In contrast to autoregressive moving average processes (ARMA) the conditional variance of these processes is not a constant but a function of preceeding observations. This makes these models more flexible for economic purposes. In the meantime several generalizations of this modeling have been proposed. An overview about the latest developments in research can be found in Pagan (1996).

In Severin/Schmid (1996) the theory of control charts for dependent observations is applied to GARCH models. Several modified control schemes were compared with each other. Here we introduce some new residual charts for GARCH models. In a computer study these control designs are compared with those discussed in Severin/Schmid (1996). Moreover, these results are used to analyze stock market returns. Several examples illustrate the utility of these methods in finance.

2 GARCH Processes

Since in the last years GARCH models are widely discussed in econometrics (e.g. Engle/Rothschild (1992), Cox et al. (1996)) we only want to give a brief introduction in this topic.

A stochastic process $\{Y_t\}$ is called a GARCH process of order (p, q) (Bollerslev (1986)) if it is a solution of the stochastic difference equation

$$Y_t = \epsilon_t \sqrt{h_t} \tag{1}$$

with

$$h_t = \alpha_0 + \sum_{i=1}^{q} \alpha_i Y_{t-i}^2 + \sum_{j=1}^{p} \beta_j h_{t-j} \tag{2}$$

for $t \in \mathbb{Z}$. To avoid a negative conditional variance it is demanded that $\alpha_0 > 0$, $\alpha_i \geq 0$ for $i = 1, .., q$ with $q \geq 0$ and $\beta_j \geq 0$ for $j = 1, .., p$ with $p \geq 0$. Usually, it is supposed that the random variables $\{\epsilon_t\}$ are independent with $E(\epsilon_t) = 0$ and $E(\epsilon_t^2) = 1$. For $p = 0$ a GARCH model reduces to an ARCH(q) process (Engle (1982)).

Statements about the existence of a weakly stationary solution of this stochastic difference equation were given by Engle (1982) and Milhøj (1985) for ARCH processes and by Bollerslev (1986) for GARCH models.

Suppose that the random variables $\{\epsilon_t\}$ are independent with $E(\epsilon_t) = 0$ and $E(\epsilon_t^2) = 1$ for $t \in \mathbb{Z}$. Furthermore let $\sum_{i=1}^{q} \alpha_i + \sum_{j=1}^{p} \beta_j < 1$. Then it follows (Bollerslev (1986)) that $\{Y_t\}$ is a weakly stationary process with

$$\begin{aligned} E(Y_t) &= 0, \\ \mathrm{Var}(Y_t) &= \frac{\alpha_0}{1 - \sum_{i=1}^{q} \alpha_i - \sum_{j=1}^{p} \beta_j} = \gamma_0 \quad \text{and} \\ \mathrm{Cov}(Y_t, Y_s) &= 0 \quad \text{for } t \neq s. \end{aligned}$$

On the other side suppose that the random variables $\{\epsilon_t\}$ are independent with $\mathrm{E}(\epsilon_t) = 0$ and $\mathrm{E}(\epsilon_t^2) = 1$ for $t \in \mathbf{Z}$ and let $\{Y_t\}$ be a stationary solution of equation (1). Then the coefficients satisfy the inequality $\sum_{i=1}^{q} \alpha_i + \sum_{j=1}^{p} \beta_j < 1$.

3 Control Charts for GARCH Processes

Let $x_1, x_2, \ldots$ denote the observed data. They are considered as a realization of the stochastic process $\{X_t\}$. The underlying target process is supposed to be a weakly stationary GARCH process and it is denoted by $\{Y_t\}$. We analyze the performance of several control schemes assuming that a deviation in the mean structure of the target process may arise i.e. we focus on the change point model

$$X_t = Y_t + a\sqrt{\gamma_0}\, I_{\{s,s+1,\ldots\}}(t) \tag{3}$$

with $a \in \mathbb{R}$ and $s \in \mathbb{N}$. γ_0 denotes the variance of the target process $\{Y_t\}$. It is said that a change point is present at position s if $a \neq 0$. Then the process $\{X_t\}$ is said to be out-of-control, else it is called in-control. Thus the target value is equal to $\mathrm{E}(Y_t) = 0$. It is important to note that a change point has only an influence on the mean but not on the variance.

A measure for the performance of a control chart is the average run length $\mathrm{E}(N)$. The run length N is equal to the first time point at which it is concluded that the process is out-of-control. To avoid a false decision the run length should be large if the process is in-control. However, if it is out-of-control, an alarm should be given immediately after the change has arisen. Frequently, it is assumed that the change occurs at position $s = 1$. The process is out-of-control even at beginning. This is only a minor restriction (cf. Kramer (1997)).

3.1 Modified Control Charts for GARCH Processes

The **modified Shewhart chart** was introduced by Vasilopoulos/Stamboulis (1978). It is concluded that the process $\{X_t\}$ is out-of-control at time t if

$$|X_t| > c\sqrt{\gamma_0}$$

with a constant $c > 0$. The run length of this scheme is equal to

$$N_S(c) = \inf\{t \in \mathbb{N} : |X_t| > c\sqrt{\gamma_0}\}.$$

Here the deviation of each observation with the target value $\mathrm{E}(Y_t) = 0$ is compared with the standard deviation of the target process. A detailed examination of this scheme can be found in Schmid (1995).

To detect deviations from a stationary process Schmid (1997a) introduced the **modified EWMA chart**. It is based on the EWMA (exponentially weighted moving average) statistic

$$Z_t = (1-\lambda)Z_{t-1} + \lambda X_t$$

for $t \geq 1$, $\lambda \in (0,1]$. The starting value is taken equal to the target value i.e. $Z_0 = E(Y_t) = 0$. Now an alarm is given at position t if $|Z_t| > c\,\sigma_e$ with $c > 0$. σ_e^2 denotes the asymptotic variance of $\{Z_t\}$. A general expression for a stationary process was derived in Schmid (1997a). For a GARCH process it follows that $\sigma_e^2 = \frac{\lambda}{2-\lambda}\,\gamma_0$. This quantity coincides with the asymptotic variance of independent variables (Roberts (1959)). This is due to the fact that the variables of a GARCH process are uncorrelated.

The run length of this scheme is

$$N_E(c,\lambda) = \inf\{t \in \mathbb{N} : |Z_t| > c\,\sigma_e\}.$$

Setting $\lambda = 1$ the modified EWMA chart reduces to the modified Shewhart scheme.

In order to protect against a small shift the smoothing parameter λ should be chosen small too. But if a large change is expected then $\lambda = 1$ is the best choice. Assuming that the size of the shift is known Severin/Schmid (1996) gave tables for an optimal choice of λ.

Modified CUSUM charts were discussed by Yashchin (1993) and Schmid (1997b). The derivation of these schemes is based on the assumption that the underlying process is Gaussian. Since up to now no characterization of the distribution of a GARCH process can be found in literature (it is not Gaussian) the sequential probability ratio can not be calculated. However, as shown in Schmid (1997b) for autoregressive processes, the loss in power by taking the **classical CUSUM chart** due to Page (1954) is not dramatical. This control design is based on the recursions

$$S_t^+ = \max\{0, S_{t-1}^+ + X_t - K\sqrt{\gamma_0}\} \quad \text{and}$$
$$S_t^- = \min\{0, S_{t-1}^- + X_t + K\sqrt{\gamma_0}\}$$

for $t \geq 1$ and $S_0^+ = S_0^- = 0$. It is concluded that the process $\{X_t\}$ is in an out-of-control state at time t if

$$\max\{S_t^+\,, -S_t^-\} \geq c\sqrt{\gamma_0}$$

with $c > 0$. The run length of this chart is given by

$$N_C(c,K) = \inf\{t \in \mathbb{N} : \max\{S_t^+\,, -S_t^-\} \geq c\sqrt{\gamma_0}\}.$$

K is called the reference value which has to be determined by the practitioner. In Severin/Schmid (1996) it was shown via simulations that for GARCH processes the best choice for K seems to be $a/2$.

Similar as in the case of independent variables it has turned out that the EWMA and CUSUM schemes are better for small shifts, else a Shewhart chart should be chosen (see Severin/Schmid (1996)).

3.2 Residual Charts for GARCH Processes

For ARMA processes residual charts are widely discussed in literature. They are based on the comparison of the observation with the best linear one-step predictor (in the L^2-sense) in terms of preceeding values. For a GARCH process we get

$$E(Y_t|Y_{t-1}, Y_{t-2}, ..) = \sqrt{h_t}\ E(\epsilon_t) = 0.$$

Thus the predictor $\widehat{X}_t$ is equal to 0 and therefore the prediction errors coincide with the observations. Following this attempt the residual charts and the modified charts are the same. This shows that in the context of GARCH processes it is not successible to follow the same argumenation as for linear processes. Here we introduce several control schemes for the residuals of a nonlinear process.

First note that for an ARCH process it follows

$$\epsilon_t = \frac{Y_t}{\sqrt{\alpha_0 + \sum\limits_{i=1}^{q} \alpha_i Y_{t-i}^2}}$$

and that the variables $\{\epsilon_t\}$ are independent. This motivates the consideration of the residuals

$$\widehat{\epsilon}_t = \frac{X_t}{\sqrt{\alpha_0 + \sum\limits_{i=1}^{q} \alpha_i X_{t-i}^2}} \qquad (4)$$

for $t \geq q+1$. Thus a starting problem is present. Therefore we replace X_v^2 in (4) by $E(X_v^2) = \mathrm{Var}(X_v) = \gamma_0$ if $v \leq 0$. We get for $t \geq 1$

$$\widehat{\epsilon}_t = \frac{X_t}{\sqrt{\alpha_0 + \sum\limits_{i=1}^{min\{q,t-1\}} \alpha_i X_{t-i}^2 + \gamma_0 \sum\limits_{i=t}^{q} \alpha_i}}$$

$$= \frac{X_t}{\sqrt{\gamma_0 + \sum\limits_{i=1}^{min\{q,t-1\}} \alpha_i (X_{t-i}^2 - \gamma_0)}}$$

since $\gamma_0 = \frac{\alpha_0}{1-\sum\limits_{i=1}^{q} \alpha_i}$. For the special case $q = 1$ it follows that

$$\widehat{\epsilon}_1 = \frac{X_1}{\sqrt{\gamma_0}}.$$

Note that the variables $\{\widehat{\epsilon}_t\}$ are not independent in the out-of-control state.

This procedure can immediately be extended to GARCH processes. Then the residuals are given by $\widehat{\epsilon}_t = \frac{X_t}{\sqrt{\widehat{h}_t}}$ with

$$\widehat{h}_t = \gamma_0 + \sum_{i=1}^{min\{q,t-1\}} \alpha_i \, (X_{t-i}^2 - \gamma_0) + \sum_{j=1}^{p} \beta_j \, \widehat{h}_{t-j}$$

for $t \geq 1$ and $\widehat{h}_t = \gamma_0$ for $t \leq 0$.

Now it is possible to apply all classical control schemes to the residuals $\{\widehat{\epsilon}_t\}$. In our study we focus on the EWMA attempt i.e. we consider the recursion

$$Z_t = (1-\lambda)\, Z_{t-1} + \lambda \, \widehat{\epsilon}_t$$

for $t \geq 1$, $\lambda \in (0,1]$ and $Z_0 = \mathrm{E}(\epsilon_t) = 0$. For an ARCH(1) process the quantities $\{\widehat{\epsilon}_t\}$ are independent with variance 1 in the in-control state and thus $\mathrm{Var}(Z_t) = \frac{\lambda}{2-\lambda}$. An alarm is given at position t if $|Z_t| > c \sqrt{\frac{\lambda}{2-\lambda}}$ where $c > 0$. The run length is denoted by $N_{E,\epsilon}(c, \lambda)$.

Bollerslev (1986) showed an interesting relation between GARCH and ARMA models. If $\{Y_t\}$ is a GARCH(p,q) process then $\{Y_t^2\}$ is an ARMA(max{p,q},p) process. This relation gives rise to the consideration of the deviations $Y_t^2 - \mathrm{E}(Y_t^2 | Y_{t-1}, Y_{t-2}, ..Y_1)$. Replacing $\{Y_t\}$ by $\{X_t\}$ we obtain another type of residuals. A recursive procedure for calculating the best linear one-step predictor for ARMA processes is given in Brockwell/Davis (1991).

For an ARCH(1) process we get $\mathrm{E}(Y_t^2 | Y_{t-1}, Y_{t-2}, .., Y_1) = \alpha_0 + \alpha_1 Y_{t-1}^2$ for $t \geq 2$ and $\mathrm{E}(Y_1^2) = \gamma_0$. This leads to the residuals

$$\widehat{\eta}_t = \begin{cases} X_1^2 - \gamma_0 & \text{for } t = 1 \\ X_t^2 - (\alpha_0 + \alpha_1 X_{t-1}^2) & \text{for } t \geq 2. \end{cases} \tag{5}$$

It is important to note that even in the in-control state the residuals $\{\widehat{\eta}_t\}$ are not independent but uncorrelated. The variance of the process $\{\widehat{\eta}_t\}$ in the in-control state is given by

$$\mathrm{Var}(\widehat{\eta}_t) = \begin{cases} \mathrm{E}(Y_1^4) - \mathrm{E}(Y_1^2)^2 & \text{for } t = 1 \\ (1-\alpha_1^2)\mathrm{E}(Y_t^4) - 2\alpha_0 \mathrm{E}(Y_t^2) + \alpha_0^2 & \text{for } t \geq 2 \end{cases} \tag{6}$$

with (cf. Bollerslev (1986))

$$\mathrm{E}(Y_t^2) = \frac{\alpha_0}{1-\alpha_1},$$

$$\mathrm{E}(Y_t^4) = \frac{3\alpha_0^2(1+\alpha_1)}{(1-\alpha_1)(1-3\alpha_1^2)}.$$

For the proof of equation (6) we refer to Appendix A. Note that the fourth moment of $\{Y_t\}$ only exists if $0 \leq \alpha_1 < 1/\sqrt{3}$.

Now we apply the modified EWMA control scheme to the quantities $\{\widehat{\eta}_t\}$. Thus

$$Z_t = (1-\lambda)Z_{t-1} + \lambda\widehat{\eta}_t$$

for $t \geq 1$, $\lambda \in (0,1]$ and Z_0 is taken as $\mathrm{E}(\widehat{\eta}_t)$ in the in-control state i.e. $Z_0 = 0$. Since the variables $\{Y_t\}$ are uncorrelated it follows that $\mathrm{Var}(Z_t) = \frac{\lambda}{2-\lambda}\,\mathrm{Var}(\widehat{\eta}_t)$. It is concluded that the process is out-of-control at time t if

$$|Z_t| > c\,\sigma_t \quad \text{with } c > 0 \text{ and } \sigma_t^2 = \frac{\lambda}{2-\lambda}\,\mathrm{Var}(\widehat{\eta}_t).$$

The run length is given by

$$N_{E,\eta}(c,\lambda) = \inf\{t \in \mathbb{N} : |Z_t| > c\,\sigma_t\}.$$

4 A Comparison of Several Control Charts

In this section we compare the control schemes of Section 3 with each other. As a measure of performance we use the average run length (ARL). Now a fair comparison of control charts is only possible if all schemes are treated the same. Therefore for each chart the quantity c was determined such that the in-control ARL is the same. Contrary to industrial applications the number of observations in finance is usually small and thus the fixed value for the in-control ARL is chosen small, here equal to 20. The reason for this choice lies in the fact that we are mainly interested in stock market returns. This value corresponds about one month at the stock exchange.

In our study the target process was chosen as an ARCH(1) modell. The distribution function of ϵ_t, $t \in \mathbb{Z}$, was taken as the standard normal distribution (symbol: Φ). Since the EWMA and the CUSUM schemes depend on an additional smoothing parameter the ARL of these charts was determined for various values of λ and K. The parameter λ was taken equal to 1.0, 0.75, 0.5, 0.25 and 0.1. For the CUSUM schemes the reference value K varies in $\{0.0, 0.25, 0.5, \ldots, 1.5, 1.75\}$.

Since the in-control ARL of these charts does not depend on α_0 we confine ourselves to $\alpha_0 = 1.0$. Due to its symmetry it is sufficient to discuss the out-of control behaviour for $a \geq 0$. The practical computation of the ARL of a control scheme for correlated variables turns out to be extremely time consuming. Only for exchangeable variables an explicit formula for the ARL is known (Schmid (1995)). In most cases the ARL is determined via simulations or an Markov chain approach. The latter method can not be used here because no Markov process is present. Consequently we used simulations. Our results are based on 100 000 repetitions for each process. The ARL was estimated by the empirical ARL.

Table 1 and Table 2 show the results for the modified EWMA chart and the classical CUSUM chart. These tables are taken from Severin/Schmid (1996) where the corresponding critical values are also listed. In each table the value of λ and K which leads to the smallest ARL for a given value of the shift is as well given. In nearly all cases the ARL of the modified EWMA chart turns out to be smaller than the ARL of the classical CUSUM procedure.

Table 1. Out-of-control ARL of the modified EWMA chart applied to an ARCH(1) process for the optimal smoothing parameter λ ($\epsilon_t \sim \Phi, \lambda \in \{0.1, 0.25, 0.5, 0.75, 1.0\}$, c such that $E_0(N_E) = 20$, above: minimal ARL over λ, below: value of λ where the minimum is attained).

$a \mid \alpha_1$	0.0	0.1	0.2	0.3	0.4	0.5	0.6	0.7	0.8	0.9
0.50	8.66	8.55	8.35	8.18	7.91	7.56	7.10	6.48	5.63	4.13
	0.1	0.1	0.1	0.1	0.1	0 1	0.1	0.1	0.1	0.25
1.00	4.17	4.13	4.05	3.94	3.80	3.64	3.40	3.08	2.61	1.88
	0.25	0 25	0.25	0.25	0.25	0.25	0.25	0.25	0.5	0.5-0.75
1.50	2.56	2.54	2.48	2.43	2.33	2.24	2.09	1.91	1.62	1.22
	0.5	0.5	0.5	0.5	0.5	0.5	0.5	0.5-0.75	0 75	1 0
2.00	1.80	1.78	1.76	1.72	1.66	1.59	1.50	1.38	1.22	1.06
	0.75	0.75	0.75	0.75	0.75	0.75	0.75	0.75	$\geq$ 0.75	$\geq$ 0.75
2.50	1.39	1.38	1.37	1.35	1.31	1.27	1.21	1.15	1.07	1.02
	0.75	0.75	0.75	$\geq$ 0.75	0.75	$\geq$ 0.75	1.0	$\geq$ 0.75	1.0	$\geq$ 0.75
3.00	1.18	1.17	1.16	1.15	1.13	1.11	1.08	1.05	1.03	1.01
	$\geq$ 0.75	$\geq$ 0.75	$\geq$ 0.75	$\geq$ 0.75	1.0	$\geq$ 0.75	1 0	1.0	$\geq$ 0.75	$\geq$ 0.5
3.50	1.07	1.06	1.06	1.06	1.05	1.04	1.03	1.02	1.01	1.01
	$\geq$ 0.75	1.0	$\geq$ 0.75	$\geq$ 0.75	1.0	1.0	1.0	1.0	1.0	$\geq$ 0.1
4.00	1.02	1.02	1.02	1.02	1.02	1.02	1.02	1.01	1.01	1.00
	$\geq$ 0.75	$\geq$ 0.75	$\geq$ 0.75	$\geq$ 0.75	$\geq$ 0.75	$\geq$ 0.5	$\geq$ 0 5	$\geq$ 0.75	$\geq$ 0.25	$\geq$ 0.5

Tables 3 and 4 refer to the residual charts introduced in this paper. In Table 3 the out-of-control ARL of $N_{E,\epsilon}$ is given while Table 4 handles with the chart based on $N_{E,\eta}$. Note that the critical values for $N_{E,\epsilon}$ are the same as in the case of independent variables. The critical values for $N_{E,\eta}$ can be found in Tabel 8 (Appendix B).

The bold typed values in both tables indicate on the smallest ARL over all considered values of λ for a fixed constellation (a, α_1). For the residual chart based on $\{\widehat{\epsilon}_t\}$ the smoothing parameter λ should be determined as follows. For a shift $a \in \{0.5, 1.0\}$ the best results are obtained for $\lambda \leq 0.25$ while for $a \in \{1.5, 2.0, 2.5\}$ λ should be taken within the intervall $[0.25, 0.75]$. $\lambda = 0.75$ is the best choice for shifts $a \in \{3.0, 3.5\}$ provided that $\alpha_1 \leq 0.7$. For extremely large shifts like $a = 4.0$ the smoothing parameter λ should be chosen greater equal to 0.5 .

Table 2. Out-of-control ARL of the classical CUSUM chart applied to an ARCH(1) process for the optimal reference value K ($\epsilon_t \sim \Phi, K \in \{0.0, 0.25, 0.5, 0.75, 1.0, 1.25, 1.5, 1.75\}$, c such that $E_0(N_C) = 20$, above: minimal ARL over K, below: value of K where the minimum is attained).

$a \mid \alpha_1$	0.0	0.1	0.2	0.3	0.4	0.5	0.6	0.7	0.8	0.9
0.50	9.53	9.44	9.28	9.12	8.83	8.52	8.05	7.35	6.38	4.68
	0.25	0.25	0.25	0.25	0.25	0.25	0 25	0 25	0.25	0.25
1.00	4.50	4.47	4.39	4.29	4.15	3.97	3.71	3.34	2.78	1.92
	0.5	0.5	0.5	0 5	0 5	0.5	0.5	0 5	0.5	0.5
1.50	2.69	2.64	2.60	2.54	2.47	2.34	2.17	1.95	1.64	1.22
	0.75	0 75	0 75	0.75	0.75	0 75	0.75	0.75	0.75	$\geq$ 0 75
2.00	1.84	1.82	1.79	1.75	1.70	1.63	1.53	1.38	1.22	1.06
	1.0	1.0	1.0	1.0	1.0	1.0	1.0	1.0	1.0	$\geq$ 0.75
2.50	1.41	1.40	1.39	1.37	1.34	1.29	1.23	1.15	1.07	1.02
	1.0	1.0	1.0	1.0	1.0	1.0	1 0	1.0	1.0	0.75-1.0
3.00	1.19	1.19	1.18	1.17	1.15	1.13	1.09	1.06	1.03	1.01
	1.5	1 5	1.5	1.5	1 5	1.5	1.5	1.5	0.75-1.25	0.5-1.0
3.50	1.08	1.08	1.07	1.07	1.06	1.05	1.04	1.03	1.02	1.01
	1.75	1.75	1.75	1.75	1.75	1 75	0.75-1.5	0.75-1.5	0.5-1.25	0.25-1.0
4.00	1.03	1.03	1.03	1.02	1.02	1.02	1.02	1.01	1.01	1.00
	$\geq$ 1.25	$\geq$ 1.25	$\geq$ 1.25	$\geq$ 1 0	$\geq$ 1.0	$\geq$ 0.75	1.25-1.5	1.0-1.5	0 5-1.25	0.5-1.0

The residual chart based on $\{\widehat{\eta}_t\}$ is only applicable for $0 \leq \alpha_1 < \frac{1}{\sqrt{3}}$. For small shifts the behaviour of the ARL of this chart differs extremely from the behaviour of the ARL of the other EWMA-typed schemes. For the shift $a = 0.5$ and $\alpha_1 \in \{0.3, 0.4, 0.5\}$ the best choice is $\lambda = 1.0$. For $a = 0.5$ and $\alpha_1 \in \{0.0, 0.1, 0.2\}$ the minimum ARL is attained for $\lambda = 0.5$. For shifts $a \in \{1.0, 1.5, \ldots, 3.0\}$ λ should be taken between 0.25 and 0.5. For changes $a \in \{3.5, 4.0\}$ the smoothing parameter $\lambda \geq 0.75$ provides good results.

A comparison of all charts shows that especially the modified EWMA chart and the chart based on $\{\widehat{\epsilon}_t\}$ provide the smallest out-of-control ARL. It is very interesting that for large values of α_1 the modified EWMA scheme seems to be the best one.

In practice we have no hint about the size of the shift. Thus, it is recommendable to use the modified EWMA chart to analyze GARCH data.

Table 3. Out-of-control ARL of the EWMA residual chart based on $\{\widehat{\epsilon}_t\}$ ($\epsilon_t \sim \Phi$, $\lambda \in \{0.1, 0.25, 0.5, 0.75, 1.0\}$, c such that $E_0(N_{E,\epsilon}) = 20$, first line: $\lambda = 0.1$, second line: $\lambda = 0.25$, third line: $\lambda = 0.5$, fourth line: $\lambda = 0.75$, fifth line: $\lambda = 1.0$).

$a \mid \alpha_1$	0.0	0.1	0.2	0.3	0.4	0.5	0.6	0.7	0.8	0.9
0.50	**8.66**	**8.54**	**8.31**	**8.11**	**7.79**	**7.40**	**6.95**	6.39	5.73	4.88
	8.93	8.81	8.61	8.36	7.99	7.53	6.98	**6.37**	**5.64**	**4.73**
	9.84	9.86	9.69	9.42	9.14	8.69	8.12	7.36	6.42	5.35
	11.12	11.15	11.12	10.94	10.57	10.11	9.57	8.75	7.75	6.61
	12.64	12.69	12.62	12.35	11.95	11.43	10.73	9.78	8.74	7.52
1.00	4.32	4.30	4.16	4.21	4.15	4.04	3.95	3.84	3.73	3.66
	4.17	**4.09**	**4.08**	**3.99**	**3.91**	**3.81**	**3.71**	**3.62**	**3.55**	**3.55**
	4.34	4.34	4.26	4.21	4.12	4.02	3.95	3.91	3.95	4.24
	4.91	4.90	5.51	4.95	4.94	4.95	4.95	5.00	5.30	6.43
	5.88	6.05	6.13	6.16	6.21	6.17	6.15	6.20	6.54	8.00
1.50	2.89	2.93	2.93	2.94	2.95	2.97	2.98	3.02	3.06	3.13
	2.65	2.66	2.65	2.66	2.67	2.69	2.70	2.83	**2.83**	**2.99**
	2.56	**2.57**	**2.57**	**2.57**	**2.59**	**2.63**	**2.69**	**2.76**	3.12	3.83
	2.70	2.73	2.77	2.82	2.94	3.10	3.37	3.85	4.87	7.86
	3.10	3.22	3.40	3.61	3.87	4.15	4.58	5.24	6.48	10.69
2.00	2.22	2.24	2.25	2.29	2.31	2.35	2.40	2.46	2.55	2.69
	1.97	1.98	2.00	2.01	2.02	2.04	2.07	2.10	2.17	**2.28**
	1.82	**1.82**	**1.84**	**1.85**	**1.87**	**1.90**	**1.92**	**1.98**	**2.11**	2.50
	1.80	1.83	1.86	1.89	1.94	2.05	2.26	2.68	3.59	6.40
	1.94	2.00	2.10	2.27	2.51	2.84	3.32	3.98	5.22	8.95
2.50	1.82	1.84	1.85	1.87	1.88	1.90	1.93	1.95	1.99	2.03
	1.59	1.60	1.61	1.61	1.62	1.62	1.63	1.62	1.62	1.60
	1.43	1.44	1.45	1.45	1.45	1.45	**1.45**	**1.43**	**1.41**	**1.34**
	1.39	**1.40**	**1.41**	**1.42**	**1.43**	**1.44**	1.49	1.60	1.83	2.28
	1.42	1.44	1.48	1.53	1.64	1.82	2.08	2.42	2.88	3.51
3.00	1.55	1.55	1.56	1.57	1.58	1.58	1.59	1.59	1.59	1.58
	1.34	1.34	1.34	1.34	1.35	1.34	1.32	1.30	1.26	1.20
	1.21	1.21	1.22	1.22	1.22	1.21	1.19	**1.17**	**1.14**	**1.09**
	1.18	**1.18**	**1.18**	**1.18**	**1.18**	**1.18**	**1.17**	**1.17**	1.17	1.22
	1.18	1.19	1.19	1.21	1.23	1.29	1.36	1.45	1.51	1.58
3.50	1.33	1.34	1.34	1.33	1.33	1.33	1.32	1.30	1.26	1.20
	1.17	1.17	1.17	1.17	1.17	1.16	1.14	1.12	1.10	1.06
	1.09	1.09	1.09	1.09	1.09	1.09	1.08	1.07	**1.05**	**1.03**
	1.07	**1.07**	**1.07**	**1.07**	**1.07**	**1.07**	**1.06**	**1.06**	**1.05**	1.05
	1.07	**1.07**	**1.07**	**1.07**	1.08	1.09	1.11	1.13	1.15	1.19
4.00	1.17	1.17	1.17	1.17	1.16	1.16	1.14	1.13	1.10	1.06
	1.07	1.07	1.07	1.07	1.07	1.06	1.06	1.05	1.04	1.03
	1.03	1.03	1.03	**1.03**	1.04	1.04	**1.03**	**1.03**	1.03	**1.02**
	1.02	**1.02**	**1.02**	**1.03**	**1.03**	**1.03**	**1.03**	**1.03**	**1.02**	**1.02**
	1.02	**1.02**	**1.02**	**1.03**	**1.03**	**1.03**	1.04	1.05	1.06	1.06

Table 4. Out-of-control ARL of the EWMA residual chart based on $\{\widehat{\eta}_t\}$ ($\epsilon_t \sim \Phi$, $\lambda \in \{0.1, 0.25, 0.5, 0.75, 1.0\}$, c such that $E_0(N_{E,\eta}) = 20$, first line: $\lambda = 0.1$, second line: $\lambda = 0.25$, third line: $\lambda = 0.5$, fourth line: $\lambda = 0.75$, fifth line: $\lambda = 1.0$).

$a \mid \alpha_1$	0.0	0.1	0.2	0.3	0.4	0.5
0.50	15.13	15.78	16.48	17.21	17.87	18.73
	12.96	12.96	13.02	13.00	12.76	12.33
	12.18	**12.36**	**12.55**	12.60	12.72	12.63
	12.44	12.62	12.64	12.65	12.63	12.50
	12.63	12.65	12.58	**12.54**	**12.38**	**12.17**
1.00	6.64	6.75	6.95	7.03	7.13	7.19
	5.60	5.53	**5.45**	**5.29**	**5.10**	**4.85**
	5.42	**5.47**	5.53	5.51	5.54	5.47
	5.68	5.73	5.75	5.74	5.72	5.60
	5.89	5.87	5.88	5.78	5.73	5.58
1.50	3.29	3.30	3.27	3.24	3.17	3.07
	2.89	**2.82**	**2.74**	**2.65**	**2.51**	**2.36**
	2.83	**2.82**	2.83	2.77	2.72	2.64
	2.97	2.97	2.96	2.92	2.89	2.80
	3.12	3.09	3.05	3.01	2.97	2.89
2.00	2.08	2.05	2.03	2.00	1.94	1.88
	1.85	1.83	**1.78**	**1.72**	**1.65**	**1.57**
	1.82	**1.82**	1.80	1.77	1.73	1.67
	1.88	1.87	1.85	1.82	1.78	1.72
	1.94	1.93	1.91	1.87	1.83	1.76
2.50	1.52	1.51	1.50	1.48	1.45	1.40
	1.39	1.38	**1.36**	**1.32**	**1.29**	**1.25**
	1.37	**1.37**	1.37	1.35	1.32	1.29
	1.40	1.40	1.39	1.37	1.34	1.30
	1.42	1.42	1.40	1.38	1.35	1.31
3.00	1.24	1.24	1.23	1.22	1.20	1.18
	1.17	1.17	**1.16**	**1.14**	**1.13**	**1.10**
	1.16	**1.16**	**1.16**	1.15	1.14	1.12
	1.17	1.17	**1.16**	1.16	1.15	1.13
	1.18	1.17	1.17	1.16	1.15	1.13
3.50	1.10	1.10	1.10	1.09	1.09	1.08
	1.07	**1.06**	**1.06**	**1.06**	**1.05**	**1.04**
	1.06	**1.06**	**1.06**	**1.06**	1.06	1.05
	1.06	**1.06**	**1.06**	**1.06**	1.06	1.05
	1.07	1.07	1.07	**1.06**	1.06	1.06
4.00	1.04	1.04	1.04	1.04	1.04	1.04
	1.02	**1.02**	**1.02**	**1.02**	**1.02**	**1.02**
	1.02	**1.02**	**1.02**	**1.02**	**1.02**	**1.02**
	1.02	**1.02**	**1.02**	**1.02**	**1.02**	**1.02**
	1.02	**1.02**	**1.02**	**1.02**	**1.02**	**1.02**

5 Empirical Examples

The intension of this section is to demonstrate the utility of control charts in finance. As examples, the daily returns of the Lufthansa and the AEG stock are analyzed. The aim is to detect deviations from a supposed process. In the present cases the target process was determined from the data of a starting block. These observations are used to fit a model including a GARCH process. In practice these data set should be chosen based on the experience of the analyst. It should reflect the characteristic behaviour of the returns at a given period. For the analysis of the data set the SAS procedures ARIMA and AUTOREG were applied. The parameters of the GARCH process were estimated by the maximum likelihood method.

5.1 The Lufthansa Stock

The starting block consists the returns of the Lufthansa stock dated from 01/03/92 to 02/27/92 (40 observations). First the test of Ljung/Box (1978) was considered to decide whether the observations are uncorrelated or not at all. Taking the number of lags h equal to 8 we obtained a p-value of 0.685. Consequently, it is concluded that there are no significant deviations from whiteness. Next the squares of the observations were determined. The McLeod and Li statistic (McLeod/Li (1983)) was calculated which is a Ljung-Box-type statistic applied to the squared returns. The number of lags was again chosen equal to 8. Now, the alternative hypothesis was accepted (p-value: 0.005), i.e. it turned out that the squared process is correlated. These two results give rise to the assumption that the data can be described by a GARCH model.

To determine the order of the GARCH(p,q) process we make use of the fact that the squared series follows an ARMA(max{p,q},p) process (cf. Bollerslev (1986)). The analysis of the partial autocorrelations and the use of Schwarz's Bayesian information criterion (SBC) as well leads to identification of the following ARCH(1) process

$$Y_t \;=\; \epsilon_t \, \sqrt{2.6 + 0.425\, Y_{t-1}^2}\,. \tag{7}$$

Next, it was examined whether the variables $\{\epsilon_t\}$ are normally distributed indeed. The normality test of Bera et al. (see SAS/ETS manual) yields a p-value of 0.777. Consequently, the normal assumption was not rejected.

For our further studies the process (7) was considered as the target process. Based on this model the returns $\{R_t\}$ of the Lufthansa stock from 02/28/92 to 07/24/92 (100 observations, see Figure 1) were analyzed. Now let $Y_t = R_t$. As usually R_t is calculated by $R_t = 100 \log(K_t/K_{t-1})$ where K_t is the value of the stock. The factor 100 is used to express the returns in percent. We made use of the modified Shewhart chart and the modified EWMA chart with $\lambda = 0.25$. The other attempts (described in Sections 3 and 4) were neglected since the EWMA procedure provides better or similar results (cf. Section 4). The critical values were taken from Severin/Schmid (1996).

In Figure 1 the control limits of the Shewhart chart are plotted. The dotted lines are control limits of the modified Shewhart chart. Here the in-control ARL is equal to 20. All observations falling outside the control limits are marked by the symbol *. It is concluded 8 times that the process is out-of-control.

Fig. 1. Returns of the Lufthansa stock ($E_0(N_S) = 20$).

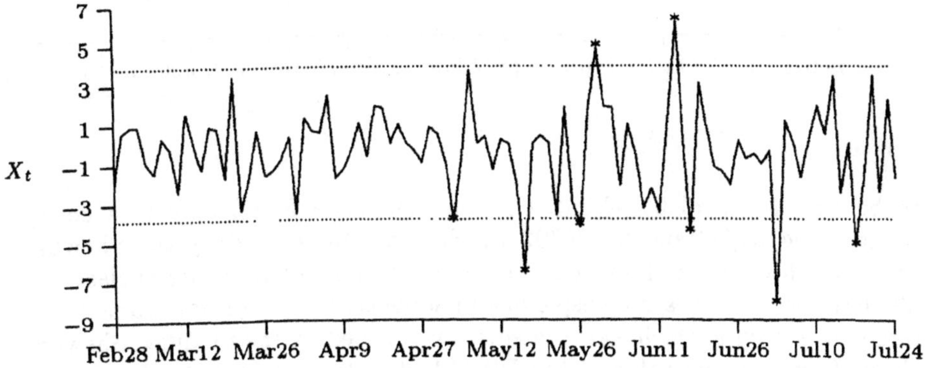

Table 5 illustrates what happens if the in-control ARL is taken equal to 40 and 60 respectively. Since the critical values are larger the number of detected positions is getting smaller. The notation $S(20)$ means that the modified Shewhart chart with an in-control ARL of 20 was applied. The date of each suspicious observation is indicated by $\times$. $S_{iid}(20)$ refers to the case that instead of the exact critical value that of the independent case is used. Since this value is larger than the exact one the number of detected change points is smaller (7 observations). The application of the EWMA scheme is a little bit more difficult. The question arises how we continue our study if an observation is classified to be out-of-control. Here we made use of a restart, i.e. for the calculation of the EWMA statistic immediately after a change point the suspicious value is replaced by the expectation of the process (here 0). Figure 2 displays the behaviour of the EWMA statistic with smoothing coefficient $\lambda = 0.25$. The dotted lines refere to the control limits of the modified EWMA chart. Again the in-control ARL is chosen equally to 20. A comparison with Figure 1 shows that the number of detected changes is the same, but not the positions.

The date of each position is listed in Table 5 (see $E(\cdot)$ and $E_{iid}(\cdot)$). While $S(20)$ gives an alarm on 05/04/92 and 07/17/92 these values are not classified

Fig. 2. Modified EWMA chart applied to the Lufthansa stock returns ($\lambda = 0.25$, $E_0(N_E) = 20$).

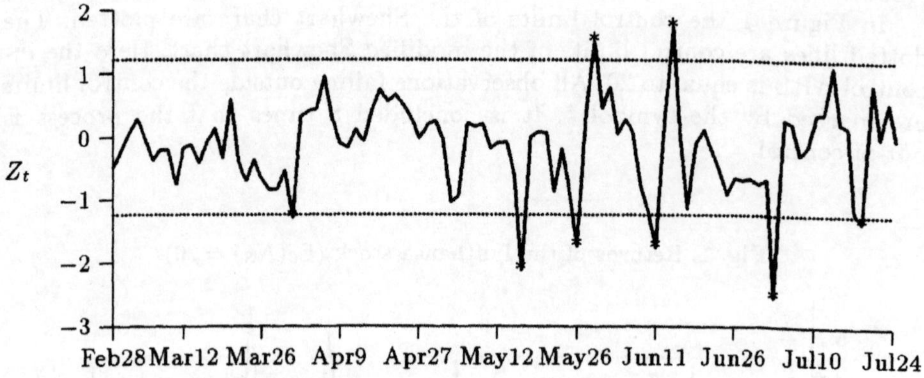

to be out-of-control by $E(20)$. On the other side $E(20)$ detects a change point on 04/01/92 and 07/20/92 which is not discovered by the Shewhart attempt. However, a closer look on Figure 1 and Figure 2 makes clear that all these observations are suspicious in some sense. On Friday, June 17th a large jump occurs which was immediately detected by the modified Shewhart chart. However, the value of the EWMA statistic is a little bit larger than the lower control limit. On the next trading day an alarm is given. This result is not surprising since, as we pointed out in Severin/Schmid (1996), the Shewhart chart is more sensible to find large shifts.

Table 5. Changes detected by the modified Shewhart and the modified EWMA chart with $\lambda = 0.25$.

date	return	$S(20)$	$S(40)$	$S(60)$	$E(20)$	$E(40)$	$E(60)$	$S_{iid}(20)$	$E_{iid}(20)$
Apr01	-3.52				×				
May04	-3.98	×							
May15	-6.65	×	×	×	×	×	×	×	×
May26	-4.25	×			×	×		×	×
May29	4.77	×	×		×			×	×
Jun02	1.80					×			
Jun11	-3.52				×	×			×
Jun15	6.10	×	×	×	×	×		×	×
Jun17	-4.54	×						×	
Jul03	-8.19	×	×	×	×	×	×	×	×
Jul17	-5.21	×	×	×				×	
Jul20	-1.80				×				

Another model to describe the behaviour of stock market returns is the Capital Asset Pricing Model (CAPM).Here it is assumed that the returns of a stock are related with the returns of an index. Now we compared the Lufthansa returns R_t dated from 01/03/92 to 02/27/92 with the returns of the DAX(30) R_t^{DAX} (from 01/03/92 to 02/27/92). The model is given by

$$R_t = \beta R_t^{DAX} + Y_t. \tag{8}$$

$\{Y_t\}$ stands for a GARCH process.

The SAS procedure AUTOREG provides $\beta = 1.4682$ and the ARCH(1) error process

$$Y_t = \epsilon_t \sqrt{0.8 + 0.865\, Y_{t-1}^2}. \tag{9}$$

Furthermore, the SBC for this model is 157.5 while for the previous model we have 176.4 . In the following the process (8) with $\{Y_t\}$ as given in (9) was assumed to be the target process. Based on this model the Lufthansa stock from 02/28/92 to 07/24/92 (100 observations, see Figure 3) was analyzed. The residual process is

$$X_t = 100 \left\{ \log\left(\frac{K_t}{K_{t-1}}\right) - 1.4682 \log\left(\frac{K_t^{DAX}}{K_{t-1}^{DAX}}\right) \right\}, \tag{10}$$

where K_t is the value of the Lufthansa stock and K_t^{DAX} is the value of the DAX(30). Again we make use of the modified Shewhart chart and the modified EWMA chart with $\lambda = 0.25$. In Figures 3 and 4 the control limits of the modified Shewhart chart and the modified EWMA chart ($\lambda = 0.25$) for an in-control ARL of 20 are dotted. All observations falling outside the control limits are signed by an asterisk.

Fig. 3. Modified Shewhart chart applied to the residuals $\{X_t\}$ of the Lufthansa stock ($E_0(N_S) = 20$).

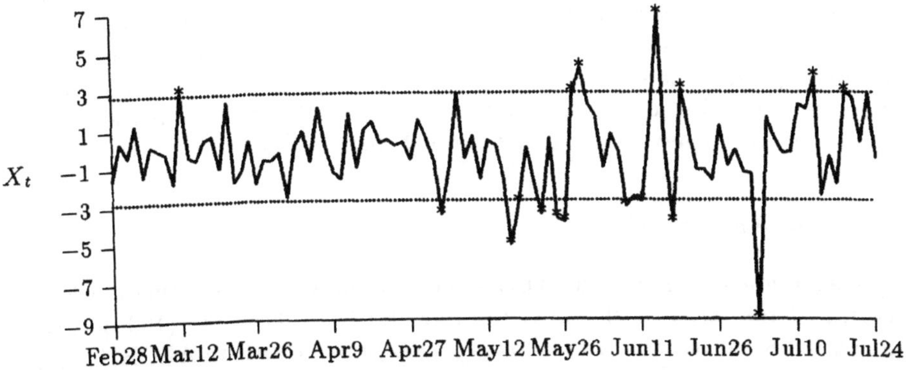

For an in-control ARL of 20 the modified Shewhart chart gave 16 alarms while the EWMA attempt led to 12 changes. An overview is given in Table 6. Note that in the present case the control schemes provide hints about shifts in the residual process $\{Y_t\}$ with $Y_t = R_t - 1.4682\, R_t^{DAX}$ and not directly about the returns $\{R_t\}$ itself. Consequently we are interested in detecting deviations from the stock return with respect to the index return. This is another standpoint as treated before. However, a comparison of Table 5 and Table 6 shows, that in a lot of cases the detected positions are the same. Based on model (9) neither the Shewhart chart (cf. Table 5, $S(20)$) nor the EWMA design (cf. Table 5, $E(20)$) detected a change point at May 18th and at May 20th. This is obvious since the returns are equal to zero. If we consider the process $\{X_t\}$ based on the CAPM the Shewhart chart signals a change point on May 18th (cf. Table 6 and Figure 3) and the EWMA attempt gives an alarm on May 20th (cf. Table 6 and Figure 4).

Fig. 4. Modified EWMA chart applied to the residuals $\{X_t\}$ of the Lufthansa stock ($\lambda = 0.25$, $\mathrm{E}_0(N_E) = 20$).

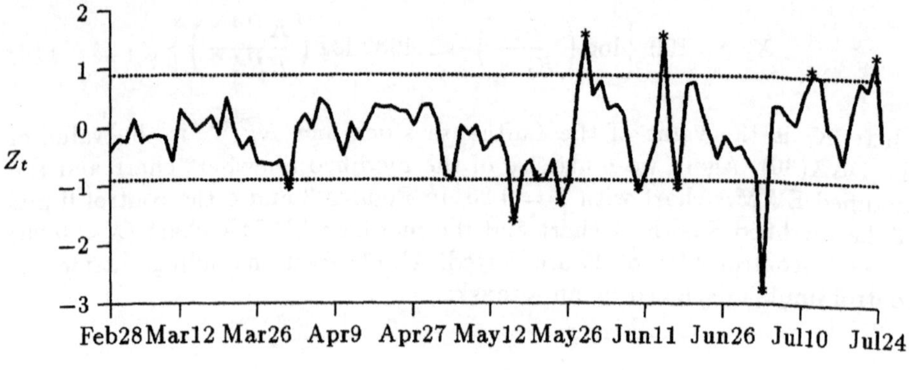

Frequently the residuals are not normally distributed. As Bollerslev (1987) pointed out, in many cases the distribution is heavy-tailed. He recommended to use a t-distribution. Note that even in this case the proposed procedure can be applied. It is only necessary to calculate the critical values for the underlying distribution. In the next example we take care about the recommendation of a heavy-tailed distribution (t-distribution).

5.2 The AEG Stock

In this section we analyze the AEG stock. We chose as the starting block the returns of the AEG stock dated from 12/05/84 to 09/11/85 (189 observations). It turned out that the following AR(2) model with an ARCH(1) error

Table 6. Changes detected by the modified Shewhart and the modified EWMA chart with $\lambda = 0.25$.

date	return	$S_{iid}(20)$	$S(20)$	$S(40)$	$S(60)$	$E(20)$
Mar12	1.59		×			
Apr01	-3.52					×
May04	-3.98		×			
May15	-6.65	×	×	×	×	×
May18	0.00		×			
May20	0.00					×
May21	-3.68		×			
May25	-2.99		×			×
May26	-4.26		×	×		×
May27	1.72		×			
May29	4.78		×	×		×
Jun09	-3.33		×			
Jun10	-2.28					×
Jun11	-3.52		×			
Jun15	6.11	×	×	×	×	×
Jun17	-4.55					×
Jun19	3.05		×			
Jul03	-8.20	×	×	×	×	×
Jul13	0.43					×
Jul14	3.36		×			
Jul20	-1.80		×			
Jul23	2.23					×

process is a suitable model to describe the AEG returns.

$$R_t = 0.238\,R_{t-1} - 0.16\,R_{t-2} + Y_t, \tag{11}$$

$$Y_t = \epsilon_t\,\sqrt{1.981 + 0.322\,Y_{t-1}^2}, \tag{12}$$

$$\epsilon_t \sim t_6. \tag{13}$$

The random variables $\{\epsilon_t\}$ are t-distributed with 6 degrees of freedom.

In the following the process (12) is considered as the target process. Based on this model the returns of the AEG stock from 09/12/85 to 11/29/85 (56 observations) were analyzed. Therefore, the residual process is

$$X_t \;=\; R_t - 0.238\,R_{t-1} + 0.16\,R_{t-2}.$$

Fig. 5. Values of the AEG stock (09/12/85 - 11/29/85).

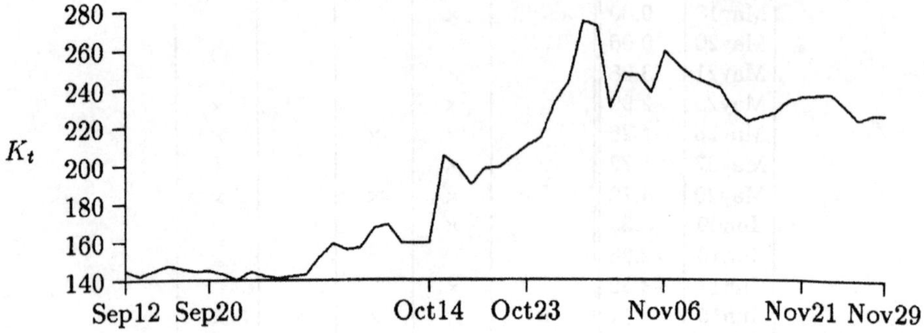

Figure 5 shows that their is an enormous increase of the stock value between 09/12/85 and 11/29/85, caused by the acquisition of Daimler Benz. At that time a chairman was suspected to profit by inside informations. Actually, he bought 700 shares between 09/12/85 and 09/20/85. The intention of the acquisition was published on 10/14/85. The chairman bought the shares before that time.

The Shewhart chart and the EWMA attempt (calibrated for an in-control ARL of 20) reflect the enormous pike at 10/15/85 (cf. Figures 6 and 7, Table 7) but also the activities of this chairman. The Shewhart chart gives an alarm at 10/02/85, 10/08/85 and 10/10/85 and the EWMA chart at 10/02/85 and 10/08/85. This fact gives rise to the suspicion that there were speculations or inside informations about the take over before the announcement of the acquisition. Within three weeks the value of the AEG stock increased by 70% after this announcement. In the period 10/15/85 to 11/13/85 the Shewhart and the EWMA chart detected 11 disturbances due to the enormous ascent of the stock value.

Fig. 6. Modified Shewhart chart applied to the AEG stock returns (09/12/85 - 11/29/85, $E_0(N_S) = 20$).

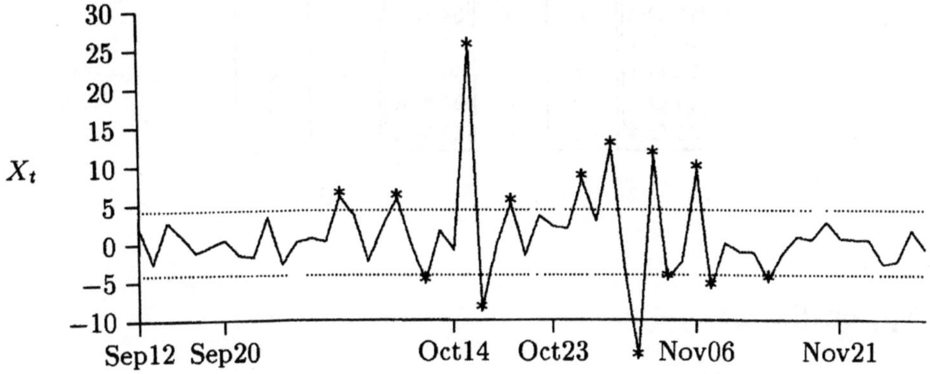

Fig. 7. Modified EWMA chart applied to the AEG stock returns (09/12/85 - 11/29/85, $\lambda = 0.25$, $E_0(N_E) = 20$).

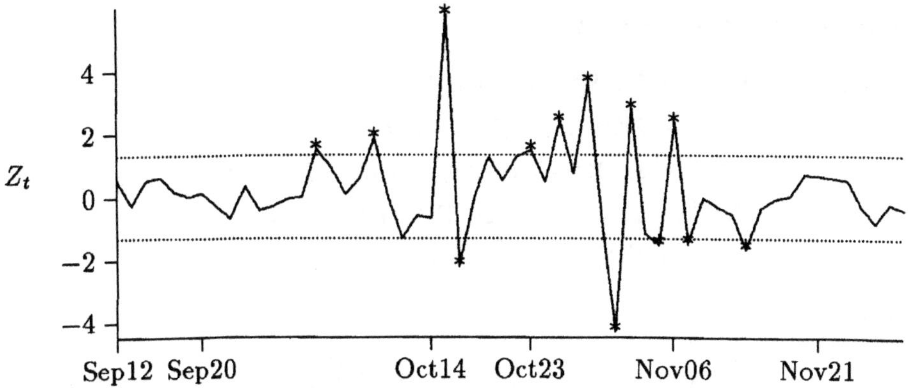

Table 7. Changes detected by the modified Shewhart and the modified EWMA chart with $\lambda = 0.25$ (AEG stock returns dated from 09/12/85 to 11/29/85).

date	$S(20)$	$E(20)$	date	$S(20)$	$E(20)$
02.Oct	×	×	29.Oct	×	×
08.Oct	×	×	31.Oct	×	×
10.Oct	×		01.Nov	×	×
15.Oct	×	×	04.Nov	×	
16.Oct	×	×	05.Nov		×
18.Oct	×		06.Nov	×	×
23.Oct		×	07.Nov	×	×
25.Oct	×	×	13.Nov	×	×

Appendix

A Proof of Equation (3.4)

In this section we prove equation (6). The process $\{\widehat{\eta}_t\}$ is given by

$$\widehat{\eta}_t = \begin{cases} Y_1^2 - \mathrm{E}(Y_1^2) & \text{for } t = 1 \\ Y_t^2 - (\alpha_0 + \alpha_1 Y_{t-1}^2) & \text{for } t \geq 2. \end{cases}$$

First note that $\mathrm{E}(\widehat{\eta}_t) = 0$ since $\gamma_0 = \frac{\alpha_0}{1-\alpha_1}$.
For $t = 1$ we obtain

$$\mathrm{Var}(\widehat{\eta}_1) = \mathrm{E}(\widehat{\eta}_1^2) = \mathrm{E}\left((Y_1^2 - \mathrm{E}(Y_1^2))^2\right) = \mathrm{E}(Y_1^4) - \mathrm{E}(Y_1^2)^2.$$

Moreover, it follows for $t \geq 2$

$$\begin{aligned} \mathrm{Var}(\widehat{\eta}_t) &= \mathrm{E}(\widehat{\eta}_t^2) = \mathrm{E}\left((Y_t^2 - (\alpha_0 + \alpha_1 Y_{t-1}^2)^2\right) \\ &= \mathrm{E}\left(Y_t^4 - 2Y_t^2(\alpha_0 + \alpha_1 Y_{t-1}^2) + (\alpha_0 + \alpha_1 Y_{t-1}^2)^2\right) \\ &= \mathrm{E}\left(Y_t^4 - 2\alpha_0 Y_t^2 - 2\alpha_1 Y_t^2 Y_{t-1}^2 + \alpha_0^2 + 2\alpha_0\alpha_1 Y_{t-1}^2 + \alpha_1^2 Y_{t-1}^4\right) \\ &= \mathrm{E}(Y_t^4) - 2\alpha_0 \mathrm{E}(Y_t^2) - 2\alpha_1 \mathrm{E}(Y_t^2 Y_{t-1}^2) + \alpha_0^2 \\ &\quad + 2\alpha_0\alpha_1 \mathrm{E}(Y_{t-1}^2) + \alpha_1^2 \mathrm{E}(Y_{t-1}^4) \end{aligned}$$

with

$$\begin{aligned} \mathrm{E}(Y_t^2 Y_{t-1}^2) &= \mathrm{E}\left(\epsilon_t^2(\alpha_0 + \alpha_1 Y_{t-1}^2) Y_{t-1}^2\right) \\ &= \mathrm{E}(\epsilon_t^2)\ \left(\alpha_0 \mathrm{E}(Y_{t-1}^2) + \alpha_1 \mathrm{E}(Y_{t-1}^4)\right). \end{aligned}$$

To guarantee the existence of the $\mathrm{Var}(\widehat{\eta}_t)$ the existenceounterparties contained within the portfolio. of the fourth moment of Y_t is required. Thus we have the restriction $0 \leq \alpha_1 < \frac{1}{\sqrt{3}}$. Because of the weak stationarity of $\{Y_t\}$ and this restriction it follows

$$\begin{aligned}\mathrm{Var}(\widehat{\eta}_t) &= \mathrm{E}(Y_t^4) - 2\alpha_0\mathrm{E}(Y_t^2) - 2\alpha_1\left(\alpha_0\mathrm{E}(Y_{t-1}^2) + \alpha_1\mathrm{E}(Y_{t-1}^4)\right) + \alpha_0^2 \\ &\quad + 2\alpha_0\alpha_1\mathrm{E}(Y_{t-1}^2) + \alpha_1^2\mathrm{E}(Y_{t-1}^4) \\ &= (1-\alpha_1^2)\mathrm{E}(Y_t^4) - 2\alpha_0\mathrm{E}(Y_t^2) + \alpha_0^2.\end{aligned}$$

B Critical Values for the EWMA Residual Chart

Table 8. Critical values c for the EWMA residual chart based on $\{\widehat{\eta}_t\}$ ($\epsilon_t \sim \Phi$, $\mathrm{E}_0(N_{E,\eta}) = 20$).

$\lambda \mid \alpha_1$	0.0	0.1	0.2	0.3	0.4	0.5
0.10	1.205	1.151	1.066	0.945	0.775	0.520
0.25	1.378	1.288	1.164	1.002	0.798	0.521
0.50	1.654	1.605	1.505	1.348	1.119	0.755
0.75	1.905	1.846	1.726	1.546	1.282	0.865
1.00	2.009	1.940	1.810	1.615	1.334	0.904

References

Alwan, L. C. and Roberts, H. V. (1988): Time-series modeling for statistical process control. Journal of Business and Economic Statistics, **6**(1), 87-95.

Bollerslev, T. (1986): Generalized autoregressive conditional heteroscedasticity. Journal of Econometrics, **31**, 307-327.

Brockwell, P. J. and Davis, R. A. (1991): Time Series: Theory and Methods. Springer, New York.

Cox, D. R., Hinkley, D. V. and Barndorff-Nielsen, O. E. (1996): Time Series Models. Chapman & Hall.

Engle, R. F. (1982): Autoregressive conditional heteroscedasticity with estimates of the variance of United Kingdom inflation. Econometrica, **50**(4), 987-1007.

Engle, R. F. and Rothschild M. (1992): ARCH Models in Finance. Journal of Econometrics, **52**.

Harris, T. J. and Ross, W. H. (1991): Statistical process control procedures for correlated observations. Canadian Journal of Chemical Engineering, **69**, 48-57.

Kramer (1997): On Control Charts for Time Series. Dissertation, Universität Ulm, Germany.

Ljung, G. M. and Box, G. E. P. (1978): On a measure of lack of fit in time series models. Biometrika, **65**(2), 297-303.

McLeod, A. I. and Li, W. K. (1983): Diagnostic checking ARMA time series models using squared-residual autocorrelations. Journal of Time Series Analysis, **4**, 269-273.

Milhøj, A. (1985): The moment structure of ARCH processes. Scand. J. Statist., **12**, 281-292.

Montgomery, D. C. and Mastrangelo, C. M. (1991): Some statistical process control methods for autocorrelated data. Journal of Quality Technology, **23**(3), 179-204.

Pagan, A. (1996): The econometrics of financial markets. Journal of Empirical Finance, **3**, 15-102.

Page, E.S. (1954): Continuous inspection schemes. Biometrika, **41**, 100-114.

Roberts, S. W. (1959): Control chart test based on geometric moving averages. Technometrics, **1**(3), 239-250.

Schmid, W. (1995): On the run length of a Shewhart chart for correlated data. Statistical Papers, **36**, 111-130.

Schmid, W. (1997a): On EWMA charts for time series. Frontiers in Statistical Quality Control. Lenz, H.-J. and Wilrich, P.-Th. (eds.), Physica Verlag, Heidelberg.

Schmid, W. (1997b): CUSUM control schemes for Gaussian processes. Statistical Papers, 38(2), 191-217.

Severin, T. and Schmid, W. (1996): Monitoring changes in GARCH models. Technical Report No. **64**, Europe-University Viadrina, Frankfurt (Oder), Germany.

Vasilopoulos, A. V. and Stamboulis, A. P. (1978): Modification of control chart limits in the presence of data correlation. Journal of Quality Technology, **10**(1), 20-30.

Yashchin, E. (1993): Performance of CUSUM control schemes for serially correlated observations. Technometrics, **35**(1), 35-52.

An Analysis of the Financing Behavior of German Stock Corporations Using Artificial Neural Networks

Manfred Steiner[1], Sebastian Schneider[1], and J. Benedict Wolf[1,2]

1 University of Augsburg, Department of Finance and Banking, Universitätsstraße 16, 86135 Augsburg, Germany
2 corresponding author: tel. +49 821 598 4119, fax: +49 821 598 4223, email: Benedict.Wolf@WiSo.Uni-Augsburg.de

Abstract. The purpose of this paper is to test the managerial and pecking order hypotheses for German stock corporations using annual reports, stock prices and other information for the years 1974 - 1992. The managerial hypothesis predicts the preference of internal to external financing if the manager's utility function diverges from the owners', to avoid additional monitoring. If internal funds are unavailable or insufficient, the management will prefer to issue debt instead of equity, because creditors usually have less chances of influencing managerial practice than the company's owners. Issuing equity is thus seen as a positive signal, as it announces the reduction of information asymmetries (contrary to the pecking order theory). In the empirical section the degree of informational asymmetry and managerial discretion is described by such variables as block ownership or market segment. Probabilistic and general regression neural networks complement the discriminant analysis, polynomial and general regression neural networks the seemingly unrelated regression.

1 Introduction

As descriptive finance developed into a theory of finance with explicit explanatory models, capital market theory has become a focus of financial theory, supplementing the theretofore dominant investment theory. The goal of the capital market theory is the simultaneous explanation of capital acquisition and employment following certain optimality criteria. While all agree on the goal, the optimality criteria are the object of discussion.

Financial actions can be classed into three general categories:

1. internal,
2. equity and
3. debt financing.

Their hierarchy can be described by various models. The starting point is *Modigliani* and *Miller's* irrelevance theorem, which does not stand up to the reality of taxes, informational asymmetries and imperfect markets.[1] Newer models partially incorporate these aspects. The pecking order theory bases on the asymmetrical information level between investors and the management of a company and assumes that the management is trying to maximize the owners' utility. As issuing stock implies the greatest tranfer of wealth away from the previous ownership, managers will prefer internal or debt financing to issuing equity. The managerial hypothesis assumes that the management is exploiting the asymmetric informational distribution in order to pursue its own interests. In this case equity finance does not transfer wealth if the investment has a positive net present value.

The goal of this paper is to analyze the financial behavior of German corporations in view of the managerial and pecking order hypotheses. The study of isolated finance forms tests the pecking order hypothesis. The managerial hypothesis by contrast is an integrated model to test causes and effects of information asymmetry. For this reason the managerial hypothesis can only be tested by simultaneously including all types of financing.

2 Financing decisions under asymmetric information

2.1 Internal vs. external funds

As long as the management of a company is solely acting in the interest of the ownership, it will only realize investment projects with a positive net present value:

NPV > I

where: NPV = net present value

I = investment outlay.

The critical state s_A identifies that state in which the ownership of a company is indifferent between investing or not. The value of the growth possibilities, depending on the probability distribution of the separate states' happening, is:

$$V(Growth) = \begin{cases} 0; s \leq s_A \\ \int_{s_A}^{\infty} qs * [V(S) - I]ds; s > s_A \end{cases}$$

By taking in external funds, less internal funds have to be provided. On the other hand dividend or interest demands in the case of equity and debt finance respectively reduce the value of the growth opportunities. In addition, assume there are informational asymmetries between investors and the creditors. While the management can estimate the cost of capital using its insider information, outside creditors must generally formulate their return demands

[1] Cf. Modigliani F., Miller, M. H. (1958), p. 261 ff.

on the basis of publically available information that was possibly manipulated by the management. If the management can issue capital for a positive NPV investment below its correctly risk-adjusted price, this increases the wealth of the current ownership. As the finance cost of nominally risk-free capital is not determined by the risk structure of a company, the information asymmetry can only be exploited by issuing risky capital. External funds always entail payment obligations, meaning the present value of the cash flows from a project must be large enough to fund all payment obligations (P) and the investment outlay:

$$V(Growth) = \begin{cases} 0; s \leq s_B \\ \int_{s_A}^{\infty} qs * [V(S) - I]ds; s > s_B \end{cases}$$

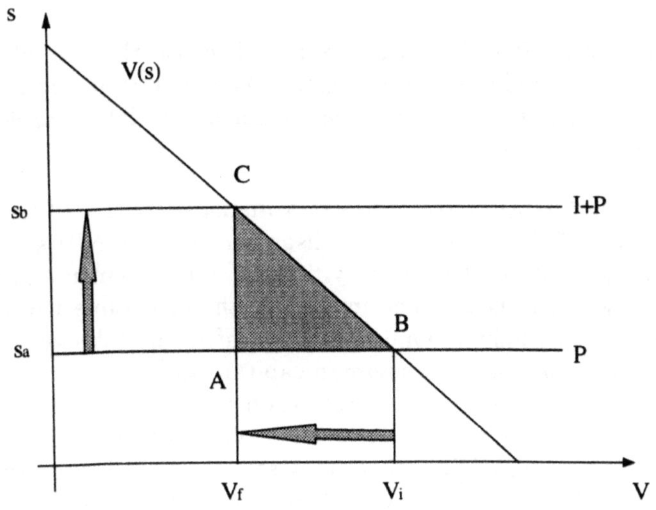

Fig. 1. External financing under asymmetric information[2]

State s_B analogously describes the situation under asymmetric information in which the management is indifferent to investing or not. As leverage can never become negative, debt financing entails a reduction in wealth to the

[2] Cf. Myers, Stewart C. (1977), p. 153

amount of the cumulated present value of all originally profitable investments (the shaded area in Figure 1) that are forgone if $s \in [s_A, s_B >$.

Equity financing does not entail any amortization obligation. That is why equity finance is preferable to debt finance if there are no information asymmetries.[3] But *Jensen* and *Meckling* point out agency problems of equity financing.[4] Both managers and owners are risk averse, as otherwise neither convex indifference curves nor capital market theories hold. Managers will be much more risk averse than the shareholders, though. The latter are able to diversify their investment capital. They only must bear the systematic portion of the risk of a certain investment. Managers on the other hand enter their non-diversifiable human capital into the firm and must watch the total risk of the company. This implies a risk asymmetry in the investors' favor as opposed to the information asymmetry in the managers' favor. Managers will want to diversify companies they control in order to reduce risk. Investments for this purpose do not necessarily have positive NPVs. Of course negative NPV projects that are financed with debt also have disadvantages for a manager:

1. Repaying the debt reduces the funds available to a manager and increase the human capital risk that the investment was originally meant to reduce.
2. The cost of capital rises with leverage because the risk of bankruptcy increases. This reduces the manager's discretionary freedom.
3. Positive NPVs of other investments must be partially transferred to the creditors.

Thus equity finance makes sense in a manager controlled firm if the advantageousness of the investment is unsure or negative NPVs are expected. Rational owners will react negatively to issue announcements in the context of the pecking order theory. The impact on the firm value will increase the probability of proxy fights, monitoring, and unfriendly takeovers, all of which negatively affect the manager's human capital risk.[5]

In owner controlled firms diversification normally is not important, as owners can more cheaply diversify on the capital markets.[6] These firms still prefer internal to external financing, because otherwise the proportion of their ownership will decrease and they will lose control of company resources.[7] In addition, the convex relation of debt to the probability of bankruptcy will restrict debt financing.[8] So the preference for internal over external funds follows from both the managerial and pecking order hypotheses in the case

[3] Cf. Myers, Stewart C.; Majluf, Nicholas P. (1984), p. 195
[4] Cf. Jensen, Michael C.; Meckling, William H. (1976), p. 315
[5] Cf. Jung, Kooyul; Kim, Yong-Cheol; Stulz, René M. (1996), p. 163
[6] Cf. Stulz, René M. (1997), p. 12
[7] Cf. Amihud, Yakov; Lev, B.; Travlos, N. G. (1990), p. 603 f.
[8] Cf. Stulz, René M. (1988), p. 25 ff.

of information asymmetry, regardless of the degree of managerial control in the company.

2.2 Substitutional effects of investments

Modigliani and *Miller's* irrelevance theorem is superseded by the discussion above. Investment and finance decisions may not be considered separately. Investment decisions are rather the basis for finance decisions in the double sense that the expected present value and the risk of the investment both affect the wealth of the the financier of the company, be it the ownership or a creditor. The literature oftentimes overlooks that positive net present values ($\frac{dV_t}{dI_t} > 0$ where V = value of company and I = investment) are not a sufficient criterion for an owner to agree to an investment. The reason for this is the substitutional relationship between the owners and creditors of a firm. The wealth effects from an investment consist of the sum of the changes in the market value of the equity andd ebt. Owner controlled firms prefer investments that increase the value of equity:

$$\frac{dE_t}{dI_t} = \frac{dV_t}{dI_t} - \frac{dD_t}{dI_t} > 0 \tag{1}$$

where E_t represents the value of the equity and D_t the value of the debt at time t. Wealth transfers at the creditors' expense increase the propensity to invest. *Myers* points out that the market value of debt can be written as a function of the corporate value and the corporate risk σ: $D_t = f(V_t, \sigma)$.[9] As decisions to invest change both the corporate value as well as the risk structure of a firm, their effect can be split into a direct effect, an indirect effect and a risk structural effect:[10]

$$\frac{dE_t}{dI_t} = \underbrace{\frac{dV_t}{dI_t}}_{direct\, effect} - \underbrace{\frac{dV_t}{dI_t}\frac{\delta f_t}{\delta I_t}}_{indirect\, effect} - \underbrace{\frac{df_t}{d\sigma_t^2}\frac{\delta\sigma_t^2}{\delta I_t}}_{risk\, structural\, effect} > 0 \tag{2}$$

When considering positive NPV investments, the direct and indirect effects are always positive, as such investments reduce the default probability of debt, $\frac{df_t}{dI_t} > 0$. On the other hand the risk structural effect will have a negative sign, due to option pricing considerations. *Merton* showed that equity can be interpreted as a European call option on the company's assets.[11] The market value of the equity increases with higher volatility of the corporate value. Rational, diversified owners maximize their wealth if: $\frac{\delta\sigma_t^2}{dI_t} > 0$. The creditors are the stillholders in this construction. The market value of the debt therefore is inversely proportional to the company risk ($\frac{df_t}{d\sigma_t^2}$). Risky investments thus

[9] Cf. Myers, Stewart C. (1984), p. 575 ff.
[10] Cf. Myers, Stewart C. (1977), p. 165
[11] Cf. Merton, R. C. (1974), p. 449 ff.

transfer wealth from the creditors to the owners. This risk structural effect is independent of the advantageousness of the investment. Negative NPV investments have negative direct and indirect effects. As manager controlled firms invest risk aversely, the risk structural effect will be positive in this case and counteract the negative NPV: $\overbrace{\underbrace{\frac{df_t}{d\sigma_t^2}}_{<0} * \underbrace{\frac{d\sigma_t^2}{dI_t}}_{<0}}^{>0}$ At the same time this implies wealth transfers to the creditors at the owners' expense which causes agency costs of equity finance. Such investments thus are more probable for owner controlled firms (Figure 2), while manager controlled firms will more often forego profitable investments. In consequence the underinvestment problem described in the literature for manager controlled firms is offset by a potential overinvestment problem of owner controlled firms.[12]

This description shows that wealth transfers between financiers of a company that are seen as the reason for agency costs are not the result of the financing decision but of the investment decision.[13] The financing decision instead determines the level of the wealth transfer and accordingly the order of preference of the financing instruments.

	owner controlled firms		manager controlled firms	
direct effect	+	-	+	-
indirect effect	-	+	-	+
risk structural effect	+	+	-	-
net present value	$\frac{dV_t}{dI_t} > 0$	$\frac{dV_t}{dI_t} < 0$	$\frac{dV_t}{dI_t} > 0$	$\frac{dV_t}{dI_t} < 0$

Fig. 2. Substitutional effects of corporate finance on investments

2.3 Debt vs. equity financing

We showed above that internal funds will always be preferred to external financing both by the ownership and the management in the case of asymmetric information. It also became clear that wealth transfers are initiated by the decision to invest, not the choice of finance method. The method of finance only determines the absolute and relative level of these transfers between investor groups.

The priority ranking of the financial instruments is a result of the level of wealth transfers and can be explained with option pricing theory. Here

[12] Cf. Cooney, John W. Jr.; Kolay, Avner (1993), p. 158

[13] Carpenter, Robert E. (1994), p. 22 comes to principally the same results.

we must differentiate between managerialism and pecking order. Both are grounded on similar assumptions with managerialism as a special case of the pecking order, as the asymmetric information condition is expanded by the assumption of managers maximizing their own utility. The pecking order theory focuses on the interests of the current ownership and ranks the financing instruments by the relative level of wealth transfers induced by new financing actions. The managerialism hypothesis results from the level of control in the company. Managers in owner controlled firms maximize the utility of the current ownership and follow the incentive structure of the pecking order hypothesis. Managers in manager controlled firms optimize their own risk-return profile.[14] The human capital risks of the managers are negatively correlated to the resources under the manager's control. That is why a manager will choose the finance instrument for an investment by the resources that will be transferred to the external financier as a compensation for receiving the investment outlay.

As companies in all cases prefer internal to external finance, they will create liquidity reserves to continuously enable internal financing. If the investment volume I exceeds the liquidity reserves S, external funds will be necessary at the amount of I - S. The money may be provided by issuing equity or debt.[15]

When assuming that the manager is maximizing the market value of the equity of the current ownership (pecking order hypothesis[16]) or that he is maximizing the value of the investment remaining in the company (managerialism), the manager will only invest if the value of the company increases through the investment. In the case of equity financing:

$$S + a \leq S + a + b - \Delta E \tag{3}$$

The left term shows the wealth of the current ownership without the investment and therefore also without a seasoned issue. It consists of the sum of the financial slack S and the future value of assets a that the management can better quantify than external investors due to the information asymmetry. The right hand side shows the wealth of the current ownership in the case of investing via an equity issue. It contains the slack, the future value of the assets and the present value of the investment b minus the transfer of resources to the new shareholders ΔE. The investment will only take place if the present value of the investment exceeds the value of the transfer to the new investors, so if $b \geq \Delta E$.

Credit finance analogously takes place if the added value of the investment is greater than the resources transferred to the creditors in the light of the changed risk structure of the company, so if $S+a \leq S+a+b-\Delta D$ and hence $b \geq \Delta D$. If a company can issue risk-free capital, the creditors cannot profit

[14] Cf. Garvey, Gerald T.; Hanka, Gordon (1997), p. 5 ff.
[15] Cf. Garvey, Gerald T.; Hanka, Gordon (1997), p. 5 ff.
[16] Cf. Myers, Stewart C., Majluf, Nicholas P. (1984), p. 207

from a change in state. In that case all investments with a positive present value will be undertaken. The finance decision is governed by the comparison of ΔE and ΔD both for the pecking order and the managerialism. With option price theory one can show that $\Delta E \geq \Delta D$.

Issuing debt is akin to selling the total assets to the creditor. At the same time the owners receive a European call option on the total assets from the creditors. The strike price is the retirement price P of the debt and the maturity date is the repayment date T for the debt. Using the *Black & Scholes* model[17], the value of the equity is:

$$E = V * N(d_1) - P * e^{r_F * t} * N(d_1) \tag{4}$$

$$d_1 = \frac{ln\frac{V}{P} + (r_F + 0,5 * \sigma^2)t}{\sigma * \sqrt{t}} \tag{5}$$

$$d_2 = d_1 - \sigma * \sqrt{t} \tag{6}$$

with: E = market value of equity
V = market value of company
σ = standard deviation of corporate profits
r_F = risk free interest rate
$N(d)$ = cumulative standard normal density function
$n(d)$ = standard normal density function
P = retirement price of debt
t = years to maturity of debt
T = maturity date of debt

Assuming that the investment changes the risk structure and the corporate value, the effects of the investment on the wealth of the shareholders or creditors can be analyzed separately. The partial derivative of the market value of equity by the variance as an indicator of corporate risk illustrates the effect of a change in risk due to an investment: $\frac{dE}{D\sigma^2} = P * e^{-r_F * t} * n(d_2) * \frac{\sqrt{T}}{\sigma} > 0$.[18] The value of the equity increases with the variance. If the investment raises the corporate value, the value of the equity also rises: $\frac{dE}{dV} = N(d_1) > 0$.

The market value of a corporation is the sum of the market values of equity and debt. The wealth of the creditors can therefore be written as: $D = V * N(-d_1) + P * e^{-r_F * t} * N(d_2)$.[19] The impact of corporate risk on the value of the debt is visible in the partial derivation of the value of debt by the variance:

$$\frac{dD}{d\sigma^2} = \overbrace{\underbrace{- P * e^{-r_F * t} * n(d_2) * \frac{\sqrt{t}}{\sigma}}_{>0}}^{<0}. \tag{7}$$

[17] Black, Fischer; Scholes, M. (1973), p. 637 f.
[18] Merton, R. C. (1974), p. 449 f.
[19] $D = V - E$ and therefore $D = V - V * N(d_1) + P * E^{-r * t} * N(d_2)$

One sees that the value of debt falls when corporate risk rises. The value of debt increases with the corporate value, as: $\frac{dD}{dV} = N(-d_1) = [1 - N(d_1)] > 0$.

By contrasting the various effects we can compare the wealth transfers to new creditors or owners (pecking order hypothesis) or their participation in the creation of added value by investing (managerialism). One has to be careful to look at the full owner or creditor groups though. The level of transfers can be determined as the new financiers are integrated into the old groups. Their gains in wealth determine the financing decisions.

The costs of the principal-agent relation are higher in equity financing than in debt financing, as the following terms are always true in the option pricing model: $\frac{dE}{dV} \geq \frac{dD}{dV}$ and $\frac{dE}{d\sigma^2} > \frac{dD}{d\sigma^2}$.[20] Therefore the wealth transfers are always greater in equity financing than in debt financing, i. e. $\Delta E \geq \Delta D$. Hence debt financing will always be preferred in both models, regardless of the control situation in the company.[21]

2.4 Hypotheses for the empirical analysis

The empirical analysis must thus address two hypotheses. In the pecking order and managerial context, companies prefer internal finance to external finance. If the internal resources are depleted, the company turns to external sources, preferring debt finance to credit finance.

ranking by preference	pecking order/managerialism
1.	internal finance
2.	debt finance
3.	equity finance

Fig. 3. Hypotheses for the empirical analysis

Long term debt is preferable to short term debt in the managerial context, if that implies that the possibility to monitor the company's situation are less frequent. Very short term debt such as liabilities for supplies and services is usually not connected to frequent revelations of corporate standing though, so the order might be ambiguous. The pecking order theory makes no reference to a maturity preference for debt.

---

20 The first condition must always hold as d_1 never can be negative and for all positive fractiles of the standard normal distribution $N(d_1) \geq 0.5$. The second inequality always holds as $\frac{dE}{d\sigma^2}$ is always positive and $\frac{dD}{d\sigma^2}$ is always negative.

21 For this reason more positive NPV investments are foregone in equity financing than with dept financing.

3 The GMDH algorithm as an instrument for the empirical analysis

Conventional mathematical-statistical analysis usually prespecifies the structure of the decision space or the instruments of analysis. The definition of optimality criteria requires a knowledge of *a priori* probabilities. Density functions are also described by a number of parameters. Implicitly the assumption is made that these parameters remain untouched by the determinants of the estimation.[22] In empirical analyses researchers are often confronted with complex problems of unknown structure. Applying traditional statistical methods means possibly incompletely describing the decision space and thus misspecifying the procedures. The results can then only be interpreted conditionally. Several techniques have been developed to solve such problems. We present one such algorithm in more detail here.

Polynomial neural networks (PoNN) are based on the 'Group Method of Data Handling' (GMDH) algorithm developed by Ivakhnenko.[23] GMDH algorithms allow a successive development of a system description. Simple mathematics can describe the system. Of a great number of variants, for this analysis "Iterative Multilayered GMDH Algorithms" proved useful.[24]

Optimization is a three-step process (cf. Figure 4).[25] In the first step all independent variables $(x_1, ..., x_n)$ are grouped pairwise $\{(x_1, x_2), (x_1, x_3), ..., (x_1, x_n), ..., (x_{n-1}, x_n)\}$. Using these variables $n(n-1)/2$ regression equations $y = A + Bx_i + Cx_j + Dx_i^2 + Ex_j^2 + Fx_ix_j$ are created. Then each regression term is verified with the complete training and test data sample $\{(x_{1i}, x_{1j}), (x_{2i}, x_{2j}), ..., (x_{mi}, x_{mj})\}$.[26] The result of the calculations for m variable realizations are put together with the estimates of the remaining regression functions into a new matrix Z. Only the best new variables as determined using the test sample are retained ('survival of the fittest').[27] The newly generated variables can be seen as improved versions of the first generation variables.

In the second optimization step the original variables x_i are replaced by the newly generated variables z_i. Those z_i are used that best describe y. For every combination of the original variables x_i and x_j the root of the mean square error is used to compute the so-called regularity criterion as a measure

22 Cf. Malone, James M., p. 69

23 Cf. Ivakhnenko, A. G. (1968)

24 For an overview over existing GMDH algorithms cf. Ivakhnenko, A. G.; Ivakhnenko, G. A. (1995), p. 572 ff.; Dolenko, S. A.; Orlov, Yu. V.; Persiantsev, I. G. (1996); Müller, J.-A.; Ivakhnenko, G. A. (1995); Müller, J.-A. (1995); Müller, J.-A.; Ivakhnenko, G. A. (1996)

25 Cf. Farlow, Stanley J. (1984), p. 4 f.

26 The dataset is divided into three samples, a training, test and holdout sample.

27 The number of variables to be retained is decided beforehand.

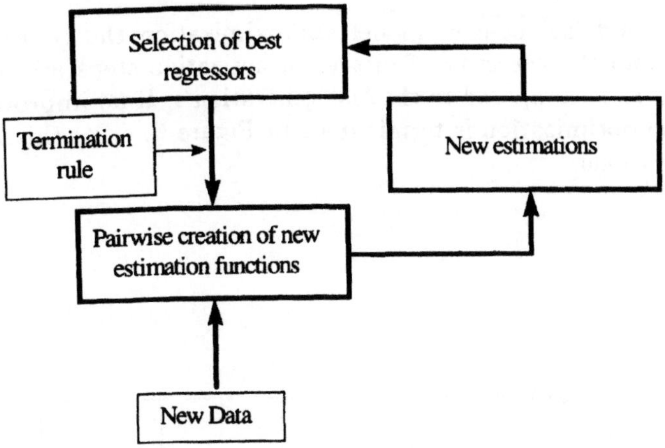

Fig. 4. GMDH algorithms

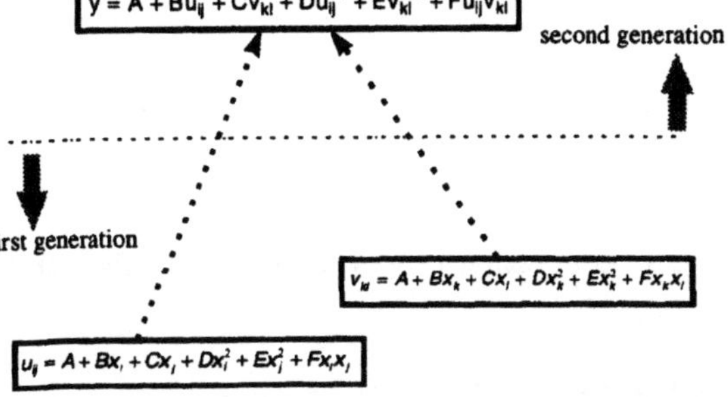

Fig. 5. Variable generation with iterative multilayered GMDH algorithms

of goodness - again via the test data.[28] All variables z_{ij} with $r_j^2 < R$ are kept as new regressors. R is prespecified.

$$r_j^2 = \frac{\sum_{i=mt+1}^{m}(y_i - z_{ij}^2)}{\sum_{i=mt+1}^{m}(y_i^2)} \quad j = 1, 2, ..., n. \tag{8}$$

In the third and last step of model approximization the goodness of the total model and the necessity of further optimization steps are established. The smallest r_j^2 is compared to the last generation's. If no improvement was achieved, the optimization is terminated. In Figure 6, e. g., this is the case after 8 generations.

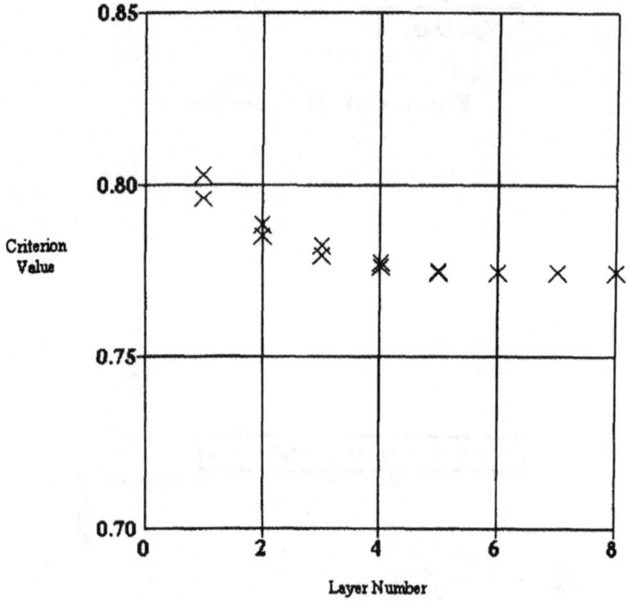

Fig. 6. Adaptation process of a GMDH algorithm

[28] In contrast to Farlow, Stanley J. (1984) all optimality criteria are computed exogenously using an holdout sample, cf. Ivakhnenko, G. A. (1997); Ivakhnenko, A. G.; Müller, J.-A. (1995), p. 93 ff.; Lemke, F.; Müller, J.-A. (1995); Müller, J.-A.; Ivakhnenko, G. A. (1997); Ivakhnenko, A. G.; Müller, J.-A. (1997); Ivakhnenko, A. G.; Ivakhnenko, G. A.; Müller, J.-A. (1994), p. 185 ff.

The final regression term consists of regressors that were created as a synthesis of various variables from earlier generations. These variables can be recursively expressed in terms of the original variables. The resulting term is called an Ivakhnenko polynomial (or Kolmogorov-Gabor polynomial).

$$y = a + \sum_{i=1}^{n} b_i x_i + \sum_{i=1}^{n}\sum_{j=1}^{n} c_{ij} x_i x_j + \sum_{i=1}^{n}\sum_{j=1}^{n}\sum_{k=1}^{n} d_{ijk} x_i x_j x_k + \ldots.. \quad (9)$$

By successively developing a system of equations to describe the problem on hand a simultaneous solution algorithm is used that is similar to genetic algorithms. It does not require the prespecification of the decision space or of the instruments of analysis.[29]

Especially the third optimization step is problematic. The regularity criterion described above is sensitive to noisy data, so it regularly fails in long term forecasts.[30] That is why other criteria are often used. These include:

- the unbiased criterion and
- the combined criterion.

The unbiased criterion is insensitive to noise. To determine it the data is subdivided into two samples A and B. The GMDH algorithm is then independently applied to each sample. The unbiased criterion is the difference in forecasting quality between both (combined) variables z_{ij}:

$$u_j^2 = \frac{\sum_{i=l}^{m}(z_{ij}^A - z_{ij}^B)^2}{\sum_{i=l}^{m}(y_i^2)} j = 1, 2, ..., n. \quad (10)$$

The variable z_{ij} is only taken over into the new optimization step if $u_j^2 < U$, with U again being a prespecified hurdle rate. Empirical research has shown that applying the unbiased criterion leads to a poorer approximization of the problem set. The combined criterion tries to avoid this deficiency by integrating the regularity criterion (high forecast goodness) and the unbiased criterion (insensitive to noise) into a new measure. After splitting the dataset into three samples A, B and C and creating the matrices Z_A and Z_B as shown above, the combined criterion is:

$$c_j = \sqrt{(r_j^2 - u_j^2)} \quad j = 1, 2, ..., n. \quad (11)$$

The variables z_{ij} are taken over into the next generation as long as $c_j^2 < C$. C again is a prespecified constant.

---

29 Cf. Steiner, Manfred; Wittkemper, Hans-Georg; Wolf, J. Benedict (1998); Wittkemper, Hans-Georg (1994), p. 66 ff. on genetic algorithms

30 Cf. Ivanhnenko, A. G.; Ivanhnenko, G. A.; Vysotsky, V. M. (1976)

Modifications of the GMDH algorithm accompanied the expansion of the GMDH method by alternative non-parametric algorithms. The original variables x_i for example are also considered as alternatives when determining the regressors. As the *a priori* data distribution dilineates the decision space by generating cluster centers, these centers signal if extrapolating or interpolating GMDH algorithms are necessary to identify the optimal solution.[31] [32] The value of the cluster generation lies in the combination of locally receptive areas with the GMDH algorithms that cover the whole decison space, which allows the decision space containing a concentration of clusters to be dilineated. This permits the isolated analysis of the structure of the data field. Clusters significantly outside of this space (noise) can be eliminated and nonstationary relationships can be identified. Nonstationary relationships are characterized by shifting datafields with high concentrations of cluster centers.[33] In addition, in the multidimensional decision space variously shifting single datafields causing autocorrelation can also be accounted for.

GMDH algorithms in PoNN are implemented following the procedures described above and are combined with genetic algorithms. Genetic algorithms ensure the optimal regressor combination to describe the model. PoNN also employ active neurons instead of genetic algorithms.[34]

As opposed to conventional passive neurons that are based on exactly defined base functions, active neurons themselves are multilayered processing units. The GMDH algorithms execute inside each neuron. Each neuron depicts a newly generated variable by independently selecting the relevant inputs necessary for an optimal model. Similar to the optimality criteria, the optimal structure of the neuron must be evaluated. Therefore the polynomials are not created by a conventional but by a doubly multilayered feed-forward network.

Next to PoNN probabilistic neural networks (PNN) and general regression neural networks (GRNN) are used that are based on radial basis functions. These functions correspond to locally receptive fields, so they can perform more detailed classifications using 'nearest-neighbor' regression techniques.[35]

---

[31] Cf. Barron, R. L. (1976), p. 25 ff.

[32] Cf. Mucciardi, A. N.; Gose, E. E. (1972), p. 247 ff.

[33] Cf. Barron, R. L.; Mucciardi, A. N.; Cook, F. J. ; Craig, J. N.; Barron, A. R. (1984), p. 38

[34] For active, GMDH algorithm-based neurons cf. Ivakhnenko, G. A. (1997) Ivakhnenko, A. G. (1997); Ivakhnenko, A. G.; Müller, J.-A. (1995), p. 93 ff.; Lemke, F.; Müller, J.-A. (1995); Müller, J.-A.; Ivakhnenko, A. G. (1996); Müller, J.-A.; Ivakhnenko, G. A. (1997); Ivakhnenko, A. G.; Ivakhnenko, G. A. (1995); Müller, J.-A. (1994), p. 185 ff.

[35] GRNN and PNN were described in detail in the literature and will not be further presented here. For more on these artificial neural network topologies cf. Specht, Donald (1991), p. 568 ff.; Wittkemper, Hans-Georg (1997), p. 27 ff.; Specht, Donald (1990), p. 110 ff.; Steiner, Manfred; Wittkemper, Hans-Georg; Wolf, J. Benedict (1998); Wittkemper, Hans-Georg; Steiner, Manfred (1996), p.

4 The dataset

The data used in our study comes from the annual report databank of the Rheinisch-Westfälische Technische Hochschule in Aachen and the stock price databank from the University of Karlsruhe. The year-end stock price was used to compute the market value of the equity. Block ownership data was collected at our department and in the case of concerns excludes the mother's ownership of its daughters. We thus disregard internal concern agency problems. There are up to 240 corporations listed at the Frankfurt stock exchange per year in the set. We do not include banks and insurance commpanies that have special accounting prerogatives. The data goes from 1973 to 1992, with ratios using lagged data making analyses possible from 1974 on. The reform of accounting laws 1986 made alternate calculations of some variables necessary, also certain values are less detailed before 1987. E. g., before 1987 dividend payments are only available together with other profit transfers. Also liabilities are not as differentiated in the pre-1987 data. We grouped the data into 15 sectors and employ 14 dummy variables to capture sectoral effects. If all sector dummies are set to 0, the company is a holding.

Sectors in the dataset	
chemicals	wood/paper/printing
plastics/rubber	leather/apparel
minerals/ceramics	food/beverages
metal and metalworking	construction
machinery	trade
automotive	utilities
electronics/precision mechanics	holdings
others	

Fig. 7. Sectors in the dataset

582. For practical applications of artificial networks in business economics cf. Steiner, Manfred; Wittkemper, Hans-Georg (1993, 1997)

5 Empirical analyses

5.1 Previous empirical analyses

The following tables list the most important studies for this analysis.[36] The papers are split into three groups. First there are those that study the effect of non-agency problems on capital structure, typically the debt ratio (cf. Figure 8). These include studies that measure the information asymmetry via firm specificity (Bradley/Jarrel/Kim 1984, Titman/Wessels 1988). R&D expenses are used as a proxy for company activity that cannot be evaluated by the ownership. Opler/ Titman (1994) link the information asymmetry to the dividend signal, firm size and previous returns reasoning that managers with informational leads will time equity issues after phases of greater stock returns (timing model). Tax shields are also regularly included (Bradley/Jarrel/Kim 1984, Titman/Wessels 1988, Homifar/Zietz/Benkato 1994). The company risk or return volatility helps explain capital structure in Marsh (1982), Titman/Wessels (1988), Chung (1993) and Homifar/Zietz/Benkato (1994). A current analysis of the German capital market is Wohlschieß (1996), who studies the debt ratio in relation to fundamental data with OLS regressions and the LISREL method.

The second group of studies listed in Figure 9 concentrates on agency problems and includes block or manager ownership in the study together other control variables. Our study belongs to this group. Most of the papers listed only analyze the debt ratio. None of the studies include short and long term debt, equity or internal financing simultaneously in their studies. Other variables included are the dividends and manager ownership, which are analyzed separately in ordinary least squares regressions (Crutchley/Hansen 1989) or in a three step least squares regression (Jensen/Solberg/Zorn 1992). In the latter study the dependent variable of one regression is entered as an independent variable in another function Control variables include tax variables (Kim/Sorensen 1986), proxies for information asymmetry (Morck/Shleifer/ Vishny 1988, Crutchley/Hansen 1989) and firm risk (Kim/Sorensen 1986, Friend/Lang 1988, Crutchley/Hansen 1989, Jensen/Solberg/Zorn 1992). The most comprehensive study as regards explanatory variables is Jung/Kim/ Stulz (1996) for whom financial slack is an important variable, as well as macroeconomic indicators.

The third group of papers (cf. Figure 10) analyzes the influence of agency problems on profitability listed. On an aggregated level return on equity or capital are put into relation with block ownership, firm size, sector and other variables. The results are inconclusive and find advantages for manager controlled firms (Ware 1975, Thonet 1977) or owner controlled firms (McEachern 1975). Kaulmann (1987) does not find any difference in returns.

[36] Further references can be found in Wohlschieß, Volker (1996), p. 151 ff.

author (year) dependent var.	independent variables	operationalisation	method country/years	results
Marsh (1982) • equity/debt issue probability	• size • firm risk • assets • capital market eff.	• log(total assets) • $\sigma_{profits}$ • fixed/total assets • $R_i - R_M$	• event study w/ logit and probit GB 1959-70	• equity class.: 62 % • debt class.: 33 %
Bradley/Jarrel/ Kim (1984) • long term debt ratio	• firm risk • non-debt tax shield • specificity	• $\sigma_{profits/total\ assets}$ • depreciation/Σ profits • R&D + advertising	• cross sectional OLS USA 1962-81	• R^2 up to 58.6 %
Titman/ Wessels (1988) • long/short term lev. • convertible bonds/debt	• fixed asset ratio • non-debt tax shield • growth • specificity • size • firm risk	• fixed/total assets • depreciation/ total assets • $R_{total\ assets}$ • R&D/sales • sales • σ_{sales}	• LISREL USA 1974-82	• inconclusive results
Chung (1993) • manager ownership • Tobin's Q • manag. pay	• firm risk • growth • fixed asset ratio • size	• R^2 from market model • price/earning ratio • fixed/total assets • total assets	• OLS • 3SLS USA 1987	• R^2 up to 59% • system-R^2 up to 65,4 %
Homaifar/Zietz /Benkato (1994) • debt ratio	• tax rate • non debt tax shield • size • growth opportun. • capital market effects • inflation • firm risk	• $(T_{reported}-T_{deferred})$ +ti/CF • deprec./total assets • total assets • market/book value of equity • $R_i - R_M$ • $\Delta\ln(CPI)$ • $\ln\sigma_{\Delta operating\ result/total\ assets}$	• pooled regress. with twice lagged variables (general autoregress. distributed lag model) USA 1978-88	• R^2 up to 91,27 %

Fig. 8. Previous emirical studies of non-agency determinants of capital structure

Opler/Titman (1996) • equity, debt issues	• target debt ratio • company outlook • wealth transfer • information asymmetry • market outlook • size	• E(debt ratio)– debt ratio debt ratio • op. res./total ass. operating loss losses brought forw./total assets • short term debt/total debt • stock return dividend signal total assets • S&P 500-return T-bond – BAA bond ret. 20-year bond ret. • dummy for large issues	• OLS • logit USA 1976-93	• R^2 up to 20 % • pseudo-R^2 up to 15 %
Wohlschieß (1996) • debt ratio	• size • security • growth opportun. • liquidity • quality • firm risk • specificity • age • sector	• total assets, sales market value of company • (fixed assets + supplies)/total assets • market/book value of equity • (profit+int. paid +taxes)/tot. ass. • operating profit/ necess. assets • profits/Øtot. ass. σ(op. profit)/ Øtot. assets • dummy = 1 if electronics/automotive/machine • dummy = 1 if listed <7 years • sector dummies	• OLS • corr • LISREL D 1980-92	• R^2 up to 58,5 % • Adj. goodness of fit index up to 54,2 %

Fig. 8. Continued

author (year) dependent variables	independent variables	operationalisation	method country/per.	results
Kim/Sorensen (1986) • debt ratio	• growth • firm risk • size • non debt tax shield • block own.	• $R_{profits}$ • $\sigma^2_{profits}$ • total assets • depreciation/total assets • block ownership	• ANOVA • OLS USA 1978-80	• R^2 up to 18 %
Morck/Shleifer/ Vishny (1988) • Tobin's Q • net cash flow/ tangible ass.	• block owner. • intangible assets • leverage • size • sector	• block ownership • R&D/total assets advertising/total assets • long term lever. • total assets • sector dummies	• OLS USA 1980	• R^2 up to 64 %
Friend/Lang (1988) • debt ratio	• assets • profitability • risk • size • manager ownership • block own.	• fixed/total assets • Øprofits*total/ fixed assets • $\sigma_{profits}$ • ln(total assets) • manager ownership • block ownership	• OLS USA 1979-83	• R^2 up to 35 %
Crutchley/Hansen (1989) • manager ownership • Ølong term lever. • Ødividends	• firm risk • specificity • diversif. • manager ownership • issue costs	• $\sigma_{operating\ result/total\ assets}$ • (advertising + R&D)/sales • Sharpe ratio Øtotal assets • manager ownership • $\sigma_{returns}$ total assets	• OLS USA 1981-85	• R^2 up to 46 %

Fig. 9. Previous empirical studies of agency determinants of the capital structure

Jensen/Solberg/ Zorn (1992) • long term debt ratio • dividends/op. profits • manager ownership	• firm risk • size • diversificat. • specificity • profitability • assets • growth	• $\sigma_{operating\ result/total\ assets}$ • log(total assets) • # of divisions • R&D/total assets • operating result/ total assets • fixed capital/total assets • sales growth	• 3SLS USA 1982, 1987	• system-R^2 up to 30 %
Jung/Kim/Stulz (1996) • equity issues	• taxes • investment opportun. • leverage information asymmetry • bankruptcy cost • financial slack • liquidity • business climate • ROE • excess return	• taxes/total assets • market/book value of equity • long term lev. total assets • return volatility • current/total assets • cash flow • macroeconomic indicators • ROE • market model excess ret.	• logistic USA 1977-84	• R^2 up to 41 %

Fig. 9. Continued

author (year) depend. var.	independent variables	operationalisation	method country/per.	results
Ware (1975) • ROE	• block ownership	• block ownership	• covariance analysis for food & bev. USA 1960-70	• $R_{manager\ contr} > R_{owner\ controlled}$
McEachern (1975) • stock return	• block ownership	• block ownership	• OLS USA 1963-72	• $R_{owner/manag.\ contr.} > R_{owner\ contr.} > R_{manager\ contr.}$
Thonet (1977) • book/ market ROE	• block ownership	• block ownership	• OLS G 1961, 1966, 1970	• $R_{manager\ contr.} > R_{owner\ controlled}$
Bühner (1984) • market model ret.	• block ownership • diversific.	• block ownership • Herfindahl-index	• OLS G 1966-81	• R^2 up to 57 %
Kaulmann (1987) • ROE • ROC	• block ownership • size • sector	• block ownership • total assets • sector dummies	• OLS G, F, GB 1975-84	• no difference
Lipka (1995) • ROE • ROC	• block ownership • size • sector • concentration • leverage • cost structure	• block ownership • sales • dummies • concentration ratios • leverage • personnel costs/sales	• OLS G 1982-88	• R^2 up to 19 %

Fig. 10. Previous empirical studies of agency determinants of returns

5.2 Variable definitions

In the following analysis the values for each method of financing are divided by the fixed assets in order to exclude size effects at this level. In the numerator we put the difference between the value of the form of financing, say long term capital to last year's. In the denominator we put last year's fixed capital. The free cash flow[37] is the profit of the year plus non-liquid expenditures and earnings minus rises in fixed and financial assets and dividends. Thus we use the following variables:

variable	calculation
d_l/ta	(change in long term debt) / lag(fixed assets)
d_s/ta	(change in short term debt) / lag(fixed assets)
d/ta	(change in total debt) / lag(fixed assets)
e/ta	(change in equity) / lag(fixed assets)
cf/ta	free cash flow / lag(fixed assets)

Fig. 11. Calculation of the finance variables

We use five agency variables. *Block ownership* is the most common of these.[38] Small investors will hardly organize to monitor the management, as the expense is too high for each individual investor. Thus the management is only controlled if there is block ownership. We assume such control if block ownership is 25 % or more. Others already take 5 % or more as the boundary.

In addition the *market segment* with its specific information requirements the company is traded in designates varying degrees of information asymmetry between owners and management. The German stock market structure is comprised of four market segments, the official continuous trade market, the official spot market, the regulated and the unregulated free trade (telephone trading). The admission criteria to the official continuous trade market are listed in §§36-44d BörsG together with the stock market admission guide; the most important requirements are a minimal combined value of DM 2.5 million of the shares to be issued, a public spread of at least 25 %, a comprehensive issue prospectus and at least one interim report per year. The regulated free trade is meant for those corporations who cannot or will not fulfill these requirements. There are no admission criteria for the unregulated market.

---

[37] Cf. Jensen, Michael C. (1986)

[38] Cf. e.g. Ware, Robert F. (1975). McEachern, William, A. (1975), Kim, Wi S.; Sorensen, Eric H. (1986), Friend, Irwin; Lang, Larry H. P. (1988), Morck, Randall; Shleifer, Andrei; Vishny, Robert W. (1989); for Germany Thonet, Peter J. (1977), Bühner, Rolf (1984), Kaulmann, Thomas (1987), Lipka, Sabine (1995).

Assuming that managers actually exploit the leeway given to them by information asymmetries, companies in free trade will have the largest agency problems, those in official trade the least. The market segment variable has the following values: official continuous trade = 3, official spot market = 2, free trade = 1. We create two dummy variables for the market segment or alternatively one, for which a one signifies the official continuous trade and zero all others. Thus higher values of the variable point to lesser agency problems.

The company *size* as measured by total assets indicates which companies are more in the spotlight of public attention and thus have less agency problems. We chose DM 100 million total assets as the boundary value to distinguish companies with greater agency problems from those with less problems. On the other hand a manager invests a great part of his human capital in a company. Bankruptcy would devaluate his company-specific knowhow. For this reason managers prefer diversified and hence larger companies to minimize their human capital risks. The empirical analysis must show which influence prevails.

Leverage as the ratio of debt to equity can be used as an agency variable because higher debt leads to higher regular payment obligations that reduce the discretionary possibilities of a manager.[39] We take an above average leverage (per sector and year) as an indicator of lesser agency problems. Leverage is computed from elements also used in the financing variables, which could cause 'wrong' signs from the managerialism point of view.

The fifth agency variable is the *dividend signal*, that operates as a corrective to the other variables.[40] In every year in which the dividend per share exceeds the previous year's dividend, the dividend signal dummy is set to 1, otherwise to 0. Managers who increase the dividend signal to investors that the prospects for future investments are good, thus reducing the information asymmetry.[41] As a dividend hike is costly, especially as subsequent reductions are hard to implement,[42] the signal cannot be copied by an ailing firm that cannot afford the expense.[43] This variable is problematic in that every company is regarded as having agency problems that didn't raise the dividend per share in that year. Even agency problem-free companies will not increase the dividends every year. Using this variable therefore should produce poor results. One solution would be to aggregate the signals over the past three to five years, but that would distort the first years' values. Tests also did

---

[39] Cf. Jensen, Michael C.; Meckling, William H. (1976), Crutchley, Claire E.; Hansen, Robert P. (1989)

[40] For the signaling theory cf. Padberg, Max (1995), p. 119 ff.

[41] Cf. Amihud, Yakov; Murgia, Maurizio (1997), Easterbrook, F. (1984); Rozeff, M. S. (1984)

[42] For dividend reductions cf. Jensen, Gerald R.; Johnson, James M. (1995)

[43] Cf. Ross, Stephen A. (1978)

not confirm better results for this approach, so we compute the variable as described for the time being.

With the exception of the debt variable that describes the discretionary possibilities in financing of the management, all other variables contain information on the level of information asymmetry between management and owners of a company. All agency variables are operationalized so that low values stand for large agency problems.

Other possible agency variables could be used if we assume that the management is maximizing sales or book values of assets. These would be the growth rate of total assets or the ratio sales to total assets.[44] This is possible if manager performance is judged by these values even if that is not in the owners' best interests. As studies in this area are more or less based on subjective experience,[45] we do not adopt any of these variables for our approach.

variable	reason for influence	agency problems assumed if:
block ownership	Determines the monitoring possibilities of external owners.	< 25 %
market segment	Trade in the market segments entails varying demands on publicity, creating different levels of information with the ownership.	< 3 (not official continuous trade)
size	As size increases, so does the public attention, effectively monitoring managers' actions. On the other hand managers prefer diversified companies in order to reduce their human capital risk.	< DM 100 million total assets
leverage	Diminishes the discretionary possibilities of the management by reducing cash flow and raising default risk.[46]	< average leverage per year and sector
dividend signal	Raising dividends signals future prospects of the firm to the owners.[47]	= 0 (no signal)

Fig. 12. Agency variables in the analysis

[44] Cf. Baumol, William (1967), Marris, Robin (1968)
[45] Cf. Ridder-Aab, Christa-Maria (1980), p. 29 f.
[46] Cf. Jensen, Michael C. (1986); p. 323 ff., Stulz, René M. (1990), p. 3 ff.
[47] Cf. Black, Fischer (1976)

5.3 Determining the explanatory power of the agency variables

Agency costs are opportunity costs and cannot be measured directly. We can only check for the relevance of agency problems and the managerial hypothesis by measuring and comparing variables that cause agency problems (Figure 12) or are caused by agency problems (financial instruments, Figure 11, pecking order variables, Figure 13). In the pecking order context the agency variables should be insignificant, as owners' and managers' financial preferences are identical.

variable	description
financial slack	Cash flow minus increase in fixed assets and dividends. This variable shows how much external funding a firm needs.[48]
return volatility	Manager controlled firms invest in less risky projects and should produce less volatile returns.
market to book value of equity	Owner controlled firms maximize shareholder value and should have higher market/book ratios of equity. The ratio also mirrors investment possibilities of a company.[49]
sector	Control variable. Sector dummies capture differences like capital-intensive production for the 15 sectors in the study. If all dummies are 0 the company is a holding.

Fig. 13. Pecking order and control variables in the empirical analysis

If the managerial hypothesis holds, the signs of the coefficients we expect are given in Figure 14.

	diminishing agency problems = rising values of the agency variables
d_l/ta	-
d_s/ta	o (-)
e/ta	+
cf/ta	- -

Fig. 14. Expected signs of the coefficients for the managerial hypothesis, by finance instrument

48 Cf. Shyam-Sunder, Lakshmi; Myers, Stewart C. (1994), p. 5
49 Cf. Smith, Clifford W.; Watts, Ross L. (1992), p. 263 ff.

The correlations of the agency variables amongst each other can tell us if they are capturing the same or different aspects of agency problems. Using the original market segment variable is not quite correct as the variable is only ordinally scaled. The results of the correlation analysis should therefore only be taken as tendencies.

	market segment	size	leverage	dividend signal
block own.	-.1362**	.0796**	.0232	-.0405**
market seg.	1.0000	.5787**	-.0722**	.0574**
size		1.0000	.0643**	.0362**
leverage			1.0000	-.0712**
dividend sig.				1.0000

dark gray or ** = significant at 5 %-level (light gray or * = 10 %)

Fig. 15. Correlations of the agency variables

Figure 15 shows that most agency variables are significantly but not strongly correlated to each other. This gives a more realistic picture of the level of agency problems in a company, as the problems are not monocausal. Especially the dividend signal has a corrective function for the other variables describing information asymmetry. For this reason one might expect a negative correlation between the dividend signal and the block ownership, market segment and size variables. This is only the case for block ownership, though. The high correlation of size to the market segment was to be expected because of the legal requirements for market segment trading.

In Figure 16 the large proportion of non-significant correlations is noteworthy, especially in view of block ownership which is often solely used in literature as a sign of agency problems. The signs for leverage are reversed, for the mathematical reasons discussed above. The signs for the equity and cash flow variables are generally other than expected.

5.4 Analysis of the financial behavior with artificial neural networks

Verification of the managerial and pecking order hypotheses We use polynomial neural networks for the separate estimation of the financial variables according to the pecking order hypothesis and general regression neural networks for their joint estimation according to managerialism.

	d_l/ta	d_s/ta	e/ta	cf/ta
expected sign	–	o	+	—
block ownership	-.0238	.0199	-.0321*	.0194
market segment	-.0335*	-.0225	-.0156	.0211
size	-.0275*	-.0028	-.0320**	.0092
leverage	.0646**	.0343**	-.0601**	-.0312**
dividend signal	-.0267*	-.0239	.0231	.0093

dark gray or ** = significant at 5 %-level (light gray or * = 10 %)

Fig. 16. Correlations between the agency and finance variables

In the first step we determine the influence of the agency and pecking order variables on financial decisions using the mean square error (MSE) as the measure of goodness. We first calculate the means of the variables which we successively enter into the training and testing phases of the artificial neural networks when we want to test their power. This forces the networks to base their estimates for the financing variable on the remaining variables without reducing the dimension of the decision space.[50] We then compare the MSE to our reference model that contains all unmodified independent variables. The importance E of variable i is computed as:

$$E_i = \frac{MSE_{GRNN,i} - MSE_{GRNN,reference}}{MSE_{GRNN,reference}} \tag{12}$$

$with\, i = \{x_1, \cdots, [x_{601}, \cdots, x_{614}], [x_{71}, x_{72}], \cdots, x_9\}$. The explanatory power of the two market segment and fourteen sector dummies is determined on an aggregate level, meaning we modify all dummies of one category simultaneously. The PoNN achieved the results shown in Figure 17 this way. The negative signs of the agency variables market segment, size, and the pecking order variable market to book value of equity are not significant. On the other hand, the R^2-value noticeably rises under exclusion of the market segment dummies and the market to book value of equity. We see that PoNNs recognize noncausal relationships and transfer these to the holdout sample.

Altogether the agency variables hold little explanatory power (E = 0.1954) for debt financing decisions as compared to the pecking order variables (E = 0.7952). The information asymmetry-producing variables explain less than

[50] Succesively reducing the number of variables would decrease the dimensions of the decision space, as fewer independent variables would be entered into the artificial neural network as information patterns.

the asymmetry-produced variables. Leverage performs best of the agency variables, which is due to its computational relation to debt financing, though. The other agency variables explain next to nothing. For the pecking order variables, again one variable dominates (financial slack, E = 0.6524) while the rest are insignificant, except for several sector dummies. As financial slack is a core variable of the pecking order theory, which states that companies only revert to debt financing if their internal reserves are exhausted, this implies a predominance of the pecking order theory over managerialism.

	reference model	**modified models**	
modified variable	---	block ownership	market segment
adj. R^2	**0.5222**	**0.5006**	**0.5693**
MSE	0.4068	0.4253	0.3667
influence E	1	0.04547689	-0.09857424
modified variable	size	leverage	dividend signal
adj. R^2	**0.5236**	**0.445**	**0.5222**
MSE	0.4057	0.4726	0.4068
influence E	-0.00270403	0.16175025	0
modified variable	financial slack	return volatility	market/book equity
adj. R^2	**0.2106**	**0.5222**	**0.5693**
MSE	0.6722	0.4068	0.3667
influence E	0.65240905	0	-0.09857424
modified variable	sector	all agency	pecking ord./sector
adj. R^2	**0.4884**	**0.4288**	**0.1424**
MSE	0.4356	0.2921	0.7303
influence E	0.07079646	0.19543773	0.79523107

Fig. 17. Isolated influence of the agency and pecking order variables on debt financing

When analyzing equity finance we basically come to the same conclusions. Figure 18 shows that the pecking order variables dominate the agency variables (E = 1.8275 vs. E = 0.024). All agency variables explain little more than naught, again with a negative impact of the market segment. The insignificance of leverage is surprising, but again the financial slack is dominant and confirms the pecking order hypothesis.[51] The sectors have only a small importance for the decision to finance by equity (E = 0.0175). The return

[51] Cf. Shyam-Sunder, Lakshmi; Myers, Stewart C. (1994), p. 10ff.

volatility probably is of small importance ($E = 0.0022$) because the preference for debt financing stems from the risk structure of the company expressed by the return volatility. As the results for internal finance are very similar, we do not show them here.

	reference model	**modified models**	
modified variable	---	block ownership	market segment
adj. R^2	**0.651**	**0.651**	**0.6681**
MSE	0.0458	0.0458	0.0435
influence E	1	0	-0.05021834
modified variable	size	leverage	dividend signal
adj. R^2	**0.651**	**0.6441**	**0.651**
MSE	0.1061	0.0467	0.0458
influence E	0	0.01965066	0
modified variable	financial slack	return volatility	market/book equity
adj. R^2	**0.1165**	**0.6502**	**0.6508**
MSE	0.1158	0.0459	0.0458
influence E	1.52838428	0.00218341	0
modified variable	sector	all agency	pecking ord./sector
adj. R^2	**0.6448**	**0.6425**	**0.0121**
MSE	0.0466	0.0469	0.1295
influence E	0.01746725	0.02401747	1.82751092

Fig. 18. Isolated influence of the agency and pecking order variables on equity finance

In sum, when using polynomial and general regression neural networks the agency-inducing variables do not influence isolated financial decisions. Financial slack is the dominant factor here. This speaks for the pecking order hypothesis and against managerialism. The advantage of managerialism is its comprehensive explanation of reasons for and results of information asymmetries, whereas the pecking order theory only explains the results of information asymmetry for single financial actions. That is why we now test the simultaneous influence of the agency variables on all financing options. A difficulty in this procedure lies in the smoothing parameter that must optimize all decision surfaces in the multidimensional decision space at the same time. For this reason we limit the number of variables to be parallely estimated in order to adequately capture the diverging cause-result-relationships of the remaining variables. Initially we simultaneously estimate the long and short term debt financing with a general regression neural network.

The results in Figure 19 show a more differentiated picture of the influence of the factors. The agency variables generally have a positive influence, except

	d/ta	d/ta
adjusted R^2	0,5292	0,5834
MSE	0,3704	0,354
variable	weight of variable in the genetic algorithm	
block ownership	0,27059	
market segment dummy 1	0,02353	
market segment dummy 2	1,45882	
size	0,17647	
leverage	2,32941	
dividend signal	0,11765	
financial slack	2	
return volatility	0,35294	
market/book equity	0,69412	
sector: chemicals	3	
sector: plastics/rubber	2,45882	
sector: minerals/ceramics	2,34118	
sector: metals	2,34118	
sector: machinery	2,81176	
sector: automotive	2,31765	
sector: electronics/fine mechanics	0,57647	
sector: wood/paper/printing	0,04706	
sector: leather/apparel	2,50588	
sector: food/beverages	1,82353	
sector: construction	0,09412	

Fig. 19. Simultaneous estimation of the long and short term finance actions with general regression neural networks

for the official continuous trade market segment dummy. This might be due to the less limited access to various financial sources of companies traded in this sector. As block ownership is the typical agency variable, its high influence (E = 0.27059) is especially interesting. Leverage and financials lack determine the simultaneous financial decisions. In Figure 20 the importance of the simultaneous estimation of debt, equity and internal financing is visible. A marked shift in variable weights is apparent. All agency and pecking order variables are much more important than before.

	debt financing	equity financing	internal financing
adj. R^2	0.4495	0.6952	0.6795
MSE	0.469	0.04	0.187
	weight of the variables in the genetic algorithm		
variable	weight	variable	weight
block ownership	0.54118	market segment 3	0.28235
market segment 2	1.02353	size	1.17647
leverage	2.95294	dividend signal	0.05882
financial slack	2.44706	return volatility	2.94118
market/book equity	1.90588		
sector: plastics/rubber	1.91765	sector: chemicals	1.91765
sector: metals	1.48235	sector: minerals/ceramics	0.01176
sector: automotive	0.23529	sector: machinery	2.42353
sector: wood/paper	0.27059	sector: electronics	0.29412
sector: food/beverages	0.76471	sector: leather/clothing	0.37647
sector: trade	1.17647	sector: construction	0.48235
sector: utilities	2.55294	sector: other	0

Fig. 20. Simultaneous estimation of debt, equity and internal financing with general regression neural networks

While the isolated analysis of the financial alternatives comes out in favor of the pecking order hypothesis, the results of the simultaneous estimation using general regression neural networks are consistent with the managerial hypothesis, without falsifying the pecking order. Financial slack remains a major factor in the simultaneous estimation.

Classification of the financial behavior A distinction between pecking order and managerialism is not possible by studying the financial behavior, as both models predict the same behavior. That is why we now try to classify the financial decisions with the agency and pecking order variables. Probabilistic neural networks (PNN) serve as the counterparts of discriminant analyses, general regression neural networks (GRNN) correspond to (logistic) regressions. In the first step we distinguish between internal and external (debt and equity) finance.

probabilistic neural network	external finance	internal finance
actually belonging to category	2464	315
classified by the network	1523	1256
correctly classified datasets	1436	228
incorrectly classified datasets	87	1028
percent of correct classifications	58.28%	72.38%
percent of incorrect classifications	27.62%	41.72%

Fig. 21. Classification of internal and external finance decisions using a probabilistic neural network

Using agency and pecking order variables we can classify internal and external financing decisions with sufficient exactness. In the next step we test for the influence of the agency variables alone. We now fully separate the external financial actions into short and long term debt and equity and use a PNN and a GRNN.

	PNN	GRNN logist.
short-term debt financing	23.08%	69.24%
long-term debt financing	54.23%	57.18%
equity financing	59.56%	76.28%
internal financing	62.66%	68.19%

Fig. 22. Classification of the financial behavior using the agency variables

Although PNNs and GRNNs are based on much the same principles, the GRNN performs much better. The PNN suggests rejecting managerialism while the GRNN confirms it. Therefore we use another neural network to test if the financial behavior can be used to classify the agency problems of a company. Using the separating boundaries of the agency variables in Figure 12 we classify the companies separately by each agency variable. The agency variables are rewritten as dummies with a 1 standing for small and

0 for large agency problems. That might not be the expected order, but it corresponds to the same tendency for the continuous variables. As the block ownership and market segment are highly correlated, we omit the market segment here. The independent variables for each of the resulting four GRNNs are $d_l/ta, d_s/ta, e/ta, cf/ta$.

value	BLOCK	MS	SIZ	LEV	DIV
0	11.7	64.1	31.5	69.8	71.6
1	88.3	35.9	68.5	30.2	28.4

Fig. 23. Frequencies of the values of the agency variables in %

percent of correctly classified datasets	
BLOCK	87.77%
SIZ	56.12%
LEV	72.93%
DIV	71.99%

Fig. 24. Classification of the agency variables with GRNNs using the financial variables

The results show an empirical connection between the financial decisions and the agency variables. Especially block ownership can be correctly classified, altogether confirming the managerial hypothesis. In the pecking order theory, the agency variables have no importance. The financial slack, return volatility and market to book values of equity variables show the collusion of interests between managers and current owners. The sector control variables are also included. We test the pecking order hypothesis with logistic GRNNs. As the pecking order does not distinguish between short and long term debt finance, we do not do so here, either. All classifications are sufficiently accurate (Figure 24).

In summary neither of the two models presented can be rejected using artificial neural networks. In describing the financial behavior the pecking order variables are superior, in classification both models perform equally well.

5.5 Analysis of the financial behavior with statistical methods

Methodology In the statistical section we use t-tests of mean differences to compare the mean level of each form of finance of companies agency problems

finance alternative	% correctly classified
debt finance	56.60%
equity finance	76.61%
internal finance	68.41%

Fig. 25. Classification of the financial behavior with GRNNs on the basis of the pecking order hypothesis

to those without or with less such problems. The financing alternatives can only be entered separately, though. The border values for the classification of the companies are given in Figure 12.

The discriminant analysis allows us to test the aggregated discriminatory power of all the financial variables together, using the same separation criteria as for the t-tests. Every agency variable has to be studied separately. We are interested in the percentage of correctly classified cases and the significance and sign of the coefficients. If the class strength is lopsided, discriminant analyses often achieve high correct classification rates by simply classifying all cases into the most populous class. To avoid this we add the initial assumption that each case belongs to either class with equal probability.

Seemingly unrelated regressions (SURE)[52] permit us to simultaneously estimate several regression functions that do not seem to be related in view of their independent variables.[53] The simultaneous estimation is necessary if the functions' error terms are correlated. Estimating them separately would mean foregoing information and computing inefficient results.[54]

Tests with discrete dependent variables

T-tests for differences in means We use the t-tests to ascertain the discriminatory power of the financial behavior of firms regarding their agency problems. The first value in Figure 26 (-.0097) e. g. means that the mean long term debt financing in relation to the fixed assets is $-.0097$ lower for firms with small agency problems (block ownership greater than 25 %) than for those with bigger agency problems. High but insignificant mean differences come about if the finance variable has a high standard deviation. The debt variable differences and the internal finance differences in part have the expected signs. Equity finance is surprisingly preferred in firms with big agency problems. The mathematical connection of leverage to debt and equity finance overpowers any agency content it has.

52 Cf. Kmenta, Jan (1980), p. 635 ff.

53 A two-stage least squares regression was not attempted due to the problem of the correct identification of the functions.

54 Cf. Kmenta, Jan (1980), p. 639

	d_l/ta	d_s/ta	e/ta	cf/ta
expected sign	–	o	+	– –
block ownership	-.0097	.2915	-.0341**	-.3517
market segment	-.0091	-1.2289	-.0102	1.7887
size	-.0119	-.3086	-.0120	1.3499
leverage	.0328**	1.3192	-.0619**	-1.5208
dividend signal	-.0111*	-1.0970	.0242	.5675

dark gray or ** = significant at 5 %-level (light gray or * = 10 %)

Fig. 26. Mean differences between groups with and without agency problems

Discriminant analysis The discriminant analysis is applied to a dataset that can be preclassified due to certain characteristics.[55] In our analysis, the preclassifying variables (active variables)[56] are the agency variables, while we test for the classifying power of the financial (passive) variables. Again we use the dummy variable classification of the agency variables (Figure 12). The goodness of classification is compared to the 50 of the coefficients also help us judge the validity of the managerial hypothesis.[57]

Taking block ownership as an example we construct the following model: $BLOCK_i = \beta_{0,i} + \beta_{d_l,i} * d_l/ta + \beta_{d_s,i} * d_s/ta + \beta_e, i * e/ta + \beta_{cf,i} * cf/ta$
where

$$\begin{aligned} BLOCK &= dummy\,variable\,for\,block\,ownership \\ d_l/ta &= (\Delta\,long\,term\,debt)/lag(fixed\,assets) \\ d_s/ta &= (\Delta\,short\,term\,debt)/lag(fixed\,assets) \\ e/ta &= (\Delta\,equity)/lag(fixed\,assets) \\ cf/ta &= free\,cash\,flow/lag(fixed\,assets) \\ \beta_{0,i} &= constant \\ \beta_{d_l,i} \quad etc. &= coefficients \end{aligned}$$

We present the results for the five separately analyzed models, each with a different active agency variable, in Figure 27. The coefficients are standardized to show the relative influence of the variable, regardless of its scale. One can see that important variables are not necessarily all significant. The model

55 Cf. Klecka, William (1980)
56 Cf. Backhaus, Klaus et al. (1990), p. 167
57 Cf. Hartung, Joachim; Erpelt, Bärbel (1986), p. 269 ff.

constant, which cannot be standardized and is also unimportant for our theory, is not reported. The canonical correlation coefficient corresponds to the square root of a linear regression's R^2.

	exp. sign	BLOCK	DIV	SIZ	MS	LEV
% corr. classif.		69.70	50.38	59.49	53.99	69.41
can. corr. coeff.		.046	.050**	.044	.116**	.152**
d_l/ta	–	.204	.690*	.748*	.400*	.573**
d_s/ta	o	.020	1.367	1.501	.948*	.760**
e/ta	+	.480*	-.355	.370	.242	-.503**
cf/ta	– –	-.693**	1.352	1.420	1.199**	.131

dark gray or ** = significant at 5 %-level (light gray or * = 10 %) BLOCK = block ownership; MS = market segment; SIZ = size; LEV = leverage; DIV = dividend signal

Fig. 27. Canonic correlation coefficients and standardised coefficients of the discriminant analysis

The dividend signal variable cannot be classified. Here we see that the variable is not meant to stand alone, but should be understood as a corrective to the other agency variables. All other variables are correctly classified way over 50% of the time. The signs of the variables though are often wrong, except for the first model with block ownership as the active variable. Here, not only the sign for the long term debt is wrong, but it is also insignificant. With its high classification result of 69.7% we can cautiously speak of a confirmation of the managerial hypothesis.

Seemingly unrelated regression (SURE) The two-class discriminant analysis and a regression outwardly resemble each other, as they put one variable in relation to one or more others. There is an important difference between the two, though. In the discriminant analysis the grouping (active) variable is unchanging while the passive variables vary stochastically. In the regression analysis on the other hand the independent variables are unchanged, so that the active variables of the discriminant analysis must become the independents of a regression so as not to reverse the causality.[58]

With this type of model construction we must include other explanatory variables that also influence the financing decision. Not including them leads

[58] Cf. Backhaus, Klaus et al. (1990), p. 180

to different coefficients, significance levels and even signs. We therefore expand the model by the pecking order and sector variables listed in Figure 13:

- financial slack,
- return volatility,
- market to book value of equity and
- sector.

The SURE method allows us to use all of the information contained in two or more regression functions that are only related via their residuals. In our case we analyze the functions of the four finance instruments together.

Using long term debt finance as an example for the four models we construct the following model:

$d_l/ta_i = \beta_{0,i} + \beta_{BLOCK,i} * BLOCK + \beta_{MSD,i} * MSD + \beta_{SIZ,i} * SIZ + \beta_{lev,i} * LEV + \beta_{DIV,i} * DIV + \beta_{FINSLA,i} * FINSLA + \beta_{RETVOL,i} * RETVOL + \beta_{MBVE,i} * MBVE + \sum_k \beta_{SECTORk,i} SECTOR_k$, where

$$
\begin{aligned}
d_l/ta &= change\,in\,long\,term\,debt/lag(fixed\,assets)\\
BLOCK &= block\,ownership\\
MSD &= market\,segment\,dummy\\
SIZ &= size\\
LEV &= leverage\\
DIV &= dividend\,signal\\
FINSLA &= financial\,slack\\
RETVOL &= return\,volatility\\
MBVE &= market\,to\,book\,value\,of\,equity\\
SECTOR &= sector\,dummies\\
\beta_{0,i} &= constant\\
\beta_{BLOCK,i}\,etc. &= coefficients
\end{aligned}
$$

Using the pecking order variables, especially the financial slack, greatly increases the explanatory power of the functions, especially the cash flow function (the cash flow is used to compute financial slack). We found the same using neural networks. The expected signs are produced for the long term debt and internal finance, but they are often insignificant. Even the leverage variable is negatively correlated to the long term debt financing, despite its mathematical construction suggesting the opposite. The mathematical connection shines through with the short term debt variable, though. Otherwise no significant influence on short term debt is apparent, as hypothesized. The often negative signs for equity do not fit into our model. For this

	d_l/ta	d_s/av	e/ta	cf/ta
adjusted R^2	0.0748**	0.4013**	0.1177**	0.9703**
adjusted R^2, only agency variables	0.0042*	0.0681**	0.0054*	0.0261**
expected sign of agency variables	–	o	+	––
block ownership	-0.0145	0.0417	0.0452*	0.0090
market segment dummy	-0.0156	-0.0466	-0.0082	-0.0014
size	-0.0022	-0.0059	0.0029	-0.0079*
leverage	-0.0031*	0.0709**	-0.0231**	-0.0027
dividend signal	-0.0239**	-0.0306	-0.0276	-0.0044
financial slack	-0.0263**	-0.2935**	-0.0639**	0.9343**
return volatility	0.2220**	-1.1723**	0.5867**	0.2687
market to book value of equity	0.0000	-0.0003**	-0.0001	-0.0002**
SECTOR: chemicals	0.0240*	0.1892**	0.0075	0.0120
SECTOR: plastics/rubber	-0.0034	0.1670	-0.0447	-0.0144
SECTOR: minerals/ceramics	0.0349**	0.1350*	-0.0159	-0.0116
SECTOR: metals	-0.0006	0.0903	-0.0052	-0.0562*
SECTOR: machinery	0.0195	0.1460**	0.0832**	0.0114
SECTOR: automotive	0.0155	0.1727*	-0.0545	0.0332
SECTOR: electronics/prec. mech.	0.0110	0.1599**	0.0313	0.0059
SECTOR: wood/paper/printing	-0.0015	0.1887*	0.0233	-0.0613
SECTOR: leather/apparel	0.0166	0.4246**	0.2290**	-0.0086
SECTOR: food/beverages	0.0023	0.1457**	-0.0229	-0.0366*
SECTOR: construction	0.0280	0.0481	0.0645	0.0004
SECTOR: trade	0.0514**	0.1753**	0.0437	0.0358
SECTOR: utilities	0.0136	0.0521	-0.0755**	-0.0270
SECTOR: others	0.0328**	0.0435	-0.0047	0.0092

dark gray or ** = significant at 5 %-level (light gray or * = 10 %)

Fig. 28. Adjusted R^2s and coefficients of the SURE

reason and because of the often insignificant coefficients, we only guardedly see the managerial hypothesis confirmed here.

6 Summary and conclusions

The statistical analysis shows that a connection between agency problems and the financial actions of the management exists, albeit tenuously. Especially the discriminant analysis showed a promising confirmation of the managerial hypothesis when using block ownership as an indicator of agency problems. The results of the neural network studies are similar. The pecking order theory is confirmed in isolated and comprehensive analyses of the financial alternatives. The latter analysis also supports managerialism. Our next goal is to expand our research in this area. One plan is to group the five agency variables into a comprehensive agency variable, especially in view of the dividend signal. We also hope to include data on managerial and bank ownership, which are important factors in agency theory. Research possibilities in this area are far from exhausted.

References

Amihud, Yakov; Lev, B.; Travlos, N. G. (1990) Corporate Control and the Choice of Investment Financing: The Case of Corporate Acquisitions, in: Journal of Finance, Vol. 45, p. 603-916.

Amihud, Yakov; Murgia, Maurizio (1997) Dividends, Taxes, and Signaling: Evidence from Germany, in: Journal of Finance, Vol. 52, p. 397-408.

Backhaus, Klaus et al. (1990) Multivariate Analysemethoden, 6. edition Berlin et al.

Barron, R. L. (1976) Three Approaches for Extrapolation with Black Box Models in Process Control Systems, in: Proceedings of the 11th Conference on Use of Digital Computers in Process Control, Louisiana State University, Baton Rouge, p. 25-27.

Barron, R. L.; Mucciardi, A. N.; Cook, F. J.; Craig, J. N.; Barron, A. R. (1984) Adaptive Learning Networks: Development and Application in the United States of Algorithms Related to GMDH, in: Farlow, Stanley J. (Hrsg.): Self Organizing Methods in Modeling, New York, p. 25-65.

Baumol, William (1967) Business Behavior, Value and Growth, 2. Aufl., New York.

Black, Fischer (1976) The Dividend Puzzle, Journal of Portfolio Management 2, p. 5-8.

Black, Fischer; Scholes, M. (1973) The Pricing of Options and Corporate Liabilities, Journal of Political Economy, Vol. 81, p. 637-654.

Bühner, Rolf (1984) Rendite-Risiko-Effekte der Trennung von Eigentum und Leistung im diversifizierten Großunternehmen, in: Zeitschrift für betriebswirtschaftliche Forschung, Vol. 36, p. 812-824.

Carpenter, Robert E. (1994) Finance Constraints or Free Cash Flow? The Impact of Asymmetric Information on Investment, Working Paper, Emory University.

Cooney, John W. Jr.; Kalay, Avner (1993) Positive Information from Equity Issue Announcements, in: Journal of Financial Economics, Vol. 33, p. 149-172.
Crutchley, Claire E.; Hansen, Robert P. (1989) A Test of the Agency Theory of Managerial Ownership, Corporate Leverage, and Corporate Dividends, in: Financial Management, Vol. 18, p. 36-46.
Dolenko, S. A.; Orlov, Yu. V.; Persiantsev, I. G. (1996) Practical Implementation and Use of Group Method of Data Handling (GMDH): Prospects and Problems, in: Proceedings of ACEDC '96, PEDC, University of Plymouth.
Donaldson, Gordon (1961) Corporate Debt Capacity, Harvard University, Boston, Mass.
Easterbrook, F. (1984) Two Agency-Cost Explanations of Dividends, in: American Economic Review, p. 650-659.
Farlow, Stanley J. (1984) The GMDH Algorithm, in: Farlow, Stanley J. (Hrsg.): Self Organizing Methods in Modeling, New York, p. 1-24.
Friend, Irwin; Lang, Larry H. P. (1988) An Empirical Test of the Impact of Managerial Self-Interest on Corporate Capital Structure, in: Journal of Finance, Vol. 43, p. 271-281.
Galai, D.; Masulis, R. W. (1976) The Option Pricing Model and the Risk Factor of Stock, in: Journal of Financial Economics, Vol. 3, p. 53-81.
Garvey, Gerald T.; Hanka, Gordon (1997) The Management of Corporate Capital Structure: Theory and Evidence, Working Paper, University of British Columbia.
Hartung, Joachim; Erpelt, Bärbel (1986) Multivariate Statistik, 2. edition, Munich/Vienna.
Ivakhnenko, A. G. (1968) The Group Method of Data Handling - A Rival of Stochastic Approximation, in: Soviet Automatic Control, Vol. 3, p. 43 ff.
Ivakhnenko, A. G.; Ivakhnenko, G. A.; Vysotsky, V. M. (1976) Long Term Prediction of Random Processes by GMDH Algorithms Using the Unbiasedness and Balance-of-Variables Criteria, in: Soviet Automatic Control, Vol. 9, p. 28 ff.
Ivakhnenko, A. G.; Ivakhnenko, G. A. (1995) The Review of Problems Solvable by Algorithms of the Group Method of Data Handling (GMDH), in: Pattern Recognition and Image Analysis, Vol. 5, No. 4, p. 527-535.
Ivakhnenko, A. G.; Müller, J.-A. (1995) Self-Organization of Nets of Active Neurons, in: System Analysis Modeling Simulation, Vol. 20, p. 93-106.
Ivakhnenko, A. G.; Müller, J.-A. (1997) Present State and New Problems of Further GMDH Development, Working Paper, Glushkov Institute of Cybernetics, Kiev.
Ivakhnenko, A. G.; Ivakhnenko, G. A.; Müller, J.-A. (1994) Self-Orgnaization of Neural Networks with Active Neurons, in: Pattern Recognition and Image Analysis, Vol. 4, No. 2, p. 185-196
Ivakhnenko, G. A. (1997) Recent Developments in the Group Method of Data Handling (GMDH) for the Modelling of Complex Objects and Neuronets Self-Organization, Working Paper, Glushkov Institute of Cybernetics, Kiev.
Jensen, Gerald R; Johnson, James M. (1995) The Dynamics of Corporate Dividend Reductions, in: Financial Management, Vol. 24, S 31-51.
Jensen, Michael C. (1986) Agency Costs of Free Cash Flow, Corporate Finance, and Takeovers, in: American Economic Review, Papers and Proceedings, Vol. 76, p. 323-329.
Jensen, Michael C.; Meckling, William H. (1976) Theory of the Firm: Managerial Behavior, Agency Costs and Ownership Structure, in: Journal of Financial Economics, Vol. 3, p. 305-360.

Jung, Kooyul; Kim, Yong-Cheol; Stulz, René M. (1996) Timing, Investment Opportunities, Managerial Discretion, and the Security Issue Decision, in: Journal of Financial Economics, Vol. 42, p. 159-185.

Kaulmann, Thomas (1987) Property Rights und Unternehmungstheorie: Stand und Weiterentwicklung der empirischen Forschung, Munich.

Kim, Wi S.; Sorensen, Eric H. (1986) Evidence on the Impact of the Agency Costs of Debt on Corporate Debt Policy, in: Journal of Financial and Quantitative Analysis, Vol. 21, p. 131-144.

Kim, Wi S.; Sorensen, Eric H. (1986) Discriminant Analysis, Beverly Hills/London.

Kmenta, Jan (1986) Elements of Econometrics, 2. edition, New York/London.

Lemke, F.; Müller, J.-A. (1995) Analyse und Vorhersage mittels statistischer lernender Netze, Working Paper, Anwendersymposium zur industriellen Anwendung der Neuro-Fuzzy-Technologien, Wittenberg.

Lipka, Sabine (1995) Managementeffizienz und Kapitalmarktkontrolle, Frankfurt.

Malone, James M. (1984) Regression without models: Directions in the search for structure, Farlow, Stanley J.: Self-Organizing Methods in Modeling, Statistics: Textbooks and Monographs, No. 54, New York.

Marris, Robin (1968) The Economic Theory of "Managerial" Capitalism, New York.

McEachern, William A. (1975) Managerial Control and Performance, Lexington.

Merton, R. C. (1974) The Pricing of Corporate Debt: The Risk Structure of Interest Rates, in: Journal of Finance, Vol. 29, p. 449-470.

Modigliani, F.; Miller, M. H. (1958) The Cost of Capital, Corporation Finance and the Theory of Investment, in: American Economic Review, Vol. 48, p. 261-297.

Morck, Randall; Shleifer, Andrei; Vishny, Robert W. (1989) Alternative Mechanisms for Corporate Control, in: American Economic Review, Vol. 79, p. 842-852.

Mucciardi, A. N.; Gose, E. E. (1972) An Automatic Clustering Algorithm and Its Properties in High-Dimensional Spaces, in: IEEE Trans. Syst. Man Cybern., p. 247-254.

Müller, J.-A.; Ivakhnenko, A. G.(1996) Self-Organizing in Analysis and Prediction of Stock Market, in: Proceedings of the ICAFS Conference, Siegen.

Müller, J.-A.; Ivakhnenko, G. A. (1997) Recent Developments of Self-Organizing in Prediction and Analysis of Stock Markets, Working Paper, HTW Dresden.

Müller, J.-A. (1995) Self-Organization of Models - Present State, in: Proceedings of EUROSIM'95 Conference, Vienna.

Myers, Stewart C. (1977) Determinants of Corporate Borrowing, in: Journal of Financial Economics, Vol. 39, p. 147-175.

Myers, Stewart C. (1984) The Capital Structure Puzzle, in: Journal of Finance, Vol. 39, p. 575-592.

Myers, Stewart C.; Majluf, Nicholas P. (1984) Corporate Financing and Investment Decisions when Firms have Information that Investors do not have, in: Journal of Financial Economics, Vol. 13, p. 187-221.

Padberg, Max (1995) Der Einfluß von Kapitalerhöhungen auf den Marktwert deutscher Aktiengesellschaften, Cologne.

Opler, Tim; Titman, Sheridan (1996) The Debt-Equity Choice, Working Paper, Ohio State University.

Ridder-Aab, Christa-Maria (1980) Die moderne Aktiengesellschaft im Lichte der Theorie der Eigentumsrechte, Frankfurt.

Ross, Stephen A. (1978) Some Notes on Financial Incentive-Signalling Models, Activity Choice, and Risk Preferences, in: Journal of Finance, Vol. 33, p. 777-794.

Rozeff, M. S. (1982) Growth, Beta and Agency Costs as Determinants of Dividend Payout Ratios, in: Journal of Financial Research, p. 249-259.

Shyam-Sunder, Lakshmi; Myers, Stewart (1994) Testing Static Trade-Off Against Pecking Order Models of Capital Structure, National Bureau of Economic Research, Working Paper Series No. 4722.

Smith, Clifford W.; Watts, Ross L. (1992) The Investment Opportunity Set and Corporate Financing, Dividend, and Compensation Policies, in: Journal of Financial Economics, Vol. 32, p. 263-292.

Specht, Donald F. (1990) Probabilistic Neural Networks, in: Neural Networks, Vol. 3, p. 109-118.

Specht, Donald F. (1991) A General Regression Neural Network, in: IEEE Transactions on Neural Networks 2, p. 568-576.

Steiner, Manfred; Wittkemper, Hans-Georg (1993) Neuronale Netze - Ein Hilfsmittel für betriebswirtschaftliche Probleme -, in: Die Betriebswirtschaft, Heft 4, p. 447-463.

Steiner, Manfred; Wittkemper, Hans-Georg (1997) Portfolio Optimization with a Neural Network Implementation of the Coherent Market Hypothesis, in: European Journal of Operational Research, Vol 100, p. 27-40.

Steiner, Manfred; Wittkemper, Hans-Georg; Wolf, J. Benedict (1998) Der Einsatz der Coherent Market Hypothesis zur Portfoliooptimierung, forthcoming in: Finanzmarkt und Portfolio Management.

Stulz, René M. (1988) Managerial Control of Voting Rights: Financial Policies and the Market for Corporate Control, in: Journal of Financial Economics, Vol. 26, p. 25-54.

Stulz, René M. (1990) Managerial Discretion and Optimal Financing Policies, in: Journal of Financial Economics, Vol. 26, p. 3-27.

Stulz, René M. (1997) Rethinking Risk Management, in: Journal of Applied Corporate Finance, Vol. 9, p. 8-24.

Thonet, Peter J. (1977) Managerialismus und Unternehmenserfolg, dissertation, Saarbrücken.

Ware, Robert F. (1975) The Performance of Manager vs. Owner Controlled Firms in the Food and Beverage Industry, in: Quarterly Review of Economics and Business, Vol. 15, p. 81-92.

Wittkemper, Hans-Georg (1994) Neuronale Netze als Hilfsmittel zur Rendite- und Risikoschätzung von Aktien, Cologne.

Wittkemper, Hans-Georg; Steiner, Manfred (1996) Using Neural Networks to Forecast the Systematic Risk of Stocks, in: European Journal of Operational Research, Vol. 90, p. 577-588.

Wohlschieß, Volker (1996) Unternehmensfinanzierung bei asymmetrischer Informationsverteilung, Wiesbaden.

Portfolio Analysis Based on the Shortfall Concept

Rainer Matthes[1] and Michael Schröder[2]

[1] Metzler Asset Management, Große Gallusstr. 18, D-60311 Frankfurt, Germany
[2] Zentrum für Europäische Wirschaftsforschung (ZEW), Dept. International Finance and Financial Management, P. O. Box 10 34 43, D-68034 Mannheim, Germany

Abstract. Investors often claim that risk is not equal to volatility. While the variance or standard deviation is an appropriate measure of volatility, the shortfall approach is a superior alternative in measuring risk. Shortfall measures define risk as earning less than a desired minimum return. Therefore only that part of the return distribution left of the minimum return is considered in calculating risk. Shortfall measures can be used to measure portfolio risk and risk adjusted performance in a very general and flexible way. This article gives a survey of the major characteristics of shortfall measures and their use in asset allocation. The second part of our article describes an investment product based on a dynamic benchmark adjustment process. The adjustment procedure is governed by the probability of falling below the investors' desired minimum return.

Keywords. Portfolio Management, Asset Allocation, Shortfall Risk, Lower Partial Moments

1 Introduction

"Risk is different from volatility". Private and institutional investors often claim that using the variance as a measure of risk does not meet their own perception of risk. Most investors interpret risk as the possibility of falling below a target return. In a recent study[1] private investors and financial analysts have been asked to list the most important characteristics of investment risk. Both groups answered that "large loss" and "return below target" are the major characteristics of what they perceive as investment risk.

In addition there are also important statistical reasons why the variance should not be considered as the best or the only measure of risk. A portfolio containing a large amount of options will typically reveal a return distribution that is skewed to the right or the left. In calculating the variance positive and negative deviations from the mean are weighted symmetrically. Only in the case of normally distributed returns the variance is the correct measure of risk. In all other cases the use of variance means an over- or underestimation

[1] Olsen, Robert A. (1997)

of risk, because deviations from the normal distribution are not taken into account.

The shortfall approach is a general concept to evaluate return distributions and the different shortfall measures can be used as flexible tools to measure investment risk and risk-adjusted performance.[2]

This article is divided into two parts. In the first part the shortfall approach is described and explained. It will be shown that the new Value at Risk-concept is interpretable as a special shortfall measure. Basic to the understanding and use of shortfall measures is their foundation in stochastic decision theory. The second part of this article describes an investment product based on a dynamic benchmark adjustment. The adjustment procedure is governed by the probability of falling below the investors desired minimum return.

2 The shortfall approach

Shortfall measures define risk as falling below a desired minimum or target return. Instead of calculating statistical moments for the total return distribution, as is common e.g. in the mean-variance approach, shortfall measures are calculated in the "left" part of the return distribution only. The meaning of "left" depends on the target return. Figure 1 illustrates the shortfall concept. The figure shows a return distribution that is skewed to the right. Such a skewed distribution is typical for a portfolio that contains a large amount of put options. The different shortfall measures are calculated over the shaded area left of the arbitrarily chosen target return z. Shortfall measures are also named lower partial moments: "partial" moments because not the total distribution is used and "lower" because only the returns left of the target return are included in the risk measures.

The following formula gives the general definition of shortfall measures or lower partial moments:[3]

$$lpm_n(z) = \int_{-\infty}^{z} (z - R)^n dF(R)$$

There are two distinct parameters - n and z - that characterize the different shortfall measures. The integral is calculated from minus infinity to z. The return distribution $F(R)$ is weighted using $(z-R)^n$. $(z-R)$ is always positive. According to the number n ($n = 0, 1, 2, ...$) we can distinguish n different shortfall measures.

If n is equal to 0, then we have the simplest shortfall measure lpm_0. In this case the weight $(z - R)^n$ is always equal to 1. Lpm_0 is therefore called

[2] Balzer, Leslie A. (1994) gives a survey of different methods to measure investment risk

[3] Bawa, Vijay S. (1978)

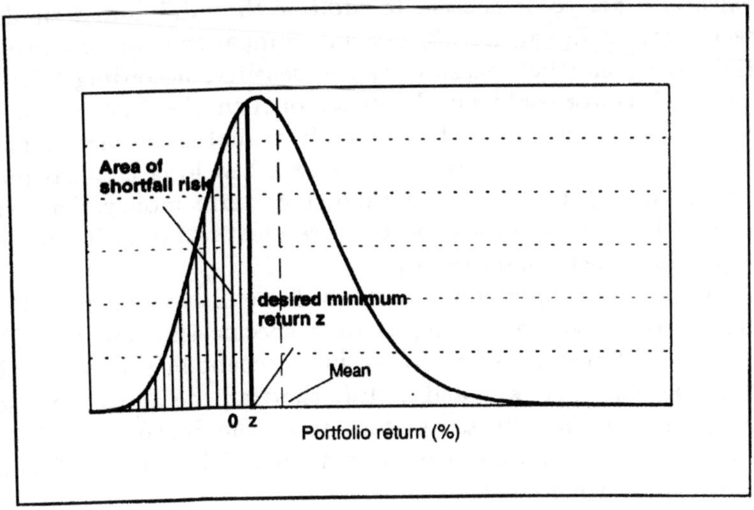

Fig. 1. Shortfall Risk

shortfall probability. It measures how often the portfolio return falls below the target return. But it does not take into account how severe the loss might be.

If n is equal to 1, then lpm_1 weighs losses with the linear weighting scheme $(z - R)$. Lpm_1 is named expected shortfall because its only difference to the common calculation of the expected return is that lpm_1 is defined only below the target return.

If n is equal to 2 we get the target semivariance lpm_2. Here the deviations from the target z are squared. Hence these deviations are punished more heavily than in the case of lpm_1. If we choose z equal to the mean of R we get the semivariance. And if the returns are normally distributed then lpm_2 is proportional to the variance.

The choice of the appropriate shortfall measure can be guided by the risk aversion of the investor.[4] Lpm_0 is the appropriate measure to evaluate a return distribution if the investor is risk-neutral. It can be shown that lpm_0 is equivalent to first order stochastic dominance. And therefore all investors with $u'(R)$, where $u(R)$ is the common von Neumann-Morgenstern utility function, can use lpm_0 to evaluate the return distribution of an investment.

[4] Bawa, Vijay S. (1975), Bawa, Vijay S. (1978) or Harlow, W. V./Rao, Ramesh K. S. (1989)

If the investor is risk averse then lower partial moments with $n > 0$ should be chosen. The more risk averse the investor is the higher should be n.[5]

To use shortfall measures correctly not only the degree of risk aversion or equivalently n has to be chosen. In addition the minimum desired return z is very important in calculating risk and comparisons among investment alternatives using shortfall measures can be sensitive according to the value of z. A private investor could e.g. choose a money market interest rate as the target rate z. The money market is a readily available alternative to equity and bond markets and can therefore serve as a benchmark to calculate the risk of an investment that is riskier than short term money. An insurance company would e. g. choose the expected invesment return of its competitors as the appropriate benchmark return.

Shortfall measures are useful tools to evaluate investment risk. Different portfolios can be ranked according to their specific shortfall risk. The investors' attitudes towards risk are taken into account in a very flexible way in choosing the parameters n and z. But shortfall measures are also very useful in comparing risk adjusted performance. The Sharpe ratio SR is the appropriate measure of risk adjusted performance if the returns of an investment are normally distributed.

$$SR = \frac{\mu - r}{\sigma}$$

In the Sharpe ratio the expected return μ minus the riskfree rate r is divided by the standard deviation σ. This is exactly the slope of the efficient frontier in the standard CAPM. In this case the Sharpe Ratio measures the risk adjusted return of the Market Portfolio.

But if the return distribution is not normal the Sharpe ratio can be misleading. If the portfolio contains put options the return distribution will be skewed to the right. In this case the portfolio without options will always be ranked higher because the standard deviation does not take into account the skewness of the distribution. And therefore the portfolio with put options seems to be too risky relative to its returns. To get a correct ranking of risk adjusted performance shortfall measures have to be used instead of the standard deviation.[6]

---

[5] The degree of the risk aversion is a characteristic of the utility function. The use of lpm_0 is appropriate for all investors with $u'(r) > 0$. The investor is risk averse if $u''(r) < 0$. Then lpm_1 is the right measure to evaluate the return distribution. Lpm_1 is equivalent to second order stochastic dominance. If the investor is still more risk averse (e. g. $u'''(r) > 0$) n should be al least equal to 2. The investor with this kind of utility function will use lpm_2. Lpm_2 is equivalent to third order stochastic dominance. See e.g. Bawa, Vijay S. (1978). In addition also higher values of n could be chosen. If n is equal to 3 the risk measure is a lower partial skewness and in case of $n = 4$ it is a lower partial kurtosis. But $n = 0, 1\, or\, 2$ are more common in practice

[6] See e.g. Zimmermann, Heinz (1994) or Albrecht Peter et al (1995)

$$SR_1 = \frac{\mu - r}{lpm_1}$$
$$SR_2 = \frac{\mu - r}{\sqrt{lpm_2}}$$

In the case of SR_1 lpm_1 is used as measure of risk, whereas SR_2 uses the square root of lpm_2. SR_2 is more appropriate when the risk aversion is relatively high because falling below the target rate z will be punished stronger.

The above described lower partial moments and the performance measure using lower partial moments are an essential part of the program MAMBA[7]. This program allows the comparison of international equity and bond portfolios containing options and futures to insure against losses.

In the recent past the Value at Risk-measure (VaR) has achieved a prominent role in the regulation of market risk of banks. VaR is defined as the potential loss that will be exceeded by only a very small probability of 1% or 5%.

$$Pr(R \leq VaR = \alpha = e.g. 5\%)$$
$$F(VaR) = \alpha$$

Hence VaR is just the α-quantile of the return distribution $F(R)$. In the case of a normal distribution VaR can be readily determined using a table of the standard normal distribution. The following figure 2 gives an illustration of Value at Risk:

It can be easily shown that VaR is equivalent to the shortfall probability lpm_0:

$$\begin{aligned} lpm_0 &= F(z) \\ &= Pr(R \leq z) \\ \text{if} \quad z = VaR & \\ &= Pr(R \leq VaR) \\ &= F(VaR) \end{aligned}$$

Value at Risk is calculated given the probability of falling below VaR ($\alpha = lpm_0$), whereas in the case of lpm_0 the target value ($z = VaR$) is given. Value

[7] MAMBA=Metzler Asset Management Benchmark Analyser. MAMBA has been developed and constructed by the Dept. International Finance and Financial Management, Zentrum für Europäische Wirtschaftsforschung (ZEW) GmbH, Mannheim.

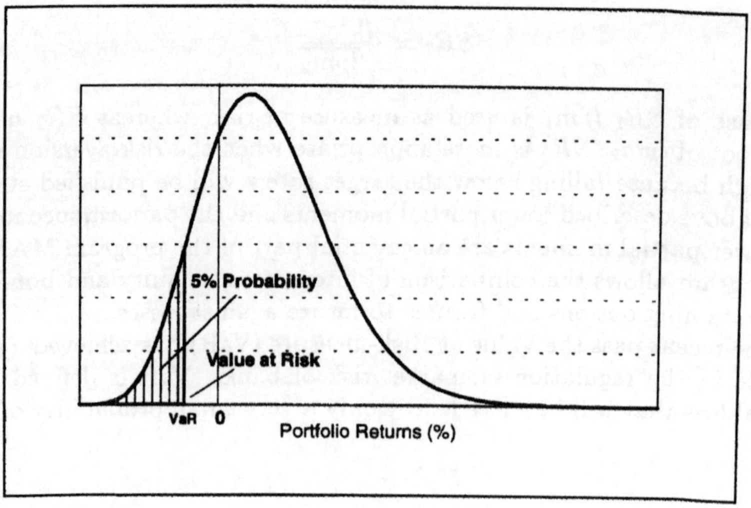

Fig. 2. Value at Risk

at Risk therefore shares the characteristics of lpm_0: VaR is appropriate for all investors with $u'(R) > 0$. Risk aversion is not covered, because losses are only included using their frequency. The severity of the losses is not taken into account.[8]

3 Dynamic benchmark adjustment

Mounting competition for capital is increasing the performance pressure on institutional investors, especially in the insurance sector (e.g. as a result of "Europeanisation") as companies need to maximise returns yet minimise risks (i.e. possible write-downs). Policyholders are generally led to expect a return of around 7.5%. However, since German insurance companies currently have only relatively low exposure to equities, it seems unrealistic to assume that they will achieve returns of this magnitude in the future. The insurance sector is thus faced with the following conflict of interest: in the medium term companies need to improve their performance by raising equity holdings but at the same time they need to avoid write-downs at year-end.

Evidently, no investment strategy can square the circle entirely. However, a strategy based on dynamic benchmark adjustment can help investors come

[8] See Schröder, Michael (1996)

close to achieving both aims. The Metzler Group's asset management staff have developed a strategy of this type based on suitable, intuitively acceptable measures of risk. This is outlined here.

3.1 The basic idea: Raising exposure to equities with a "safety margin"

The basic idea is that the risk profile of a portfolio should be kept fairly constant over time at a level defined by the client. To achieve this, asset class allocation is altered gradually in line with the hidden reserves built up over time. The aim is to achieve the portfolio structure that gives the highest possible performance at the defined level of risk. This can basically be broken down into three phases (see Figure 3).

In the initial phase, a decision has to be made on the proportion of the fund's assets that will be invested in bonds and equities when it is set up. There is a clear trade-off between the lowest possible shortfall risk (high proportion of bonds) and rapid creation of hidden reserves (high proportion of equities). It would seem sensible to accept a somewhat higher shortfall risk as this provides the only real chance of building up significant equity holdings.

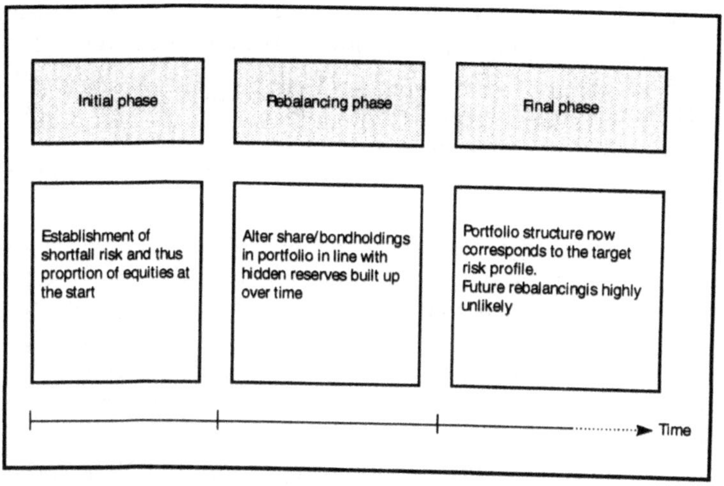

Fig. 3. Phases in the dynamic benchmark adjustment process

In the second phase, the portfolio's equity exposure is raised. The aim is to minimise the probability of falling below the minimum return set by the investor or to ensure that it does not fall below a predefined limit (for example 5%). The hidden reserves accrued by the time the portfolio is rebalanced should be included in this calculation. In other words, if hidden reserves have been built up, the proportion of equities in the fund can be increased without increasing the risk of failing to achieve the minimum return. The frequency with which the portfolio is rebalanced can be set individually.

At the start of the final phase, the fund contains the proportion of equities and bonds corresponding to the target risk profile. Exactly when this final phase starts evidently depends on market trends in preceding years and the hidden reserves built up as a result. As a guide, it takes about five years to reach this point. If market trends are negative in this phase and hidden reserves decline, equity holdings can be reduced to ensure that the shortfall risk and minimum return meet the investor's criteria. Over time it becomes increasingly unlikely that the equity ratio will have to be reduced because the longer the investment period, the more evident it becomes that equities generate a higher return. By then, sufficient hidden reserves will have been built up to cushion temporary market downturns and ensure that the shortfall risk can be kept at the required level without reducing the proportion of the fund's assets invested in equities. Figure 4 shows how dynamic benchmarking functions, using a fictitious example to depict the rise in hidden reserves. It is assumed that the investor wishes to ensure a shortfall risk of 5% or less.

The table in Figure 4 shows that the minimum shortfall risk for a minimum yield (*Rmin*) of 0% and a one-year investment horizon is 4.3%. Even a fund invested solely in bonds - the 0/100 Model Portfolio (0% equities/100% bonds) - has a higher shortfall risk (5%). In other words, the investor must select the portfolio structure with the lowest risk, even though a significant proportion is invested in equities. In our example, this is a split of 20% equities and 80% bonds in the initial phase (Year 1). This gives a higher return than a portfolio invested solely in German bonds but the risk profile is comparable. In other words, the structure of the portfolio does not increase the risk in the initial phase.

We assume that hidden reserves of 2% are built up by the end of Year 1, even though dividends and interest income are paid out. As a result, in Year 2 the minimum return that has to be obtained by the portfolio is 2 percentage points lower. It is thus possible to switch to a portfolio structure with a shortfall risk of 5% or less based on a minimum return of -2%. Figure 4 shows that in our example the equity holding can be increased to 30%, giving a shortfall risk of 4.1%.

As a result of good market conditions, by the end of Year 2, the fund's hidden reserves have grown to 8thus drops to -8% and shareholdings can be raised to 70%. The shortfall risk is now 4.4%. However, since a downtrend on the market in Year 3 reduced cumulative hidden reserves to 6%, bondholdings have to be raised to 40% again to ensure that the portfolio does not exceed the set shortfall risk. The benchmark structure is adjusted dynamically in the same way in subsequent years. Given the development of hidden reserves assumed in our example, at the start of Year 7, the entire portfolio would be invested in equities.

The shortfall risks illustrated in Figure 4 are average values. A 5% shortfall risk means that taking a statistical average, an investor would only have to fear dropping below the minimum return once in twenty years. However, no-one can guarantee that this will not occur in Year 1. The risks can be highlighted by simulated scenarios.

A simulation of this type was worked through for the 20/80 Model Portfolio (20% equities, 80% bonds) used as the basis for the initial phase in our example. Table 1 shows the potential performance of the portfolio in its first year for various alternative interest rate and share price scenarios. To simplify matters, the calculations are based on a portfolio duration of 4.5 years, parallel shifts in the yield curve, an average coupon of 6% p.a. and an average dividend yield of 2.5% p.a. These parameters are assumed to be the same for all countries represented in the portfolio.

Admittedly, this is an extremely simplified example, but it should give an idea of the portfolio risk in the first year.

The shaded area shows all combinations of equity and bond market trends that would give the required minimum return of around 0% or above. For example, if interest rates do not change, the equity markets can drop by around 25%. Conversely, if the equity markets rise by 10%, interest rates can rise by more than 2 percentage points. Other scenarios can be inferred from the table.

3.2 An example

The portfolio structure illustrated in Figure 5 refers to a portfolio set up in 1980, invested exclusively in German assets and managed on the basis of the dynamic benchmark adjustment strategy (all data are year-end figures). The minimum return set for this portfolio is 0%. In other words, the aim is to avoid write-downs if at all possible.

By 1984 the portfolio was already fully invested in equities. By then, the cumulative hidden reserves (see Figure 6) had reached a level that obviated the need for rebalancing (in this case a reduction in the percentage of equities to maintain the risk profile), even in phases when the market dropped sharply (1987 and 1990). Since 1985 the fund (DBA 100) has performed in line with the equity market. There was no need for write-downs at year-end at any time in the period shown here (1980 - 1996).

A fund managed on the basis of these criteria would have given a return of 11.3% p.a. over this period. By contrast, a portfolio (10E/90B) with a constant level of investment in bonds and equities (10% equities, 90% bonds)

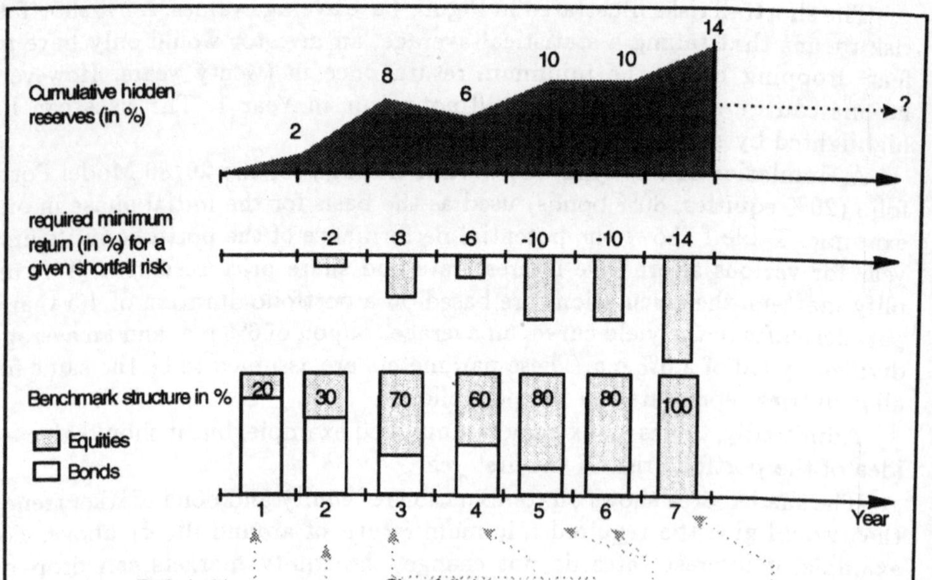

Equities/Bonds in % (rows) / Rmin in % for shortfall risk (columns)	+4	+2	0	-2	-4	-6	-8	-10	-12	-14	-16
100 / 0	26,5	23,4	20,2	16,2	13,8	10,6	8,2	6,6	5,4	4,7	3,8
90 / 10	25,7	22,2	17,9	14,9	11,5	8,9	7,0	5,5	4,4	4,0	3,6
80 / 20	24,9	19,9	16,5	12,5	9,2	7,0	5,6	4,4	4,0	3,4	2,5
70 / 30	22,9	17,9	13,9	10,1	7,4	6,1	4,4	3,8	3,2	2,6	1,6
60 / 40	21,3	15,1	11,5	8,6	6,2	4,7	4,0	3,2	2,6	1,3	0,8
50 / 50	18,5	13,5	9,1	6,6	4,7	3,6	3,1	2,0	1,2	0,5	0,4
40 / 60	16,7	11,0	7,7	5,2	3,8	2,6	1,8	0,6	0,4	0,2	0,1
30 / 70	15,9	9,7	5,8	4,1	2,3	1,3	0,5	0,2	0,1	0,1	0,0
20 / 80	15,0	8,0	4,3	2,3	1,1	0,2	0,2	0,1	0,0	0,0	0,0
10 / 90	16,0	8,9	4,1	1,8	0,2	0,2	0,0	0,0	0,0	0,0	0,0
0 / 100	19,1	10,9	5,0	1,9	0,5	0,1	0,0	0,0	0,0	0,0	0,0

Fig. 4. Hidden reserves as the key determinant of benchmark adjustment

Change in interest rates in percentage points

Change in share prices in %	3	2,5	2	1,5	1	0.5	0	-0,5	-1	-1,5	-2	-2,5	-3
-30	-11,5	-9,7	-7,9	-6,1	-4,3	-2,5	-0,7	1,1	2,9	4,7	6,5	8,3	10,1
-25	-10,5	-8,7	-6,9	-5,1	-3,3	-1,5	0,3	2,1	3,9	5,7	7,5	9,3	11,1
-20	-9,5	-7,7	-5,9	-4,1	-2,3	-0,5	1,3	3,1	4,9	6,7	8,5	10,3	12,1
-15	-8,5	-6,7	-4,9	-3,1	-1,3	0,5	2,3	4,1	5,9	7,7	9,5	11,3	13,1
-10	-7,5	-5,7	-3,9	-2,1	-0,3	1,5	3,3	5,1	6,9	8,7	10,5	12,3	14,1
-5	-6,5	-4,7	-2,9	-1,1	0,7	2,5	4,3	6,1	7,9	9,7	11,5	13,3	15,1
0	-5,5	-3,7	-1,9	-0,1	1,7	3,5	5,3	7,1	8,9	10,7	12,5	14,3	16,1
5	-4,5	-2,7	-0,9	0,9	2,7	4,5	6,3	8,1	9,9	11,7	13,5	15,3	17,1
10	-3,5	-1,7	0,1	1,9	3,7	5,5	7,3	9,1	10,9	12,7	14,5	16,3	18,1
15	-2,5	-0,7	1,1	2,9	4,7	6,5	8,3	10,1	11,9	13,7	15,5	17,3	19,1
20	-1,5	0,3	2,1	3,9	5,7	7,5	9,3	11,1	12,9	14,7	16,5	18,3	20,1
25	-0,5	1,3	3,1	4,9	6,7	8,5	10,3	12,1	13,9	15,7	17,5	19,3	21,1
30	0,5	2,3	4,1	5,9	7,7	9,5	11,3	13,1	14,9	16,7	18,5	20,3	22,1

Table 1. Risk scenarios for the 20/80 Model Portfolio in the first year

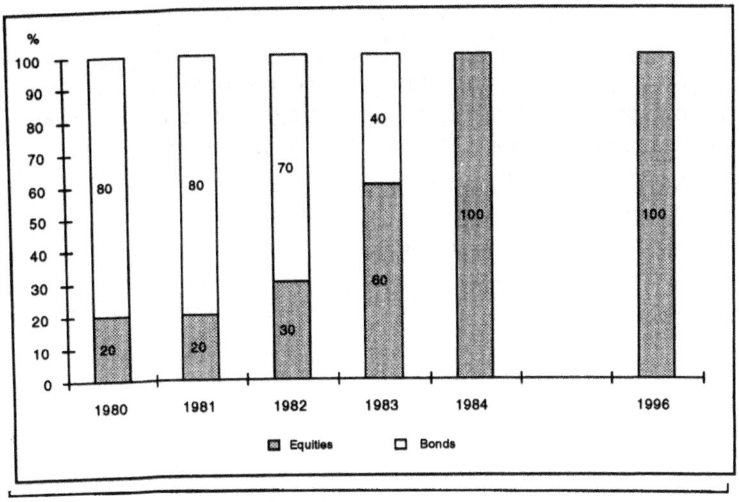

Fig. 5. Change in the structure of the portfolio over time

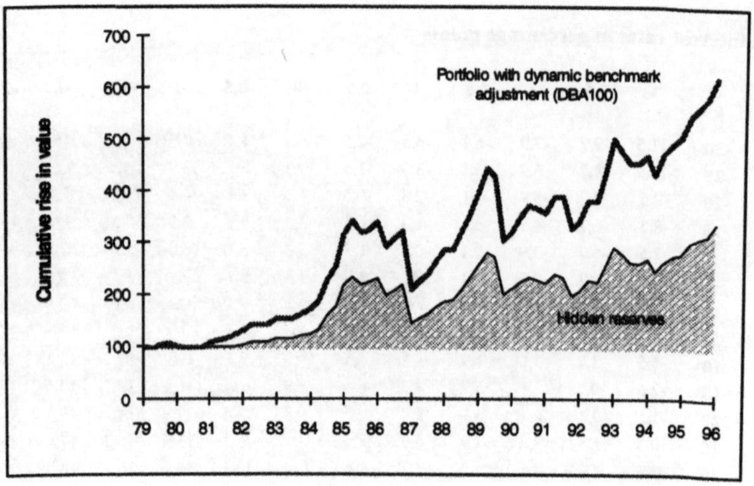

Fig. 6. Increase in the value of the fund and hidden reserves

- a split that is common in the German insurance sector-would have yielded a return of just 9.0% p.a.

In view of the legislative constraints placed on German insurance companies, we also show a portfolio using a dynamic benchmark adjustment with a maximum shareholding of 30% (DBA30). Even this greatly reduced form of benchmark adjustment would have yielded a return of 10.1% - a full percentage point above the return on the average portfolio structure used by German insurance companies. Figure 7 shows the increase in the cumulative value of all three portfolios.

In this example, it is assumed that hidden reserves are built up solely through rising equity prices and that dividends and interest income are paid out. The portfolio was rebalanced once a quarter. Management of the two sub-portfolios (German equities and German bonds) was assumed to be based on index-tracking.

4 Conclusion

The shortfall concept offers a very general and flexbile way to measure investment risk and performance. The idea behind shortfall measures is what most private and institutional investors perceive as investment risk. But the shortfall concept also has a sound foundation in stochastic decision theory.

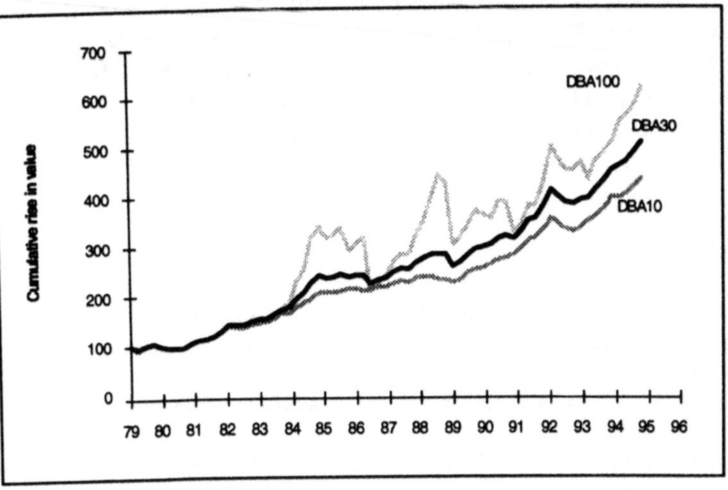

Fig. 7. Comparison of rise in value of three portfolios

Hence shortfall meaures offer a superior alternative to the common Markowitz approach.

One example for the successful application of shortfall measures is the dynamic benchmark adjustment process described in chapter 3.

The key elements in the dynamic benchmark adjustment approach are:

- Method of increasing equity exposure while maintaining a pre-defined, low level of risk
- Exposure to equities is managed through hidden reserves.
- This risk structure is analysed using measures of downside risk and realistic distribution for returns. In other words, it is not simply based on volatility and the arbitrary assumption of a normal distribution. The risk profile deduced from this is geared to investors' requirements, makes sense on an intuitive level and reflects reality.
- There is no guarantee that the value of assets will be maintained or that a minimum return will be earned, although there is a high likelihood of this. However, unlike "guaranteed products" the portfolio participates fully in market trends. Similarly it is not necessary to use derivatives, although there is no reason why they should not be used.

Dynamic benchmarking is thus particularly suitable for investors with low-risk portfolios (e.g. 100% bonds) who wish to increase their risk exposure cautiously in order to raise their earnings potential. The advantage of this strategy is that it largely eliminates the risk of write-downs.

Dynamic benchmark adjustment is thus an attractive option for insurance companies, especially as increasing globalisation and the related competitive pressure mean that bond yields are unlikely to be as high as they have been in the past. Moreover, low structural inflation will keep interest rates low, regardless of cyclical trends. The higher returns earned by investing in equities will also play an increasing role given the growing need for private pension provision and the increasing focus on shareholder value.

References

Albrecht Peter, Maurer Raimond, Stephan Thomas G. (1995): Shortfall-Performance rollierender Wertsicherungsstrategien, Finanzmarkt und Portfolio Management, vol. 9, 197-209.

Balzer, Leslie A. (1994): Measuring Investment Risk: A Review, Journal of Investing, Fall 1994, 47-58.

Bawa, Vijay S. (1975): Optimal rules for ordering uncertain prospects, in: Journal of Financical Economics, vol. 2, 95-121.

Bawa, Vijay S. (1978): Safety First, Stochastic Dominance, and Optimal Portfolio Choice, in: Journal of Financial and Quantitative Analysis, vol. 13, 255-271.

Bawa, Vijay S./Lindenberg, Eric B. (1977): Capital Market Equilibrium in a Mean-Lower Partial Moment Framework, in: Journal of Financial Economics, vol. 3, 189-200.

Fishburn, Peter C. (1977): Mean-Risk Analysis with Risk Associated with Below-Target Returns, American Economic Review, vol. 67, 116-126.

Harlow, W. V. (1991): Asset Allocation in a Downside-Risk Framework, in: Financial Analysts Journal, Sept./Oct. 1991, 28-40.

Harlow, W. V./Rao, Ramesh K. S. (1989): Asset Pricing in a Generalized Mean-Lower Partial Moment Framework: Theory and Evidence, in: Journal of Financial and Quantitative Analysis, vol. 24, No. 3, 285-311.

Lee Wayne Y.,Rao Ramesh K. S. (1988): Mean Lower Partial Moment Valuation and lognormally distributed Returns, in: Management Science, vol. 34, No. 4, 446-453.

Olsen, Robert A. (1997): Investment Risk: The Experts' Perspective, Financial Analysts Journal, March/April 1997, 62-66.

Schröder, Michael (1996): Value at Risk: Proposals on a Generalization, in: Peter Albrecht (Hrsg.): Aktuarielle Ansätze für Finanz-Risiken, Beiträge zum 6. Internationalen AFIR-Colloquium, Karlsruhe, 151-169.

Zimmermann, Heinz (1994): Editorial: Reward-to-Risk, in: Finanzmarkt und Portfolio Management, Jg. 8, Nr.1, 1-6.

Basics of Statistical VaR-Estimation

Thomas Ridder

SGZ-Bank AG, Frankfurt a.M. / Karlsruhe

Abstract. Quantifying risk exposure has become a key issue for financial institutions in recent years. The idea of Value at Risk (VaR) reflects the industry's efforts to develop new methods in financial risk management that take into account available knowledge in financial engineering, mathematics and statistics.
This paper gives a survey of common procedures to analyze risk exposure by VaR and studies basic statistical properties of various VaR-estimators. An application to foreign exchange rate risk gives insight into the accuracy of those VaR-strategies.

1 Introduction

As international financial business is perceived to become riskier nearly every year, it is not only banking regulators that agree upon the necessity of adequate strategies in internal risk controlling in financial institutions. Thus, quantifying, controlling and limiting the firm wide risk exposure on a daily basis has become a standard activity for most banks. But systematic risk management can not only protect individual firms from losses it might never have anticipated before. Risk management in general helps to allocate financial resources more efficiently as risk adjusted comparisons of different investments become feasible. The concept of Value at Risk (VaR) reflects industry's state of the art efforts in finding a reasonable definition of risk.

The idea of VaR is that of reserve capital. Uncertainty about future prices in financial markets always generates a positive probability of adverse developments. A certain amount of capital should be kept aside to cover potential losses resulting from trading activities. But the question is, how much risk capital should we hold instead of allocating it to more profitable opportunities. VaR gives the answer. As VaR is denominated in currency units it tells us, how much capital we need to be able to cover losses up to a certain degree. VaR can be computed for individual trading positions, for portfolios or with regard to the whole bank. Additionally, VaR is defined to cover potential losses occuring over a given holding period with given probabilty.

Evidently, a measure of risk showing these features can only be constructed by means of statistical models, since - in statistical terms - VaR can readily be recognized as a quantile of a loss distribution. Thus analyzing VaR means computing quantiles. But in most applications the true loss distribution of the risk position under concern is not known. So VaR can only be estimated from samples of loss data. Several methods of estimating VaR have become popular in recent years. All of them are standard techniques in statistical quantile estimation but currently they pass through a revival within the context of VaR. See JP Morgan [1996]. *Historical simulation* estimates quantiles from the empirical distribution of available sample data. *Variance-covariance models* assume linear pricing rules and normality of the sample to estimate VaR. *Monte Carlo simulations* generate synthetic sample paths for financial markets and evaluate empirical distributions of positions priced along these simulated market movements. An empirical comparison can be found in Bühler/Korn/Schmidt [1997] for example. Problems in applying standard VaR-techniques are focused in Leong [1996], Bode/Mohr[1994] or Johanning [1996]. The drawbacks can be summarized by the use of unrealistic assumptions and the question of whether VaR is a reasonable definition of risk at all. We do not focus on the last issue here. Lacking subadditivity is an important drawback to VaR. See Artzner/Delbean/Eber/Heath [1996] on this topic. Nevertheless, we assume in the following that the decision has been taken to use VaR as our measure of risk. Additionally, we will not generalize common assumptions on the stochastic structure of the underlying model. Instead we will introduce some less known quantile estimation procedures with a view to VaR. Thus, all properties are studied in an ideal statistical world.

Several survey articles on statistical quantile estimation are available. Parrish [1987], Hughey [1991] and Dielmann/Lowry/Pfaffenberger [1994] give broad insight into the subject. As their results show, gains with respect to the accuracy of estimates can be expected, if more sophisticated techniques are applied. This becomes especially important when investigating quantiles in the upper right tail of the loss distribution. If we really want to know about large or extreme quantiles, we have to apply the adequate tool kit (Bassi/Embrechts/Kafetzaki [1997]).

This paper collects results from various sources. Most results are known but some of them have not yet been applied within the context of VaR. Section 2 gives basic definitions and assumptions on the stochastic structure of our model. Section 3 introduces several VaR-estimators and studies their statistical properties. In Section 4 an application to foreign exchange rate risk compares the performance of different ways to estimate VaR. In Section 5 conclusions are drawn.

2 The P&L-Process

Consider an open financial risk position (e.g. an asset) in a discrete time world and let V_t be its market price at time t . The way the market creates this price can be described by a continuous valuation function $v : \mathbb{R}_+^k \to \mathbb{R}$ mapping $k-$vectors $R_t = [R_{t,1}, R_{t,2}, \dots, R_{t,k}]'$ of position specific risk factors into real numbers. Each period we have $V_t = v(R_t)$. As we include derivative positions, market prices can be negative. Potential risk factors are exchange rates, stock indices, interest rates, commodity prices etc. Every risk factor is assumed to exhibit *i.i.d.* one period log-returns. Different risk factors are allowed to have different return distributions. The rates of log-return of all risk factors at time t are pooled in $k-$vectors

$$X_t = [X_{t,1}, \ldots, X_{t,k}]' = \Big[\ln(R_{t+1,1}/R_{t,1}), \ldots, \ln(R_{t+1,k}/R_{t,k}) \Big]'$$

having an identical $k-$dimensional distribution each period. Independence of $X_{t,j}$ and $R_{1,j}$ for every j and t ensures independence of $X_{t,j}$ and $R_{t,j}$ since for $t \geq 2$ we have $\ln R_{t,j} = \ln R_{1,j} + \sum_{\tau=1}^{t-1} X_{\tau,j}$. The risk of the bank can be described by potential changes in the market value of the risk position resulting from adverse movements of risk factors. The one period loss function is

$$L(R_t, X_t) = v(R_t) - v(R_{t+1}) = v(R_t) - v(R_t * \exp(X_t)) \tag{1}$$

where element by element operation $R_t * \exp(X_t)$ is defined by

$$R_t * \exp(X_t) := \Big[R_{t,1} \exp(X_{t,1}), \ldots, R_{t,k} \exp(X_{t,k}) \Big]' .$$

Definition (1) always leads to positive values when the market price at time $t+1$ is lower than at time t. Periods of losses are thus characterized by positive realizations of L. As profits are negative losses, the distribution of L_t is called the P&L-distribution.

Assuming time stationarity of the valuation function v we do not consider effects resulting from the reduced time to maturity of the risk position between t and $t+1$. If the risk position being analyzed consists of a portfolio of different assets, identical portfolio structures at times t and $t+1$ are assumed. This should be acceptable since we are only interested in one period risk analysis and the time unit should be defined by one day. Additionally, we assume that trading in the asset takes place with prices computed by the fair value function v without any bid-ask spreads. Despite these apparent inconsistencies with reality, incorporating those facts into the model would only complicate notation without changing the main arguments. Hull [1993] develops pricing rules for most financial instruments.

The goal of VaR-analysis is to compute at time t an upper limit for potential losses occurring between dates t and $t+1$. This limit should be exceeded by L_t with given probabilty $1-p$ and is called the *Value at Risk* (VaR) at time t of the financial position under concern. Typically small values like 0.01 or 0.05 are used for $1-p$ and usually these probabilities are set with respect to the loss distribution conditioned on observed risk factors up to time t. Thus the calculations are being repeated every period, taking into account all available information up to time t.

Another - more statistical - formulation of the VaR-problem is to find the $p-$quantile of the conditional P&L-distribution where conditioning takes place with respect to the information set (or sigma algebra) generated by the

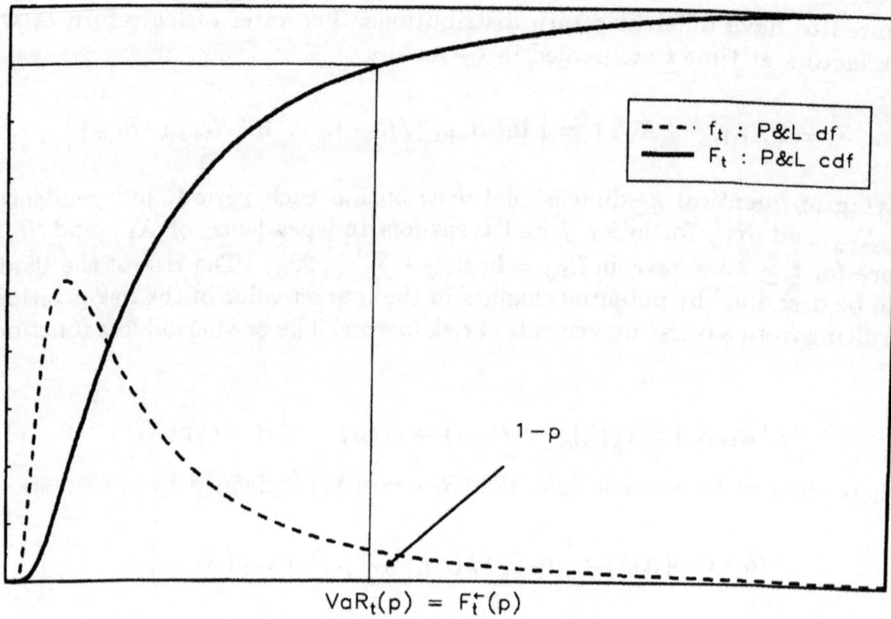

Fig. 1. Graphical representation of VaR

variables $\{R_1, R_2, \ldots, R_t\}$. Figure 1 shows the typical graphical representation of VaR, given P&L-density and distribution function at time t.

Analyzing VaR in models with general stochastic structures usually demands rather technical arguments. But the assumptions stated above restrict stochastic dependency in our model to a minimum degree. Consequently, because of independence of R_t and X_t we have for every strictly positive k−vector r

$$P\{\,L(R_t, X_t) \leq y \,|\, R_t = r\,\} = P\{\,L(r, X_t) \leq y\,\}\ ,\ t = 1, 2, \ldots$$

where $P\{A\}$ stands for probability of event A. If r_t is the realization (observation) of the random vector R_t in period t the VaR-process $\{\,\mathrm{VaR}_t(p)\,\}_{t=1}^{\infty}$ can be defined implicitly by

$$P\{\,L(R_t, X_t) \leq \mathrm{VaR}_t(p) \,|\, R_t = r_t\,\} = P\{\,L(r_t, X_t) \leq \mathrm{VaR}_t(p)\,\} = p \qquad (2)$$

where $0 < p < 1$. Indicating the conditional P&L-distribution function at time t by F_t a formal definition of VaR is

$$\mathrm{VaR}_t(p) = F_t^{\leftarrow}(p) = \inf\{x \in \mathbb{R} : F_t(x) \geq p\} \qquad (3)$$

where $F_t^{\leftarrow}$ denotes inverse P&L-distribution function.

If F_t is known, $\mathrm{VaR}_t(p)$ can be computed exactly by analytical calculations or can be approximated to a high degree by numerical methods. But this is hardly the typical situation for real world risk controllers. Usually $\mathrm{VaR}_t(p)$ has to be estimated from observations of risk factors or position values. Using (2) we see that

$$\{ L(r_t, X_{t-i}) \, , \; i = 1, 2, \ldots, n \}$$

can be regarded as a sample of size n from distribution F_t . Based on r_t and $\{X_{t-i} \, , \; i = 1, 2, \ldots, n\}$ for $n < t$ a corresponding sample of loss variables can be represented by

$$L(r_t, X_{t-i}) = v(r_t) - v(\, r_t * \exp(X_{t-i}) \,) \, , \quad i = 1, 2, \ldots, n \, .$$

So even if we do not know the true P&L-distribution function, we are able to construct samples from it as long as we know the valuation function v and as long as we can observe the evolution of relevant risk factors.

3 VaR-Estimators

We assume the true distribution function F_t to be unknown. Thus the VaR of the risk position under review has to be estimated. This can be done by various methods. All estimation procedures discussed here are based on n constructed P&L-values $\{ L(r_t, X_{t-i}) \, , \; i = 1, 2, \ldots, n \}$. These data are considered as a sample of random variables whose distribution is F_t. The VaR-estimator of the risk position under review is an estimator of the $p-$quantile of this distribution function. For ease of notation we drop time index t in the following, keeping in mind that if we want to analyze VaR at different times, the sample data have to be completely recalculated every period, since in general $r_t \neq r_{t'}$ for $t \neq t'$. We write $L_i = L(r_t, X_{t-i})$ for $i = 1, 2, \ldots, n$.

3.1 Nonparametric Approaches

Nonparametric methods of VaR-estimation dispense with an explicit specification of the class of distribution functions that F belongs to. Usually this class is not known to the risk controller. So any assumption of a specific family of distributions bears the danger of severe misspecification producing systematic estimation bias.

Minimal Assumptions: Historical Simulation. To reduce specification errors VaR(p) is often estimated by the $p-$quantile of the empirical distribution function constructed from the sample of loss data available. As the sample is the result of simulating the effects of observed historical factor returns on the current risk position, this method is usually called *historical simulation*. Let $\mathbf{1}_A(x)$ be the function

Table 1. VaR-estimators by historical simulation

p	$n = 100$	$n = 260$	$n = 520$	$n = 780$
0.9000	$L_{90:100}$	$L_{234:260}$	$L_{468:520}$	$L_{702:780}$
0.9500	$L_{95:100}$	$L_{247:260}$	$L_{494:520}$	$L_{741:780}$
0.9900	$L_{99:100}$	$L_{258:260}$	$L_{515:520}$	$L_{773:780}$
0.9950	$L_{100:100}$	$L_{259:260}$	$L_{518:520}$	$L_{777:780}$

$$\mathbf{1}_A(x) = \begin{cases} 1 & \text{if} \quad x \in A \\ 0 & \text{else}, \end{cases}$$

where $A \subset \mathbb{R}$ and x is real valued. The empirical distribution of the sample variables is given by

$$\hat{F}(x) = n^{-1} \sum_{i=1}^{n} \mathbf{1}_{(-\infty,x]}(L_i)\,.$$

Let $L_{1:n} \le L_{2:n} \le \cdots \le L_{n:n}$ be the ordered sample variables. As a non-parametric estimator of VaR(p) we use the inverse of $\hat{F}$

$$\boxed{\widehat{\mathrm{VaR}}_{\mathrm{HS}}(p) := L^*_{j:n} \quad \text{where} \quad \frac{j^*-1}{n} < p \le \frac{j^*}{n}, \qquad j^* \in \{1, 2, \ldots, n\}} \tag{4}$$

Table 1 shows estimators given by historical simulation for some selected values of p and n . Regulatary requirements call for a sample size large enough to cover the evolution of financial markets of at least one year. Thus most risk reports are based on samples consisting of 250 or 260 of the most recent daily observations at hand. But even for this sample size $\widehat{\mathrm{VaR}}_{\mathrm{HS}}(p)$ exhibits a rather discontinuous behavior in p especially for large $p-$values as 0.95 or even 0.99 since only few observations in the sample can be expected to belong to the extreme right tail of the distribution. As the estimators in Table 1 indicate, the functions $\widehat{\mathrm{VaR}}_{\mathrm{HS}}(p)$ will not be smoothed very much even if we use observations belonging to the last 3 years. Expanding the sample period to several years is usually considered dangerous, as it gets harder to justify the assumption of stationarity of the economic time series. Using historical simulation we always have $\widehat{\mathrm{VaR}}_{\mathrm{HS}}(p) \le L_{n:n}$ for all n and all $p-$values between 0 and 1. This method will never estimate the true VaR to be larger than the worst loss observed in the sample at hand.

The exact distribution function of ordered loss statistics can be computed by standard arguments from combinatorics (Büning/Trenkler [1994]). Obviously

$$\begin{aligned} P\{\, L_{k:n} \le x \,\} &= P\{\, F(L_{k:n}) \le F(x) \,\} \\ &= P\{\, U_{k:n} \le F(x) \,\} \\ &= \sum_{j=k}^{n} \binom{n}{j} F(x)^j \,(1 - F(x))^{n-j} \end{aligned} \tag{5}$$

where $U_{k:n}$ is the k–th order statistic of n independent uniform variables defined on $[0,1]$. As $\widehat{\mathrm{VaR}}_{\mathrm{HS}}(p) = L^*_{j^*:n}$ with $(j^*-1)/n < p \le j^*/n$, we get the cumulative distribution function of $\widehat{\mathrm{VaR}}_{\mathrm{HS}}(p)$ by

$$P\left\{\, \widehat{\mathrm{VaR}}_{\mathrm{HS}}(p) \le x \,\right\} = \sum_{j=j^*}^{n} \binom{n}{j} F(x)^j \,(1 - F(x))^{n-j} .$$

The density of $L_{k:n}$ - or that of $\widehat{\mathrm{VaR}}_{\mathrm{HS}}(p)$ - can be derived by differentiation of the right hand side in (5) to yield

$$f_{L_{k:n}}(x) = k \binom{n}{k} F(x)^{k-1} \,(1 - F(x))^{n-k} f(x) . \tag{6}$$

Given large sample sizes, arguments can be based on asymptotic results as the distribution of

$$n^{1/2} \, \frac{\widehat{\mathrm{VaR}}_{\mathrm{HS}}(p) - \mathrm{VaR}(p)}{[f(\mathrm{VaR}(p))]^{-1} \sqrt{p\,(1-p)}} . \tag{7}$$

converges to the unit normal (Eubanks [1985]). A first assessment of the accuracy of $\widehat{\mathrm{VaR}}_{\mathrm{HS}}(p)$ can thus be derived from asymptotic confidence intervals. Given confidence level $1-\alpha$ we get an asymptotic confidence interval using asymptotic normality of (7) as

$$\left[\, \widehat{\mathrm{VaR}}_{\mathrm{HS}}(p) \;\pm\; z_{1-\alpha/2} \,[f(\mathrm{VaR}(p))]^{-1} \sqrt{p(1-p)/n} \,\right]$$

where $z_{1-\alpha/2} = \Phi^{\leftarrow}(1-\alpha/2)$ and Φ denotes the unit normal distribution. If f is known, these intervals can be evaluated for any p and large enough n. See Jorion [1996,1997] for some examples. But in general f is not known. Using the asymptotically equivalent interval $[\ \hat{F}^{\leftarrow}(k/n)\ ,\ \hat{F}^{\leftarrow}(l/n)\ ]$ where integers k and l are chosen to ensure

$$k \approx np - z_{1-\alpha/2} \sqrt{np(1-p)}$$

and

$$l \approx np + z_{1-\alpha/2} \sqrt{np(1-p)}$$

(see Eubanks [1985]) is no solution to the problem since the approximation performs rather badly for large p–values and a sample size of just a few hundred. To compute exact small sample confidence intervals we try to find variables $L_{k:n}$ and $L_{l:n}$ where $k < l$ such that

$$P\{\, L_{k:n} \le \mathrm{VaR}(p) \le L_{l:n} \,\} = 1 - \alpha\,. \tag{8}$$

Let Y be the number of sample variables smaller than the true VaR,

$$Y = \sum_{i=1}^{n} \mathbf{1}_{(-\infty,\mathrm{VaR}(p))]}(L_i)$$

then Y is a binomial variate with parameters p and n. Instead of event $\{\, L_{k:n} \le \mathrm{VaR}(p) \le L_{l:n} \,\}$ in (8) we can obviously use the equivalent event $\{k \le Y \le l-1\}$, hence

$$\begin{aligned} P\{\, L_{k:n} \le \mathrm{VaR}_t(p) \le L_{l:n} \,\} &= P\{\, k \le Y \le l-1 \,\} \\ &= \sum_{j=k}^{l-1} \binom{n}{j} p^j\,(1-p)^{n-j} \\ &= 1-\alpha\,. \end{aligned}$$

In general it is impossible to find integers k and l such that the last equation holds exactly. So the usual textbook advice is to look for integers such that

1. $l - k$ minimal
2. $\sum_{j=k}^{l-1} \binom{n}{j} p^j\,(1-p)^{n-j} \approx 1-\alpha\,.$

Table 2 shows some solutions for selected values of n and p. Confidence intervals with exact confidence level $1-\alpha$ can be computed from the respective statistics of orders k and l. When using a spreadsheet programm to compute binomial probabilities we might encounter overflow problems. If this happens, approximating the binomial by Poisson variates should work as long as $1-p < 0.1$ and $n(1-p) < 10$ (Evans/Hastings/Peacock [1993]). The distribution of $n-Y$ is approximated by a Poisson distribution with mean $n(1-p)$ leading to

$$\sum_{j=k}^{l-1} \binom{n}{j} p^j\,(1-p)^{n-j} \approx \mathrm{e}^{-n(1-p)} \sum_{j=n-l+1}^{n-k} \frac{[n(1-p)]^j}{j\,!}$$

because events $\{k \le Y \le l-1\}$ and $\{n-l+1 \le n-Y \le n-k\}$ are equivalent.

3.2 Smoothing Empirical Quantiles

Instead of smoothing the empirical quantile function by larger sample sizes, its discontinuity could be removed by averaging. Additionally, if the true quantile function behaves smooth enough we can expect weighted sums of

Table 2. Exact confidence intervals from historical simulation

	p	confidence level $1-\alpha$	order of lower bound k	order of upper bound l
$n = 100$	0.95	0.7583	93	98
		0.8998	92	99
	0.99	0.6340	93	100
$n = 260$	0.95	0.7471	244	252
		0.8998	238	252
		0.9471	237	253
	0.99	0.7342	249	259
		0.9101	254	260
$n = 520$	0.95	0.7591	490	502
		0.8926	487	503
		0.9550	485	505
	0.99	0.7383	513	518
		0.8914	507	518
		0.9492	510	519
$n = 780$	0.95	0.7505	735	749
		0.8996	732	752
		0.9508	730	754
	0.99	0.7258	770	776
		0.8986	768	777
		0.9563	767	778

VaR-estimates in neighborhoods of p to give more reliable estimates of the true VaR at p by reducing the estimator's variance. Simple approaches suggest weights like

$$w_{i,n}(p) = \begin{cases} 1-(n+1)p+i & i = [(n+1)p] \\ (n+1)p-i & i = [(n+1)p]+1 \\ 0 & \text{else} \end{cases} \tag{9}$$

or similar schemes where $[np]$ denotes the largest integer less than or equal to np. Properties of these estimators have been studied by Parrish [1990] or Dielman/Lowry/Pfaffenberger [1994] for example. As their results show that these simple estimators can be outperformed by others especially when estimating extreme quantiles they are not studied here any further. Nevertheless, VaR-estimators created from (9) can be regarded as a first step to more stable results and can be computed without any difficulties when using historical simulation. So Table 3 gives numerical values of some weights for selected parameter sets. Thus, for example if the sample period is two years, VaR(0.99) is estimated according to (9) and Table 3 by

Table 3. Weights from (9)

	p	$i = [(n+1)p]$	$w_{i,n}(p)$	$w_{i+1,n}(p)$
$n = 100$	0.95	95	0.0500	0.9500
	0.99	99	0.0100	0.9900
$n = 260$	0.95	247	0.0500	0.9500
	0.99	258	0.6100	0.3900
$n = 520$	0.95	494	0.0500	0.9500
	0.99	515	0.2100	0.7900
$n = 780$	0.95	741	0.0500	0.9500
	0.99	773	0.8100	0.1900

$$\widehat{\text{VaR}}_t(0.99) = 0.21\, L_{515:520} + 0.79\, L_{516:520}\,.$$

In general averaging is done by real valued kernel functions $k_n(p,x)$ defined on $(0,1)\times(0,1)$ for integers n and $\int_0^1 k_n(p,x)\,dx = 1$. The kernel quantile estimator of VaR(p) is defined by

$$\widehat{\text{VaR}}_{\text{KQ}}(p) = \int_0^1 k_n(p,x)\,\hat{F}^{\leftarrow}(x)\,dx = \sum_{i=1}^{n} w_{i,n}(p)\,L_{i:n} \tag{10}$$

where weights $w_{i,n}$ can be written as

$$w_{i,n}(p) = \int_{(i-1)/n}^{i/n} k_n(p,x)\,dx$$

and hence $\sum_{i=1}^{n} w_{i,n}(p) = 1$. As equation (10) shows, $\widehat{\text{VaR}}_{\text{KQ}}(p)$ can be understood as a linear function of order statistics created by the sample variables. Like (9), these weights do not depend on the sample, so they can be computed, saved and used easily for any $n-$ sample of losses whose VaR has to be estimated.

Reasonable kernel functions assign positive weight to those order statistics that correspond to VaR-estimators for probabilities in a small neighborhood of p. Kaigh/Lachenbruch [1982], Harrel/Davis [1982] and Reiss [1989] give examples for special kernel functions. As the Harrel/Davis estimator has proved to be successful in Parish [1990] and Dielman/Lowry/Pfaffenberger [1995], we will have a closer look at it. Based on (6), the mean of $L_{k:n}$ is given by

$$E(L_{k:n}) = \frac{1}{B(k, n-k+1)} \int_{-\infty}^{\infty} y\,F(y)^{k-1}\,(1-F(y))^{n-k}\,f(y)\,dy$$

where $B(a,b)$ denotes the beta function. Integration with respect to $x = F(y)$ yields

$$E(L_{k:n}) = \frac{1}{B(k, n-k+1)} \int_0^1 F^{\leftarrow}(x)\, x^{k-1} \,(1-x)^{n-k}\, dx\,.$$

Thus Harrel/Davis [1982] suggest as an estimator of $F^{\leftarrow}(p) = \mathrm{VaR}(p)$

$$\boxed{\widehat{\mathrm{VaR}}_{\mathrm{HD}}(p) = \frac{\int_0^1 x^{(n+1)p-1}\,(1-x)^{(n+1)(1-p)-1}\,\hat{F}^{\leftarrow}(x)\, dx}{B(\,(n+1)p, (n+1)(1-p)\,)}}$$

regardless of $(n+1)p$ being an integer or not. It is easy to see, that $\widehat{\mathrm{VaR}}_{\mathrm{HD}}(p)$ is a kernel quantile estimator with kernel function

$$k_n(p,x) = \frac{x^{(n+1)p-1}\,(1-x)^{(n+1)(1-p)-1}}{B(\,(n+1)p, (n+1)(1-p)\,)}\,.$$

Implied weights are

$$\begin{aligned} w_{i,n}(p) &= \frac{\int_{(i-1)/n}^{i/n} x^{(n+1)p-1}\,(1-x)^{(n+1)(1-p)-1}\, dx}{B(\,(n+1)p, (n+1)(1-p)\,)} \\ &= I_{i/n}\Big(p(n+1), (1-p)(n+1)\Big) - I_{(i-1)/n}\Big(p(n+1), (1-p)(n+1)\Big) \end{aligned}$$

where $I_x(a,b)$ denotes the incomplete beta function. As most statistical computing environments provide access to the incomplete beta function, the weights of the Harrel/Davis estimator can be computed without difficulty. Other kernel functions are defined by

$$k_n(p,x) = \frac{1}{h_n}\, u\left(\frac{p-x}{h_n}\right)$$

where $\int u(y)\, dy = 1$ und $u(y) = 0$ for $|y| > 1$. See Reiss [1989] for detailed discussion. If the window width h_n is chosen such that inequality $h_n(p) \le \min\{p, 1-p\}$ always holds, the amount of smoothing is reduced when estimating quantiles near the boundaries of the interval $[0,1]$. Under these assumptions the first equation of (10) can be transformed to

$$\widehat{\mathrm{VaR}}_{\mathrm{KQ}}(p) = \int_{-1}^1 \hat{F}^{\leftarrow}(\,p - y\, h_n(p)\,)\, u(y)\; dy \tag{11}$$

which is equivalent to

$$\boxed{\widehat{\mathrm{VaR}}_{\mathrm{KQ}}(p) = \sum_{i=1}^n \left[U\left(\frac{p-(i-1)/n}{h_n(p)}\right) - U\left(\frac{p-i/n}{h_n(p)}\right)\right] L_{i:n}} \tag{12}$$

where $U(z) = \int_{-1}^{z} u(y)\,dy$. An admissible window function is given by

$$h_n(p) = \begin{cases} p - p^2/4\beta_n & 0 < p < 2\beta_n \\ \beta_n & 2\beta_n \le p \le 1 - 2\beta_n \\ (1-p) - (1-p)^2/4\beta_n & 1 - 2\beta_n < p < 1, \end{cases}$$

where $\beta_n \le 1/4$ for any n. See Reiss [1989] for this and other examples. Combined with the Epanechnikov kernel $u(y) = 0.75\,(1-y^2)\,\mathbf{1}_{[-1,1]}(y)$ which transformes to

$$U(y) = \begin{cases} 0 & y < -1 \\ \frac{1}{2} + \frac{3}{4}y - \frac{1}{4}y^3 & -1 \le y \le 1 \\ 1 & y > 1 \end{cases}$$

the coefficients needed in (12) can be computed. A survey of alternative kernel functions defined on $[-1,1]$ can be found in Härdle [1991]. If all assumptions on h_n and $u(y)$ hold, sample variables of orders $i \le n(p - h_n(p))$ and $i \ge n(p + h_n(p)) + 1$ will have zero weight. This shows, that only a small number of extreme observations is effectively used when estimating the value at risk for high values of p or $1-p$. Most of the sample variables will be assigned zero weight.

Confidence intervals of VaR(p) could be deduced from the asymptotic distribution of the kernel quantile estimators. Falk [1985] gives barely restrictive assumptions on the kernel function and the window width, which ensure

$$n^{1/2}\,\frac{\widehat{\mathrm{VaR}}_{\mathrm{KQ}}(p) - \mathrm{VaR}(p)}{[f(F^{\leftarrow}(p))]^{-1}\sqrt{p(1-p)}} \tag{13}$$

to be asymptotically distributed normal with mean 0 and variance 1. Here $h_n(p) \to 0$ as $n \to \infty$. This asymptotic result corresponds to that of the VaR-estimator from historical simulation. But as Falk shows, the kernel quantile estimator should be preferred with regard to the concept of deficiency as it uses available data more efficiently than empirical quantiles . But as f is usually unknown, this result cannot be used to evaluate confidence intervals in most applications. So Harrel/Davis [1982] suggest a jackknife procedure to estimate the variance of $\widehat{\mathrm{VaR}}_{\mathrm{KQ}}(p)$. Let $\widehat{\mathrm{VaR}}_{\mathrm{KQ}}^{(j)}(p)$ be the kernel quantile estimator of VaR(p) where order statistic j has been removed from the sample. Then

$$S_{KQ}^2 = \frac{n-1}{n}\sum_{j=1}^{n}\left(\widehat{\mathrm{VaR}}_{\mathrm{KQ}}^{(j)}(p) - n^{-1}\sum_{j=1}^{n}\widehat{\mathrm{VaR}}_{\mathrm{KQ}}^{(j)}(p)\right)^2 \tag{14}$$

gives an estimate of the variance of $\widehat{\mathrm{VaR}}_{\mathrm{KQ}}(p)$ for n large enough (Miller [1974]). From (13) a confidence interval with approximate confidence level $1-\alpha$ can then be constructed as

$$\left[\widehat{\text{VaR}}_{\text{KQ}}(p) \pm z_{1-\alpha/2} S_{KQ}\right].$$

Realizations of the boundaries of the interval can easily be computed from the sample.

3.3 Parametric Models

If the parametric family of distribution functions that F belongs to is known, the problem of VaR-estimation typically reduces to the estimation of unknown parameters. Estimation procedures for most common parametric distributions can be found in the literature. If the true parameters are replaced by their estimates an estimator of any $p-$quantile - or VaR(p) - can be obtained by inverting the distribution function. Several families of distribution functions have been applied to financial data, including the normal, stable or recently the hyperbolic model (Eberlein/Keller/Prause [1997]). Special focus is given here on the location and scale parameter model, where

$$F(x) = F_0\left(\frac{x-b}{a}\right)$$

and F_0 is a known distribution. Here b is called the location parameter and a denotes the scale parameter. Important examples are the normal, the Student-t and the exponential distribution. The location and scale parameter model leads for $0 < p < 1$ to

$$\text{VaR}(p) = F^{\leftarrow}(p) = b + a\,F_0^{\leftarrow}(p)\,, \tag{15}$$

so any VaR can be estimated by a linear function of parameter estimates $\hat{a}$ and $\hat{b}$. Quantiles $F_0^{\leftarrow}(p)$ of F_0 can be obtained for many distributions from tables, spreadsheet programms or from statistical software.

Normally Distributed Losses. Under the assumption of normally distributed loss variables L_i $(i = 1, 2, \ldots, n)$ model (15) writes as

$$\text{VaR}(p) = \mu + \sigma\,\Phi^{\leftarrow}(p)\,, \quad 0 < p < 1\,, \tag{16}$$

where the location and scale parameters can be shown to be mean μ and standard deviation σ of L_i. Plugging in estimators $\hat{\mu} = n^{-1}\sum_{i=1}^{n} L_i$ and $\hat{\sigma}^2 = (n-1)^{-1}\sum_{i=1}^{n}(L_i - \hat{\mu})^2$ we obtain the VaR-estimator

$$\boxed{\widehat{\text{VaR}}_{\text{ND}}(p) = \hat{\mu} + \hat{\sigma}\,\Phi^{\leftarrow}(p)} \tag{17}$$

Due to its simple structure and the familiarity of most risk controllers with the normal distribution, this estimator can be found in many applications. But like most VaR-estimators, it is not unbiased. Nevertheless, mild assumptions ensure favourable asymptotic behavior of $\widehat{\text{VaR}}_{\text{ND}}(p)$.

Estimators $\hat{\mu}$ and $\hat{\sigma}$ are independent random variables for any sample size in this model (see Fisz [1980]). Their densities are

$$f_{\hat{\mu}}(x|\mu,\sigma,n) = \frac{\sqrt{n}}{\sigma\sqrt{2\pi}} \exp\left(-\frac{n\,(x-\mu)^2}{2\,\sigma^2}\right) \tag{18}$$

and

$$f_{\hat{\sigma}}(s|\sigma,n) = \frac{n^{\frac{n-1}{2}}\, s^{n-2} \exp\left(-\frac{n\,s^2}{2\sigma^2}\right)}{2^{\frac{n-3}{2}}\, \Gamma\left(\frac{n-1}{2}\right)\sigma^{n-1}}. \tag{19}$$

Hence, the joint density of $\hat{\mu}$ and $\hat{\sigma}$ is the product of the marginal densities. (18) is a normal density with mean μ and variance σ^2/n. Given linearity of (17) the density of $\widehat{\text{VaR}}_{\text{ND}}(p)$ can be derived as

$$f_{\widehat{\text{VaR}}_{\text{ND}}}(q|\mu,\sigma,n,p) = \int_{-\infty}^{\infty} f_{\hat{\mu}}\Big(q - s\,\Phi^{\leftarrow}(p)\,|\,\mu,\sigma,n\Big)\, f_{\hat{\sigma}}(s\,|\,\sigma,n)\, ds\,.$$

In the case of large sample sizes an approximation can be used. Because $(n-1)\,\hat{\sigma}^2/\sigma^2$ is distributed χ^2 with $n-1$ degrees of freedom the variables

$$\sqrt{2(n-1)}\,\hat{\sigma}/\sigma - \sqrt{2(n-1)-1}$$

converge to unit normal variates in distribution as $n \to \infty$ (Fisz [1980, S.401]). Hence the distribution of

$$\sqrt{n}\,\Phi^{\leftarrow}(p)\,\frac{\hat{\sigma}-\sigma}{\sigma}$$

converges to a normal distribution with zero mean and variance $[\Phi^{\leftarrow}(p)]^2/2$ as $n \to \infty$. Asymptotically, we have

$$\frac{\widehat{\text{VaR}}_{\text{ND}}(p) - \text{VaR}(p)}{\sigma/\sqrt{n}} = \sqrt{n}\,\frac{\hat{\mu}-\mu}{\sigma} - \sqrt{n}\,\Phi^{\leftarrow}(p)\,\frac{\hat{\sigma}-\sigma}{\sigma}$$

as a normal variate with zero mean and variance $1 + \frac{[\Phi^{\leftarrow}(p)]^2}{2}$. Thus for $0 < \alpha < 1$ the following equation holds

$$\lim_{n\to\infty} P\left\{\left|\frac{\widehat{\text{VaR}}_{\text{ND}}(p) - \text{VaR}(p)}{\sigma/\sqrt{n}}\right| \le z_{1-\alpha/2}\sqrt{1+\frac{[\Phi^{\leftarrow}(p)]^2}{2}}\right\} = 1-\alpha\,.$$

Finally, as $\hat{\sigma}$ is a consistent estimator of σ, the interval

$$\left[\quad \widehat{\text{VaR}}_{\text{ND}}(p) \quad \pm \quad z_{1-\alpha/2}\,\frac{\hat{\sigma}}{\sqrt{n}}\sqrt{1+\frac{[\Phi^{\leftarrow}(p)]^2}{2}}\quad\right] \tag{20}$$

gives an asymptotic confidence interval of VaR(p) with confidence level $1-\alpha$. The function $\Phi^{\leftarrow}$ is not bounded on $(0,1)$. So the length of the confidence interval increases as $p \to 1$ or $p \to 0$, showing decreasing accuracy of the VaR-estimator as p tends to the boundaries of $(0,1)$.

Sometimes loss can be shown to be a linear function of risk factor changes or loss can be approximated with high degree by a linear function of X_t. The coefficients of these linear functions are called sensitivities and are derived from the partial derivatives of the value function v (delta approximations). Wilson [1996] gives a survey of these VaR-approaches. Sensitivities usually depend on the current level of risk factors R_t leading to

$$L(R_t, X_t) = \sum_{j=1}^{k} \delta_j(R_t)\, X_{t,j} = \delta(R_t)' X_t$$

where $\delta(R_t) = [\,\delta_1(R_t), \ldots, \delta_k(R_t)\,]'$. Joint normality of $[X_{t,1}, \ldots, X_{t,k}]$ with regular covariance matrix Σ and mean vector ν entails that the distribution of L - conditioned on $\{R_t = r_t\}$ - is normal, too. Dropping time index t again, the moments are $\mu = \delta(r)'\,\nu$ and $\sigma^2 = \delta(r)'\;\Sigma\;\delta(r)$. VaR can then be computed as

$$\mathrm{VaR}(p) = \underbrace{\delta(r)'\,\nu}_{=\mu} + \underbrace{\sqrt{\delta(r)'\;\Sigma\;\delta(r)}}_{=\sigma}\;\Phi^{\leftarrow}(p)\,.$$

Models of this structure are called *variance covariance models* in the risk management community. VaR can be estimated once again by replacing unknown parameters by their estimates. Let

$$\hat{\nu} = (\,\hat{\nu}_1, \ldots, \hat{\nu}_k)' = n^{-1} \sum_{i=1}^{n} X_{t-i} \qquad \text{and} \qquad \hat{\Sigma} = n^{-1}\,\tilde{X}_t'\,\tilde{X}_t\,,$$

where

$$\tilde{X}_t = \begin{bmatrix} X_{t-1,1} - \hat{\nu}_1 & X_{t-1,2} - \hat{\nu}_2 & \cdots & X_{t-1,k} - \hat{\nu}_k \\ X_{t-2,1} - \hat{\nu}_1 & X_{t-2,2} - \hat{\nu}_2 & \cdots & X_{t-2,k} - \hat{\nu}_k \\ \vdots & \vdots & \ddots & \vdots \\ X_{t-n,1} - \hat{\nu}_1 & X_{t-n,2} - \hat{\nu}_2 & \cdots & X_{t-n,k} - \hat{\nu}_k \end{bmatrix},$$

the estimator of VaR(p) is given by

$$\widehat{\mathrm{VaR}}_{\mathrm{ND}}(p) = \underbrace{\delta(r)'\,\hat{\nu}}_{=\hat{\mu}} + \underbrace{\sqrt{\delta(r)'\;\hat{\Sigma}\;\delta(r)}}_{=\hat{\sigma}}\;\Phi^{\leftarrow}(p)\,.$$

Extreme Value Models. Probability p is typically chosen to be at least 0.95. To analyze the risk exposure of an open financial position we focus on those adverse events that have small probability but cause large losses. Hence it should be a reasonable strategy to have a closer look at VaR-estimators that are derived from careful modelling of the upper right tail of the loss distribution . Statistical extreme value models have been developed for this special purpose and should be regarded as a promising approach to risk management in financial institutions (see Embrechts/Resnick/Samorodnitsky [1997]). Basic results on the statistics of extreme value models can be found in Leadbetter/Lindgren/Rootzén [1983].

Let $L_{n:n}$ be the maximum in a sample of size n drawn independently from distribution F . The distribution of this maximum can be written as $F^n(x)$. If there are real sequences $\{c_n\}$ und $\{d_n\}$ and a continuous distribution function G such that

$$\sup_{x \in \mathbb{R}} |F^n(c_n + d_n\, x) - G(x)| \to 0 \quad , \quad n \to \infty \, ,$$

it can be shown for $\xi > 0$

$$G(x) = \begin{cases} G_{1,\xi}(x) := \exp(-x^{-\xi}) \, , \; x > 0 & \text{(Frechét)} \\ \quad \text{or} & \\ G_{2,\xi}(x) := \exp(-(-x)^{\xi}) \, , \; x \leq 0 & \text{(Weibull)} \\ \quad \text{or} & \\ G_3(x) := \exp(-e^{-x}) \, , \; x \in \mathbb{R} & \text{(Gumbel)} \, . \end{cases} \qquad (21)$$

Thus, if normed maxima converge in distribution to a continiuous distribution, the limit distribution must be one of those given in (21). In this case F lies in the domain of attraction of the respective extreme value distribution. Distributions like the normal, the log-normal, the Student-t or the exponential that are typically used in financial market analysis are characterized by this property. If F lies in the domain of attraction of an extreme value distribution, Pickands [1975] shows that its upper right tail can be approximated by the generalized Pareto distribution

$$\begin{array}{lll} W_{1,\xi}(x) := 1 - x^{-\xi} \, , \; x \geq 1 & & \text{(Pareto)} \\ W_{2,\xi}(x) := 1 - (-x)^{\xi} \, , \; -1 \leq x \leq 0 & & \text{(type II)} \\ W_3(x) \;\; := 1 - \exp(-x) \, , \; x \geq 0 & & \text{(exponential)} \end{array}$$

where $\xi > 0$. The relationship between loss distribution F , extreme value distribution and generalized Pareto distribution can be summarized by the following result (see Reiss [1989, theorem 5.1.1] for a proof). Let $G = G_{i,\xi}$ and $W = W_{i,\xi}$ for $i \in \{1, 2, 3\}$ and $\xi > 0$. Then the following statements are equivalent.

$$F^n(c_n + d_n\, x) \to G(x) \, , \quad n \to \infty, \; x \in \mathbb{R}$$

$$[\, 1 - F(c_n + d_n\, x)\,]/[\, 1 - W(b_n + a_n\, x)\,] \to 1 \, , \quad n \to \infty$$

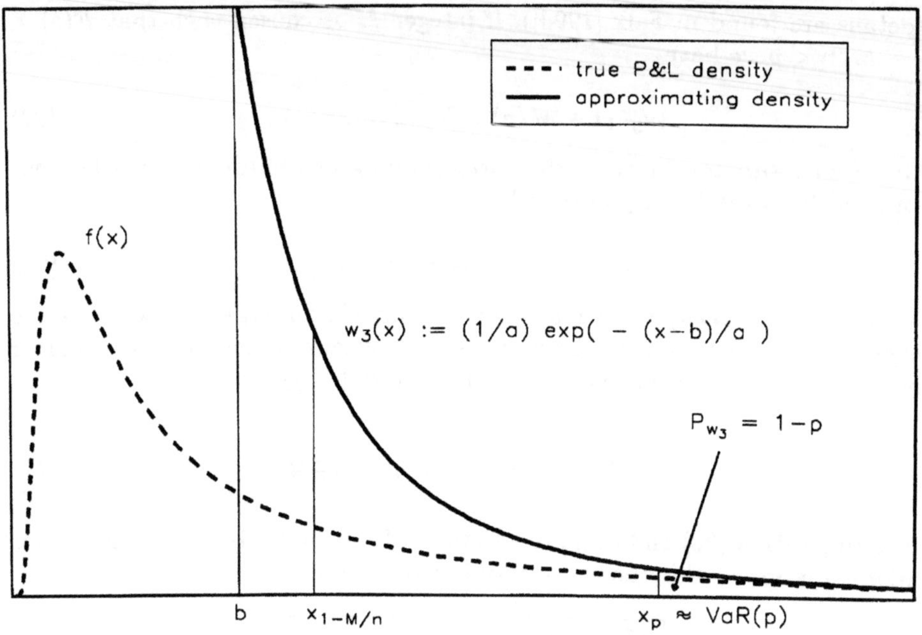

Fig. 2. Approximating the upper right tail

$$[1 - F(c_n + d_n\, x)]/[1 - G(b_n + a_n\, x)] \to 1\,, \quad n \to \infty$$

where

$$b_n = \begin{cases} 0 & i = 1,2 \\ \ln n & i = 3 \end{cases} \qquad \text{and} \qquad a_n = \begin{cases} n^{1/\xi} & i = 1 \\ n^{-1/\xi} & i = 2 \\ 1 & i = 3 \end{cases}$$

Thus $W_{1,\xi}$, $W_{2,\xi}$ und W_3 lie in the domain of attraction of $G_{1,\xi}, G_{2,\xi}$ and G_3, too. Let F be such that its upper right tail can be approximated by the exponential (case $i = 3$). See Figure 2. Then there are location and scale parameters b and a leading to

$$F(x) \approx W_3\left(\frac{x-b}{a}\right) = 1 - \exp\left(-\frac{x-b}{a}\right)$$

for $x > b$. Looking at inverse distributions we have

$$\text{VaR}(p) \approx W_3^{\leftarrow}(p) = b - a\,\ln(1-p)$$

which gives another example of the location and scale parameter model (15) as $\ln(1-p)$ is the p–quantile of the exponential with $b = 0$ and $a = 1$. Estimators of unknown parameters a and b can be constructed as follows (details are found in Falk [1994]). If integer M is chosen such that $F(b) < 1 - M/n < p$ we have

$$W_3^{\leftarrow}(1 - M/n) = b - a \ln(M/n) . \tag{22}$$

Given an estimator $\hat{a}_3$ of a, the corresponding estimator of b can be constructed from the last equation as

$$\hat{b}_3 = L_{n-M:n} + \hat{a}_3 \ln(M/n) \tag{23}$$

where $F^{\leftarrow}(1 - M/n)$ is estimated by the sample quantile $L_{n-M:n}$. As the exponential distribution can be characterized by lack of memory, an estimator of a can be derived from the losses that exceed $L_{n-M:n}$ as

$$\hat{a}_3 = M^{-1} \sum_{m=1}^{M} (L_{n-m+1:n} - L_{n-M:n}) . \tag{24}$$

Inserting this in (23) and (22) leads to the VaR-estimator based on the partial approximation of F by an exponential distribution

$$\boxed{\widehat{\text{VaR}}_{\text{ED}}(p) = \hat{b}_3 - \hat{a}_3 \ln(1 - p)} \tag{25}$$

Case $i = 1$, where F is approximated by a Pareto distribution can be referred to case $i = 3$. This preserves the structure of the location and scale parameter model. Let the $M + 1$ largest losses be positive, $L_i > 0$ for $i \geq n - M$. If these are distributed approximatly according to a Pareto distribution, the approximating distribution function is

$$W_{1,\xi}(x/c) = 1 - \left(\frac{x}{c}\right)^{-\xi} \quad , \quad (x \geq c > 0) .$$

for some positive ξ. As $P\{\ln L_i \leq x\} = P\{L_i \leq \exp(x)\}$ we get the approximating distribution of log-losses by $F(x) = 1 - (\exp(x)/c)^{-\xi}$ which reduces to $F(x) = 1 - \exp(-\xi(x - \ln c))$. But this is an exponential distribution function with location parameter $\ln c$ and scale parameter $1/\xi$. Thus in case $i = 1$, the VaR of log-losses can be estimated by the estimation procedure of case $i = 3$ where log-losses replace the original variables. The resulting VaR-estimator of the original loss variables is given by

$$\boxed{\widehat{\text{VaR}}_{\text{PD}}(p) = \exp\left(\hat{b}_1 - \hat{a}_1 \ln(1 - p)\right)} \tag{26}$$

where

$$\hat{a}_1 = M^{-1} \sum_{m=1}^{M} (\ln L_{n-m+1:n} - \ln L_{n-M:n})$$

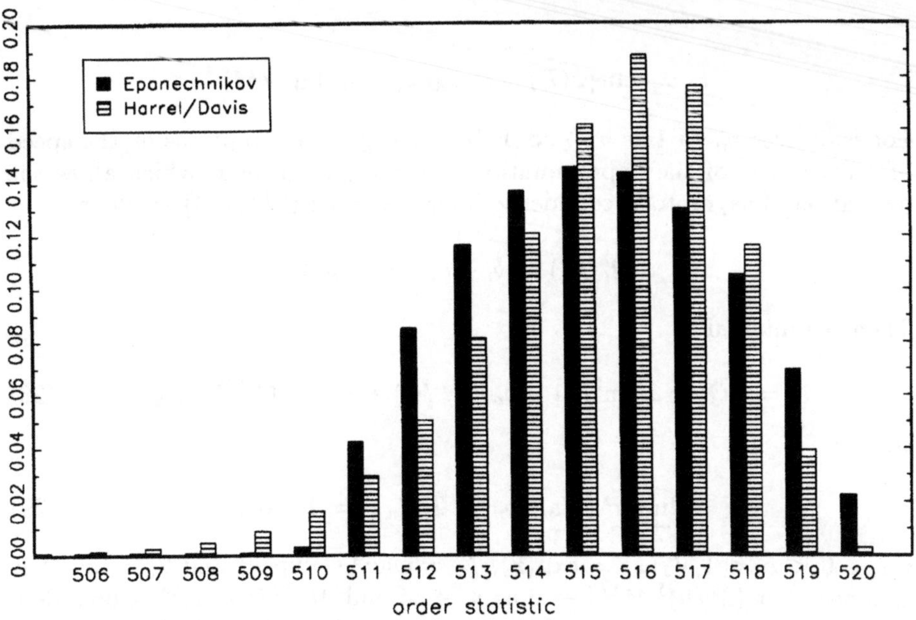

Fig. 3. Weights for kernel quantile estimators, $p = 0.99$, $n = 520$

and

$$\hat{b}_1 = \ln L_{n-M:n} - \hat{a}_1 \ \ln(n/M) \ .$$

Hosking/Wallis [1987] study alternative ways to estimate quantiles from Pareto distributions. Case $i = 2$ can be referred to case $i = 3$, too. Let $\omega(F)$ denote the right endpoint of the support of F. Then the upper right tail of F can be approximated by $W_{2,\xi}(x) = 1 - (-x/c)^{\xi}$ for some positive c. As $P\{-\ln(\omega(F) - L_i) \leq x\} = P\{L_i \leq \omega(F) - \exp(-x)\}$, we have $W_{2,\xi}(x) = 1 - \exp(-\xi[x - (-\ln c)])$, which is an exponential distribution function with scale parameter $1/\xi$ and location parameter $-\ln c$. Thus VaR can be estimated in this case by

$$\boxed{\widehat{\text{VaR}}_{\text{II}}(p) = \omega(F) - \exp\left(-\left[\hat{b}_2 - \hat{a}_2 \ \ln(1-p)\right]\right)} \tag{27}$$

where

$$\hat{a}_2 = \ln\left[\omega(F) - L_{n-M:n}\right] - M^{-1} \sum_{m=1}^{M} \ln\left[\omega(F) - L_{n-m+1:n}\right]$$

and

$$\hat{b}_2 = -\ln\left[\omega(F) - L_{n-M:n}\right] - \hat{a}_2 \,\ln(n/M)\,.$$

For sequences $p_n \to 1$ as $n \to \infty$, Falk [1994] gives assumptions on the speed of convergence of the approximation to the right tail of F which allow the derivation of asymptotic confidence intervals. For $i \in \{1, 2, 3\}$ we define

$$Q_i^*(p_n) := \hat{b}_i - \hat{a}_i \,\ln(1 - p_n)\,.$$

Then the interval

$$I_n = \left[\, Q_i^* \pm \hat{a}_i [\ln^2((1 - p_n)\, n \,/\, M\,) + 1]^{1/2} M^{-1/2}\, z_{1-\alpha/2} \,\right] \qquad (28)$$

ensures

$$\lim_{n\to\infty} P\,\{\,\mathrm{VaR}(p_n) \in T_i(I_n)\,\} = 1 - \alpha\,, \qquad (29)$$

where $0 < \alpha < 1$, $T_1(x) = \exp(x)$, $T_2 = \omega(F) - \exp(x)$ and $T_3 = x$. It is assumed that $(M/n)^\delta\, M^{1/2} \to 0$ as $n \to \infty$ and $M^{1/2}\,(1 - p_n)^\delta$ is bounded, where $\delta > 0$ controls the speed of approximation on the right tail of F (see Falk [1994]). Monotonicity of T_i $(i = 1, 2, 3)$ leads to confidence intervals in cases 1 to 3 just by transforming the boundaries of (28). If it is not known, which case holds for a given sample, the decision can be based on the extreme value index defined by

$$\hat{\gamma}_n = \ln(2)^{-1}\, \ln\left(\frac{L_{n-m+1:n} - L_{n-2m+1:n}}{L_{n-2m+1:n} - L_{n-4m+1:n}}\right)\,.$$

Dekkers/de Haan [1987] show that $\hat{\gamma}_n$ converges almost sure

$$\hat{\gamma}_n \to \begin{cases} 1/\alpha & i = 1 \\ -1/\alpha & i = 2 \\ 0 & i = 3 \end{cases}$$

where $n \to \infty$, $m \to \infty$ und $m/n \to 0$. Thus positive values of the estimator argue in favor of case $i = 1$, negative values indicate case $i = 2$. If $\hat{\gamma}_n$ tends around zero, case $i = 3$ could be assumed. Statistical tests about the true value of γ can be constructed from the asymptotic distribution of $\hat{\gamma}_n$ (Falk [1994]).

4 An Application

To compare the performance of different VaR-estimators we apply the methods of the last section to a simple problem. Let the risk position under concern be the amount of USD 100. Risk is to be denominated in DEM. So the only

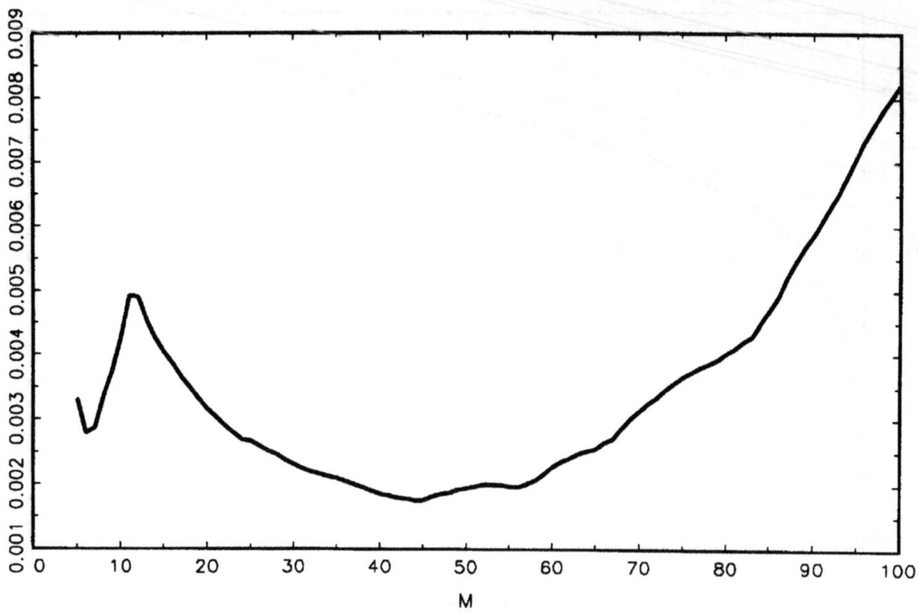

Fig. 4. MSE for different values of M

risk factor is the DEM/USD-exchange rate. The simple structure of the problem allows us to focus on differences in statistical performance of the various estimators. Under more complex valuation functions v, it could be difficult to single out effects of valuation from those of estimation.

The sample consists of 520 daily log-returns of the USD up to December 6, 1996, reflecting a sample period of about two years. According to section 2, these log-returns are applied to the USD-level of 12/6/1996 which is given by 1.5338 DEM. The resulting sample of $n = 520$ potential P&L-data is assumed to be drawn from distribution F whose $p-$quantile is to be estimated. We compute estimators by historical simulation (HS), kernel quantile VaR with Epanechnikov kernel (KQ) and Harrel/Davis kernel (HD), by assuming normality (ND) and by one of the extreme value models.

With regard to KQ we use a window width computed with parameter $\beta = 0.15$. The resulting weights for large order statistics used in methods HD and KQ are shown in Figure 3. The decision which extreme value model to use could be based on the tail index $\hat{\gamma}_n$ as indicated in Section 3. But the convergence usually performs rather slow, so no clear conclusion can be drawn from a sample size of almost 500.

But as the results of Boothe/Glassman [1987] or Koedjik/Schafgans/de Vries [1990] show, assuming a Pareto approximation to be valid can be regarded as a sensible start. This assumption leads to a linear relationship

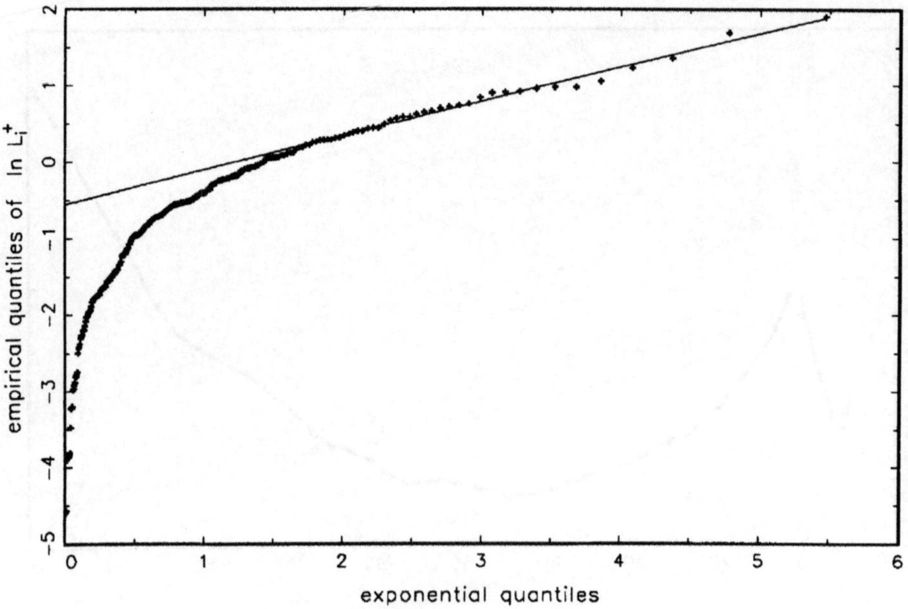

Fig. 5. QQ-plot of log-returns

between large quantiles of log-losses and the quantiles of exponential variates - see the last section. Thus we test this linearity heuristically by fitting a straight line into the QQ-plot. Fitting with different numbers M of large losses gives mean square errors as shown in Figure 4. We choose $M = 45$, which minimizes MSE in this example. The resulting line, fitted into the QQ-plot is given in Figure 5. As we regard the linear fit to be sufficient, a Pareto approximation leads to the VaR-estimator (case $i = 1$).

Estimation results for $0.95 \leq p < 1$ are shown in Figure 6. No visual differences can be detected for p smaller than 0.975 . But as p grows, the normal model yields extremely low estimators. This result corresponds with the well known failure of the normal distribution to model heavy tails. The kernel quantile estimators follow the empirical quantile function closely, while the extreme value model indicates large VaR-estimates as p tends to one.

The accuracy of the estimators can be assessed from confidence intervals. Table 4 shows results for p—values of 0.95, 0.99 and 0.995. The two rows beneath the VaR-estimate in each block give lower (CI_{lo}) and upper (CI_{up}) limits of the confidence intervals. All intervals are computed by the methods given in Section 3 where asymptotic variances of the kernel quantile estimators are estimated by the respective jackknife procedures. Confidence level $1 - \alpha = 0.95$ is chosen with the exception of method HS with $p = 0.995$. Here, no order statistics $L_{k:520}$ and $L_{l:520}$ can be found that cover the true

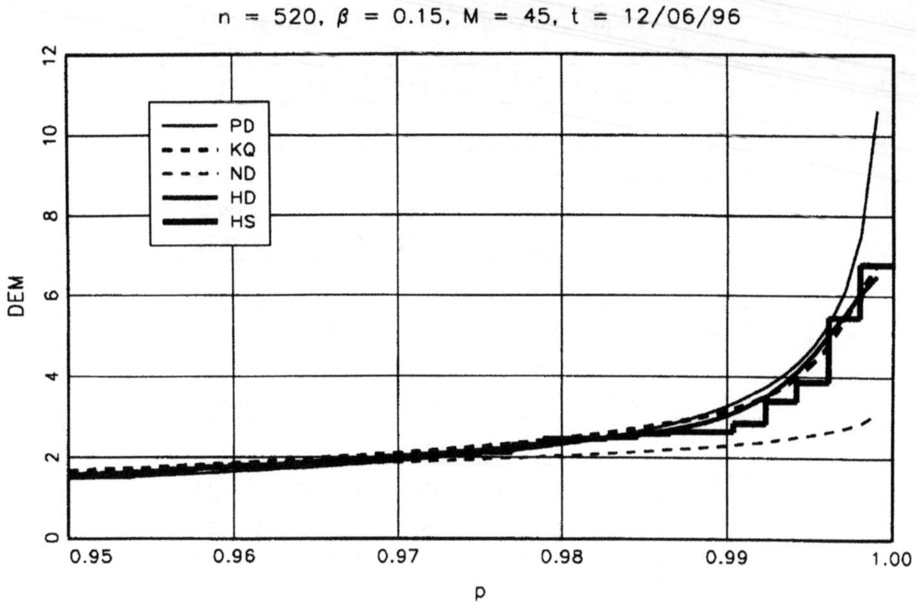

Fig. 6. 100 USD: VaR-estimators

VaR with probabiltiy as large as 0.95. Thus for method HS we use confidence level 0.9210 for $p = 0.995$.

The wide range of VaR-estimators for each $p-$value in Table 4 indicates a high sensitivity of the results with regard to the model we assume to be adequate. Within each row, estimated VaR-values differ by factors up to 1.75, which evidently has severe economic implications if these models are used to calculate risk capital. Estimation errors are seen to grow very fast as p increases. Figure 7 compares length and position of the estimated confidence intervals.

Additional to differences in the level of estimates, the accuracy of the methods differs largely and is reduced to a rather foggy degree for extreme quantiles. Thus, in contrast to the original intuition, of feeling more secure when choosing a high $p-$value, we might end up with totally wrong VaR-figures because of estimation errors when probabilities larger than 0.95 are used.

Despite the fact, that assuming normally distributed P&L-variables leads to the smallest VaR-estimates and the shortest confidence interval, we prefer the other methods to the normal model. The reason is, that the existence of heavy tails in financial series can safely be regarded as a law. But the normal model is just not able to cover this stylized fact. Thus the results

Table 4. $\widehat{\text{VaR}}$ and confidence intervals (α=0.05)

	ND	HS	KQ	HD	PD
			$p = 0.9950$		
$\widehat{\text{VaR}}$	2.5527	3.8684	4.2605	4.4459	4.3278
CI_{lo}	2.3761	2.5893	2.1242	2.2507	2.9856
CI_{up}	2.7293	6.7683	6.3967	6.6411	7.1888
			$p = 0.9900$		
$\widehat{\text{VaR}}$	2.3060	2.6507	3.1637	3.0524	3.2811
CI_{lo}	2.1424	2.4470	2.2340	2.1270	2.3217
CI_{up}	2.4696	5.4533	4.0934	3.9777	4.6369
			$p = 0.9500$		
$\widehat{\text{VaR}}$	1.6321	1.5284	1.6591	1.5687	1.4728
CI_{lo}	1.5018	1.3232	1.3805	1.2755	1.2477
CI_{up}	1.7625	1.9988	1.9377	1.8618	1.7384

of the normal model in our example should reflect rather specification error instead of more precision.

5 Conclusions

Basic properties of common VaR-estimators were deduced. Historical simulation can be studied by using combinatorics. In the case of normal loss data the distribution of the estimator can be deduced and allows construction of confidence intervals. This case does not need to be analyzed via empirical quantiles as in Jorion [1996]. Kernel quantile VaR-estimators are more flexible than standard parametric models and yield more stable results than historical simulation. As for given n and p the linear coefficients have to be calculated only once, this method can be applied without major difficulties. Extreme value models are proved to be tractable despite their complex theoretical background. Estimating VaR via approximation by the generalized Pareto distribution ends up in computing simple expressions that are easily coded even with spreadsheet programms.

But theoretical results and practical experience show that the results of VaR-estimation are far from being easy to use black box figures. Analyzing VaR incorporates econometric model building in its full range. But in addition to the usual problems in econometrics, we soon encounter new problems because if we are interested in the upper right tail of the P&L-distribution, we work in the realm of extreme value theory. Statistical precision in this

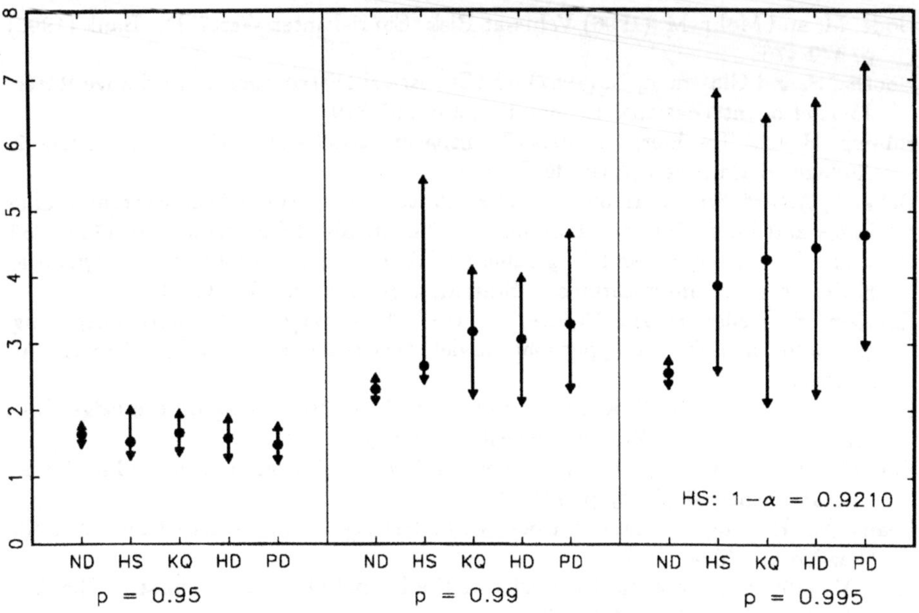

Fig. 7. 100 USD: Confidence intervals of VaR(p). ($\alpha = 0.05$)

region can be very low. We have to live with it, if we want to look at the upper distributional tail.

All results in this study are derived in an ideal world with respect to statistical and valuation assumptions. There is no reason to believe accuracy to be much higher when working in more complex worlds. So we can expect VaR-figures calculated from real portfolios with optional instruments, driven by risk factors being distributed heavy tailed or non stationary, to reflect only a limited degree of accuracy, also. This conclusion is drawn with respect to estimation risk. Model risk - the risk of severe misspecification of the VaR-model - was not analyzed in this study in detail. Estimation bias resulting from using the wrong model of the P&L-process leads to the topic of backtesting. However, up to now, only few results on this problem are available.

References

Artzner, P., Delbaen, F., Eber, J-M. and Heath, D. (1996) A Characterization of Measures of Risk, Discussion Paper, Universite Strasbourg

Bühler, W., Korn, O. and Schmidt, A. (1997) Ermittlung von Eigenkapitalanforderungen mit internen Modellen, Discussion Paper, Universität Mannheim

Bassi, F., Embrechts, P. and Kafetzaki, M. (1997) A Survival Kit on Quantile Estimation, Discussion Paper, Department of Mathematics, ETH Zürich

Bode, M. and Mohr, M. (1996) Value-at-Risk. Ein riskanter Wert?, Die Bank (1996) p. 470-476

Boothe, P. and Glassman, D. (1987) The Statistical Distribution of Exchange Rates, Journal of International Economics 22 p. 297-319

Büning, H. and Trenkler, G. (1994) Nichtparametrische statistische Methoden. 2. Auflage, Berlin, New York, de Gruyter

Dekkers, A.L.M. and de Haan, L. (1989) On the Estimation of the Extreme-Value Index and Large Quantile Estimation, The Annals of Statistics 17 p. 1795-1832

Dielman, T., Lowry, C. and Pfaffenberger, R. (1994) A Comparison of Quantile Estimators, Communications in Statistics: Simulation 23 p.355-371

Eberlein, E., Keller, U. and Prause, K. (1997) New insights into smile, mispricing and value at risk: the hyperbolic model, Discussion Paper, FDM, Universität Freiburg

Embrechts, P., Resnick, S. and Samorodnitsky, G. (1997) Living on the Edge, Discussion Paper, ETH Zürich, Cornell University

Eubanks, R.L. (1985) Quantiles, S. Kotz and N.L. Johnson, Encyclopedia of Statistical Science, 6, Wiley p. 424-432

Evans, M., Hastings, N. and Peacock, B. (1993) Statistical Distributions. 2nd ed. New York, Wiley

Falk, M. (1985) Asymptotic Normality of the Kernel Quantile Estimator The Annals of Statistics 13 p. 428-433

Falk, M. (1994) Extreme Quantile Estimation in δ-neighborhoods of Generalized Pareto Distributions, Statistics and Probability Letters 20 p. 9-21

Fisz, M. (1980) Wahrscheinlichkeitsrechnung and Mathematische Statistik, 10.Auflage, Berlin, Deutscher Verlag der Wissenschaften

Härdle, W. (1991) Smooothing Techniques. With Implementations in S, Berlin, Heidelberg, New York, Springer

Harrel, F.E. and Davis, C.E. (1982) A New Distribution-Free Quantile Estimator, Biometrika 69 p. 635-640

Hosking, J.R.M. and Wallis, J.R. (1987) Parameter and Quantile Estimation for the Generalized Pareto Distribution, Technometrics 29 p. 339-349

Hughey, R.L. (1991) A Survey and Comparison of Methods for Estimating Extreme Right Tail-Area Quantiles, Communications in Statistics: Theory and Methods 20 p. 1463-1496

Hull, J.C. (1993) Options, Futures, and other Derivative Securities. 2nd ed., Englewood Cliffs, Prentice Hall

Johanning, L. (1996) Value-at-Risk Modelle zur Ermittlung der bankaufsichtlichen Eigenkapitalunterlegung beim Marktrisiko im Handelsbereich, Zeitschrift für Bankrecht and Bankwirtschaft 8 p. 287-303

Jorion, P. (1996) Risk^2: Measuring the Risk in Value at Risk, Financial Analysts Journal November/December p. 47-56

Jorion, P. (1997) Value at Risk. The New benchmark for Controling Derivatives Risk, Chicago, Irwin

JP Morgan (1996) RiskMetrics - Technical Document. 4th ed., New York, JP Morgan

Kaigh, W.D. and Lachenbruch, P.A. (1982) A Generalized Quantile Estimator, Communications in Statistics: Theory and Methods 11 p. 2217-2238

Koedijk, K.G., Schafgans, M.A.M. and de Vries, C.G. (1990) The Tail Index of Exchange Rate Returns, Journal of International Economics 29 p. 93-108

Leadbetter, M.R., Lindgren, G. and Rootzen, H. (1983) Extremes and Related Properties of Random Sequences and Processes, New York, Berlin, Heidelberg, Springer

Leong, K. (1996) The Right Approach, Risk 9 (No 6: Value-At-Risk Supplement) p. 9-14

Miller, R.G. (1974) The Jackknife - A Review, Biometrika 61 p. 1-15

Parrish, R.S. (1990) Comparison of Quantile Estimators in Normal Sampling, Biometrics 46, p. 247-257

Pickands, J. (1975) Statistical Inference Using Extreme Order Statistics, The Annals of Statistics 3, p. 119-131

Reiss, R.D. (1989) Approximate Distributions of Order Statistics. With Applications to Nonparametric Statistics, New York, Berlin, Heidelberg, Springer

Wilson, T.C. (1996) Calculating Risk Capital, Alexander, Carol, The Handbook of Risk Management and Analysis, Chichester, Wiley p. 193-232

On the Accuracy of VaR Estimates Based on the Variance-Covariance Approach

Rakhal D. Davé[1] and Gerhard Stahl[2]

[1] Olsen & Associates Research Institute for Applied Economics, Seefeldstrasse 233, 8008 Zürich, Switzerland (email: rakhal@olsen.ch)
[2] Federal Banking Supervisory Office, Gardeschützenweg 71-101, 12203 Berlin, Germany (The views expressed herein should not be construed as being endorsed by the Federal Banking Supervisory Office)

Abstract. We present a thorough empirical study (based on over 8 years of daily data) of candidate models for forecasting losses in relation to positions held against individual risk factors as well as losses in relation to a portfolio of risk factors. As part of the study, we also define various measures and visualization techniques to evaluate the performance of the candidate models in the context of risk management and introduce two innovations: 1) tail emphasized model optimization and 2) implied covariance forecasting. Finally, we highlight the important issue of the estimation error of the covariance matrix in relation to its dimension and the number of datum from which it is estimated and outline a framework for handling this problem.

1 Introduction

To measure the market risk of a portfolio of traded assets, many banks are increasingly employing internal models based on a methodology called Value-at-Risk (VaR). In an important regulatory innovation Basle (1996), the Basle Committee on Banking Supervision has proposed in 1996 that such models may be used in the determination of the capital requirements that banks must fulfill to back their trading activities. In addition, the Bank for International Settlements – Fisher Report Basle (1994) urged financial intermediaries to publicly disclose Values-at-Risk. The Group of Thirty Group of Thirty (1993) and ISDA also advocate the VaR concept as a general measure of market risk.

The motivation behind the use of these models is to quantify risk for internal decision making and capital adequacy purposes. The quantification is in terms of the upper limit for loss inherent to a portfolio position with a given pre-specified probability, called *level of confidence*, and over a fixed period of time in the future *holding period*. An in depth description of the theory of VaR models is given in RiskMetrics (1996), Jorion (1997), Alexander (1996) and Beckström and Campbell (1995).

VaR models have to deal with 4 mathematical components:

1. Models – for predicting conditional distributions of returns for the underlying risk factors.
2. Stripping – of instruments so that they are expressed in terms of their underlying risk factors.
3. Value function of the portfolio – which may be non-linear with respect to the returns of the underlying.
4. Construction of profit and loss forecast distribution – based on the value function and class of conditional distribution of risk factors.

VaR models may differ from each other in different ways with respect to one or more of the above components. Components 3 and 4 are often mathematically interlinked – and together may be broadly addressed with the following popular methodologies:

- Variance-Covariance – or Δ approach.
- The $\Delta - \Gamma$ approach.
- Monte-Carlo simulation.
- Scenario simulation.

One of the objectives of this paper is to compare different models for predicting market risk. For this purpose we have chosen a candidate portfolio of spot foreign exchange and metal rates. Since the candidate portfolio is linear – the Variance-Covariance approach for the portfolio function is more than adequate for the purpose of this paper.

There is a considerable amount of work related to VaR models from academics, practitioners and regulators. Traditionally, a large volume of work from academics has been in the field of stochastic modeling of financial time series. This huge volume of work – review article Ghysels et al. (1995) containing 230 references – while being extremely relevant is not directly presented in the context of risk management. Academic work directly addressing the field of risk management is less common but includes a broad range of topics of practical and theoretical relevance such as the formal properties and deficiencies of the VaR concept Artzner et al. (1997), the problem of actual versus risk neutral distributions Eberlein et al. (1997), the discrepancies between the diversity of implementations of the same model – system risk Marshall and Siegel (1996), the determination of capital requirements with internal models Bühler et al. (1997), the appropriateness of VaR models as a management tool Beder (1995) and techniques for forecasting variances and covariances Alexander (1994), Alexander (1995), Alexander and Chidumba (1995).

Important contributions to VaR models from practitioners include the breakthrough work of J. P. Morgan giving a complete description of VaR models, criticisms, assumptions and methods RiskMetrics (1996), methods to judge the VaR model forecast quality Crnkovic and Drachman (1996), concepts that analyze the risk exposure of a portfolio by measuring the risk of the basic blocks of the total portfolio Chen et al. (1996) and the probabilistic foundation of scenario simulation models Jamshidian and Zhu (1997).

The contributions by regulators in this field primarily focus on issues related to the integration of VaR models within BIS recommendations and the attendant problems faced in practical implementation of these recommendations. Consequently work by regulators focus on backtesting Kupiec (1995), Lopez (1997) and on issues that relate VaR model performance to the BIS parameters Hendricks (1995), Jackson et al. (1995), Mahoney (1995) and Drudi et al. (1996).

This paper covers a cross section of issues focused on by academics, practitioners and regulators. From the academic perspective, we offer some innovations in stochastic modeling of financial time series - and offer a stringent testing framework for new models. From the regulators perspective, we introduce and discuss various measures for evaluating model performance in the context of risk management. From the practitioners perspective we carry out the entire evaluation using *real data* and try to construct performance measures which strike a balance between risk over-estimation and risk-under estimation.

In sections 2 through 6 of this paper, we evaluate the absolute and relative performance of VaR estimates based on historical simulation, the rectangular moving average, the exponential moving average, and the conditional variance forecasted by the GARCH process. In dealing with this issue we also discuss the meaning of *performance* in terms of different techniques to evaluate the success of the VaR model. In section 7, we discuss the issue of stochastic error in covariance matrix estimation and the inter-relation of this problem to the following issues:

- Possible singularities in the covariance matrix due to rank defects caused when the sample size is smaller than the dimension of the correlation matrix (which is generally true when using the popular figure of 250 days history).
- Systematic underestimation of risk arising out of the re-optimization of the portfolio (to minimize VaR) based on the estimated covariance matrix.

We end the discussion with some speculation on the innovations necessary to handle this problem. Although section 7 is directly unrelated to the earlier sections – we believe that it is a crucial point which must be emphasized for the completeness of this paper. We end the paper with a summary of innovations and recommendations for better overall handling of risk measurement.

2 General overview

Below, we present a general non-mathematical overview of this paper. The detailed mathematical description of the models and performance measures is in sections 3 and 4.

2.1 Data

This presentation is based on 2252 daily prices for the value of 1 United States Dollar (USD) expressed in the following 10 units:

- Swiss Francs (CHF)
- German Marks (DEM)
- French Francs (FRF)
- British Pounds (GBP)
- Italian Lira (ITL)
- Japanese Yen (JPY)
- Netherland Guilder (NLG)
- Swedish Kroner (SEK)
- Gold Ounces (XAU)
- Silver Ounces (XAG)

The 2252 prices $w_t^{(k)}$ corresponding to each series k are constructed by linearly interpolating the middle prices, constructed as a geometric mean of bid and ask Müller et al. (1990), from the 2 nearest quotes which bracket 16:00 UKT. Of course, as a corollary to the linear interpolation – if a quote falls exactly on 16:00 UKT then the middle price from this quote is used. Only quotes which have been filtered using O&A filtering technology – and from an original high frequency database of O&A collected quotes from 3 real-time data vendors – Knight Ridder, Reuters and Telerate are used. The daily prices do not include Saturdays, Sundays, Christmas days, December 26 and New years day. The first price in this data set belongs to February 23, 1988 and the last price belongs to October 31, 1996.

2.2 Portfolio specification

Introduction: The performance measures of the risk models considered in this paper are presented in 2 ways:

Univariate: As an average performance over the 10 trivial portfolios that are defined by the 10 risk factors k – as being the sole asset in the portfolio.

Multivariate: In conjunction with a candidate portfolio P of 10 risk factors itemized in section 2.1 and having equal weights.

We shall hereafter use *univariate* and *multivariate* to describe the context of the performance measures as defined above.

The entire analysis is done assuming zero mean price change – but to account for any bias due to long term increases or decreases in one or more of the time series – the performance information is symmetrized by taking the mean performance over the 2 portfolio position vectors defined by:

Long on the US dollar with equal investment in each risk factor.

Short on the US dollar with equal investment in each risk factor.

The above *symmetrization* procedure applies equally to the *univariate* as well as the *multivariate* presentations.

Univariate context: Taking into account the *long* and *short* positions, in the *univariate* case we introduce notation $\mu^{\pm(k)}$ to denote a generic measure, μ, associated with a specific portfolio and position. Using this notation we can than construct the mean measure, μ_u, in the *univariate* (u) context as

$$\mu_u = \frac{1}{20}\left(\sum_{k=1}^{10}\mu^{+(k)} + \sum_{k=1}^{10}\mu^{-(k)}\right) \tag{1}$$

where each of the $\mu^{\pm(k)}$ is an entirely univariate construct related to the 20 distinct *univariate* portfolios with respect to which loss is estimated by a candidate model.

Multivariate context: Taking into account the *long* and *short* positions, in the *multivariate* case we introduce notation $\mu^{\pm(P)}$ to denote a generic measure, μ, associated with a specific portfolio and position. Using this notation, we can then construct the mean measure, μ_m, in the *multivariate* (m) context as

$$\mu_m = \frac{1}{2}\left(\mu^{+(P)} + \mu^{-(P)}\right) \tag{2}$$

where each $\mu^{\pm(P)}$ is an intrinsically multivariate construct related to the 2 distinct portfolios with respect to which loss is estimated by a candidate model.

Comment: The performance measures we present in this paper are all computed as an average over candidate portfolios. Such averaging is necessary so that the performance of the model is not biased by the choice of the portfolio. In particular, long term fall or rise in prices, can by chance make a fixed portfolio consistently profitable or lossy. To prevent bias of this nature we always average over pairs of portfolios with opposing positions.

2.3 Methodology

The 2252 prices $w_t^{(k)}$ for each of the risk factor series[1] k are then log differenced to produce 2251 consecutive log differenced price changes

$$x_{t+1}^{(k)} = \ln(w_{t+1}^{(k)}) - \ln(w_t^{(k)}). \tag{3}$$

In the *univariate* context, the primary focus of each candidate model is to predict the log difference series $x_{t+1}^{(k)}$ in terms of a conditional probability density function denoted by $p_t^{(k)}$. In the *multivariate* context, however, the primary focus of each candidate model is to predict $x_t^{(P)}$ defined by

$$x_t^{(P)} = \frac{1}{10}\sum_{k=1}^{10} x_t^{(k)} \tag{4}$$

[1] Data set RISK97 is available from Olsen & Associates – http://www.olsen.ch

in terms of a conditional probability density function denoted by $p_t^{(P)}$.

It is convenient at this point to establish the following nomenclature and notational guidelines.

- Price changes are numbered from -1249 to 1001. Specific, price change events are described as a subset of $t \in [-1249, 1001]$.
- The notations (k) and (P) are used as a superscript to reference the *univariate* portfolios k and the *multivariate* portfolio P respectively.
- The notation (k, P) is used as a superscript in an expression valid in both contexts.
- The superscripts $-$ and $+$ are used to reference the *long* and *short* positions of a portfolio.
- The superscript $\pm$ is used in an expression valid for both positions.
- The nomenclature *prediction-realization pair* refers to the ordered mathematical object $(p_t^{(k,P)}, x_{t+1}^{(k,P)})$. It should be kept in mind that by definition – $p_t^{(k,P)}$ are the forecast distributions for realizations $x_{t+1}^{(k,P)}$ and depend only on data until t.

The first 1250 of the price changes $x_t^{(k,P)}$, ($t \in [-1249, 0]$), constitute the in-sample period of those models requiring buildup and/or optimization. The final 1001 data points, $t \in [1, 1001]$ are used to construct 1000 out-of-sample *prediction-realization pairs.* This latter set of 1000 pairs are then analyzed in different ways to evaluate the performance of the models.

2.4 Zoo of models

We study the following 5 different models for generating the forecast distribution p_t over a *holding period* of 1 day:

1. Historical simulation with 250 day memory
2. 250 day rectangular moving average
3. J. P. Morgan RiskMetrics – Exponential moving average
4. GARCH(1, 1)
5. Tail emphasized GARCH(1, 1)

Model 1 is non-parametric and models 2 and 3 are pre-defined and do not require any optimization. Models 4 and 5 are optimized in an in-sample period of 1250 days (250 days for build up and 1000 days for optimization). The latter in-sample period precedes a 1000 day out-of-sample in which the performance of the models is gauged in the context of risk management.

The determination of the conditional distribution $p_t^{(k)}$ forecast by a model for each of the series k is of course an entirely univariate exercise. The multivariate aspect comes through the covariance matrix Σ_t which is determined by bivariate analysis – for all but model 1. In the case of models 2 and 3 the covariance is measured in a straightforward - intuitively obvious – bivariate generalization of the variance computation (details in sections 3.2 and 3.3).

In models 4 and 5 however – a more sophisticated approach is presented - wherein the dynamics of the bivariate sum are also fit to the candidate model and the implied covariance is used (details in section 3.4). Another innovative feature (described in section 3.5) is introduced in model 5 – wherein the optimization of the model is performed using a specialized method which emphasizes optimization of the residual tails. This optimization methodology turns out to have dramatic impact when using these models in the context of risk-management – as the performance measures will show.

2.5 Performance measures

The purpose of this paper in not simply the study of the different candidate models itemized in section 2.4 but also to present the different subtleties in evaluation of models in the context of risk management. For better visualization of performance, the above measures are plotted for each model as a function of increasingly extreme classes of realizations. To construct the classes of extreme realizations we use two different parameters such that – as they increase – they define a class of more and more extreme observations. These are:

Volatility percentile: The candidate measure is plotted against the percentile level of $|x_t^{(k,P)}|$ – and for a given level the measure is evaluated as a count or average over those realizations which correspond to a higher percentile level.

Confidence level: The candidate measure is plotted against a confidence level – and for a given level the measure is evaluated as a count or average over those realizations which are outside the confidence level.

There is a subtle difference between the two kinds of plots in terms of how the X-axis refers to an increasingly extreme set of points. The first plot directly evaluates the performance of the model for large price changes. The second evaluates the performance of the model for movements which are regarded as increasingly improbable by the model itself.

From the latter, it should be clear that – when comparing plots for different models against the volatility percentile – a given point on the X-axis will always refer to the same data. On the other hand, for plots against the confidence level – a point on the X-axis will refer to differing data sets since this discrimination depends on the model itself.

The performance measures presented in this paper are:

1. Observed/Predicted exceedence ratio against confidence level
2. BIS Red, Yellow and Green Zone frequency against confidence level
3. Observed/Predicted serial exceedence against confidence level
4. Mean log likelihood against confidence level
5. Mean log likelihood against volatility percentile

As already mentioned in section 2.2, we have one further variation in the presentation of performance measures – the *univariate* context based on 10000 *prediction-realization pairs* and the *multivariate* context based on 1000 *prediction-realization pairs.* The two contexts allow us to distinguish between the performance of a model for forecasting a risk factor with the performance of the model for forecasting the portfolio. Knowledge of the relative performance of the model in the *univariate* and *multivariate* context highlights an important distinction. It is usually the case that the methodology used for the multivariate computation $p_t^{(P)}$ as prescribed by a given model can be refined without affecting the univariate computation $p_t^{(k)}$. (For all the models presented in this paper – barring historical simulation – this means that it is possible to modify the methodology for updating the off diagonal elements of the covariance matrix Σ_t without changing the prescription for the diagonal elements.) The relative performance of a given model in the *univariate* and *multivariate* context can indicate if the point of weakness for the model lies in the computation of $p_t^{(P)}$ and can therefore motivate further research in rectifying the point of weakness.

3 Model specification

Presented below are the detailed specifications for the construction of the forecast probability density $p_t^{(k)}$ and $p_t^{(P)}$ as prescribed by the different models included in this paper.

3.1 Historical simulation

Historical simulation is the method of prediction in terms of which the forecast probability density depends directly on the past empirical distribution. In general the length of the past used for forecasting can be a parameter – which is adjusted for best performance in the in-sample period. The Bank of International Settlements in Basle (BIS), Switzerland, has however recommended Basle (1996) the use of at least 250 day memory. In view of this we present this model with a 250 day memory since the comparative performance of the BIS recommendation is likely to be of general topical interest.

Since no optimization is involved, the 10000 *univariate* and 1000 *multivariate prediction realization pairs* are constructed using the last 1250 data points $t \in [-248, 1001]$ in the 10 log differenced price change series k.

To construct $p_t^{(k,P)}$ we first define

$$X_t^{(k,P)} = \{x_t^{(k,P)}, x_{t-1}^{(k,P)}, \dots, x_{t-249}^{(k,P)}\} \tag{5}$$

as the set of 250 points to be used in constructing $p_t^{(k,P)}$ at time t. Then we introduce ranking notation using parenthetical subscripts (to be distinguished from $x_t^{(k,P)}$) such that

$$x_{(i,t)}^{(k,P)} \le x_{(i+1,t)}^{(k,P)}, \quad i \in [1, 250], \quad x_{(i,t)}^{(k,P)} \in X_t^{(k,P)}. \tag{6}$$

Thus $x_{(1,t)}^{(k,P)}$ is the minimum and $x_{(250,t)}^{(k,P)}$ is the maximum element in $X_t^{(k,P)}$.

We then construct $p_t^{(k,P)}$ to be the probability density function such that:

$$\int_{-\infty}^{x_{(i,t)}^{(k,P)}} p_t^{(k,P)}(x)dx = \frac{i-\frac{1}{2}}{250} \quad \forall i \in [1,250]. \tag{7}$$

The above density function is assumed to remain constant between $x_{(i,t)}^{(k,P)}$ and $x_{(i+1,t)}^{(k,P)}$ so that:

$$\int_{x_{(i,t)}^{(k,P)}}^{x_{(i+1,t)}^{(k,P)}} p_t^{(k,P)}(x)dx = \frac{1}{250}. \tag{8}$$

If $x_{(1,t)}^{(k,P)} < x_{t+1}^{(k,P)} < x_{(250,t)}^{(k,P)}$ then the probability density and fractile value associated with $x_{t+1}^{(k,P)}$ is well defined. If however $x_{t+1}^{(k,P)} < x_{(1,t)}^{(k,P)}$ or $x_{t+1}^{(k,P)} > x_{(250,t)}^{(k,P)}$ the probability density function is not defined and the fractile value of such an event cannot be determined. To resolve this – we extend this model by asserting that $p_t^{(k,P)}$ has Gaussian tails and hence the form implied by a normal distribution with mean $\bar{x} = \bar{x}_t^{(k,P)}$ estimated from set $X_t^{(k,P)}$ and where $\sigma = {}^{l}\sigma_t^{(k,P)}$ for the left side of the distribution which satisfies

$$\int_{-\infty}^{x_{(1,t)}^{(k,P)}} \frac{1}{\sqrt{2\pi}\, {}^{l}\sigma_t^{(k,P)}} \exp\left(-\frac{1}{2}\left[\frac{x-\bar{x}}{{}^{l}\sigma_t^{(k,P)}}\right]^2\right) dx = \frac{1}{500} \tag{9}$$

and with $\sigma = {}^{r}\sigma_t^{(k,P)}$ for the right side of the distribution which satisfies

$$\int_{x_{(250,t)}^{(k,P)}}^{\infty} \frac{1}{\sqrt{2\pi}\, {}^{r}\sigma_t^{(k,P)}} \exp\left(-\frac{1}{2}\left[\frac{x-\bar{x}}{{}^{r}\sigma_t^{(k,P)}}\right]^2\right) dx = \frac{1}{500} \tag{10}$$

This completely defines the construction of the forecast distribution $p_t^{(k,P)}$ and through this the out-of-sample *prediction-realization pairs* that form the starting point for performance assessment of this model.

3.2 Rectangular moving average

Since the horizon is already fixed by the data (and the *holding period*) at 1 day, the only remaining parameter one can associate with this method is the size of the memory. Once again – because of the commonly used 250 day memory (consistent with the BIS recommendation) we do not undertake any optimization and use the model with this pre-defined value. As in the case of the historical simulation model – the 1000 *univariate* and *multivariate prediction realization pairs* are constructed using the last 1251 data points in the 10 log differenced price change series k.

Univariate context: In this model, as in all subsequent models described below, the conditional distribution $p_t^{(k)}$ is predefined to be a Gaussian distribution with variance $\sigma_t^{(k)\,2}$. The difference between this and subsequent models lies only in the manner $\sigma_t^{(k)}$ is generated from the price change history of a given series. For this model:

$$\sigma_t^{(k)\,2} = \frac{1}{250} \sum_{i=0}^{249} x_{t-i}^{(k)\,2} \tag{11}$$

and this along with $x_t^{(k)}$ constitutes the full specification of the *prediction-realization pairs* that form the starting point of the *univariate* performance analysis of the model.

Multivariate context: In this case we define a multivariate object, the covariance matrix Σ_t, determined from the moving past 250 day history of the 10 series k. The elements of Σ_t are determined by:

$$s_t^{(jk)} = s_t^{(kj)} = \frac{1}{250} \sum_{i=0}^{249} x_{t-i}^{(j)} x_{t-i}^{(k)}. \tag{12}$$

Note that the diagonal elements ($j = k$) are exactly the variances $\sigma_t^{(k)\,2}$ obtained in the *univariate context*.

Then, using standard multivariate probability theory the forecast distribution $p_t^{(P)}$ for $x_{t+1}^{(P)}$ is the Gaussian distribution with variance:

$$\sigma_t^{(P)\,2} = \frac{1}{100} \left(\sum_{k=1}^{10} s_t^{(kk)} + 2 \sum_{k=1}^{10} \sum_{j<k}^{10} s_t^{(jk)} \right). \tag{13}$$

The determination of $\sigma_t^{(P)}$ as above along with $x_{t+1}^{(P)}$ constitutes the full specification of the *prediction-realization pairs* that form the starting point of the *multivariate* performance analysis of the model.

3.3 J. P. Morgan RiskMetrics

This model is basically a non-stationary GARCH(1, 1) model. The GARCH coefficients are pre-defined by the RiskMetrics technology – and so no optimization is necessary. To be consistent with the previous 2 models – we choose a build up period of 250 days – although the effective model memory is much shorter due to the exponentially decaying weight of the moving average that the GARCH(1, 1) process emulates. As in the previous 2 models – the analysis is based on the 1000 *prediction-realization pairs* constructed from the last 1251 data points in the 10 log differenced price change series k. For convenience we define $t \in [-249, 0]$ as representing the build up period.

Univariate context: In this model the variance of the Gaussian distribution $p_t^{(k)}$ is given by the recursive formula:

$$\sigma_t^{(k)\,2} = 0.94\sigma_{t-1}^{(k)\,2} + 0.06x_t^{(k)2}, \quad \sigma_{-249}^{(k)} = |x_{-249}^{(k)}|. \tag{14}$$

The value of $\sigma_t^{(k)}$ is not sensitive to the seed value $\sigma_{-249}^{(k)}$ once the recursion has been applied 250 times or more. Of course, for the latter to be true, the seed value must be a reasonable value like the one used in expression 14. The repeated application of the recursive expression 14 will allow the specification of 1000 *prediction-realization pairs* starting with $(p_1^{(k)}, x_2^{(k)})$ and ending with $(p_{1000}^{(k)}, x_{1001}^{(k)})$. These pairs are the starting point of the *univariate* performance analysis of the model.

Multivariate context: As in the previous model, here again we construct the covariance matrix Σ_t. The matrix elements are generated with the intuitively obvious bivariate generalization of the univariate case:

$$s_t^{(jk)} = s_t^{(kj)} = 0.94s_{t-1}^{(jk)} + 0.06x_t^{(j)}x_t^{(k)}. \tag{15}$$

Then, using standard multivariate probability theory the forecast distribution $p_t^{(P)}$ for $x_{t+1}^{(P)}$ is constructed from Σ_t in the same manner as expressed at the end of section 3.2 (expression 13).

3.4 GARCH(1,1)

The GARCH process is a popular stochastic process which has been fairly successful in modeling financial time series Bollerslev et al. (1992). In general the GARCH(p, q) process has $p + q + 1$ parameters which must be fit to the data. GARCH$(1, 1)$ is the simplest of this class with 3 parameters.

The GARCH$(1, 1)$ based risk model is the first model we present for which optimization is involved – and we make full use of all 2251 data points available to us from each of the 10 log differenced price change series k. For convenience we define $t \in [-1249, -1001]$ as representing the build up period, $t \in [-1000, 0]$ as representing the in-sample period and $t \in [1, 1001]$ as the period over which we construct the 1000 *prediction-realization pairs* associated with this model.

Univariate context: In this model the variance of the Gaussian distribution p_t is given by the recursive formula:

$$\sigma_t^{(k)\,2} = \alpha_0^{(k)} + \alpha_1^{(k)}x_t^{(k)2} + \beta_1^{(k)}\sigma_{t-1}^{(k)\,2}, \quad \sigma_{-249}^{(k)} = |x_{-249}^{(k)}|. \tag{16}$$

The recursion is initialized using the same seed value as for the J. P. Morgan RiskMetrics case. As discussed in section 3.3, we emphasize that the built up values $\sigma_t^{(k)}$ for $t \in [1, 1000]$ are not sensitive to this arbitrary but reasonable choice.

We have already stated that the J. P. Morgan RiskMetrics model is a GARCH(1, 1) model. What is different here, however, is that we have separately optimized the GARCH(1, 1) for each of the k risk-factors in the in-sample period. Unlike the J. P. Morgan prescription of using $\alpha_0^{(k)} = 0$, $\alpha_1^{(k)} = 0.06$ and $\beta_1^{(k)} = 0.94$ for all k risk factors, in our case we used the values reported in Table 1.

The optimization – reported in Table 1 – was done with the objective of maximizing the average log-likelihood of the 1000 *prediction-realization pairs* in the in-sample period – corresponding to the events $t \in [-1000, 0]$. Mathematically, the average log-likelihood may be expressed as:

$$\bar{\mathcal{L}}^{(k)} = \frac{1}{1000} \sum_{t=-1000}^{-1} \mathcal{L}_t^{(k)} = \frac{1}{1000} \sum_{t=-1000}^{-1} \ln(p_t^{(k)}(x_{t+1}^{(k)})) \tag{17}$$

where, $p_t^{(k)}(x_{t+1}^{(k)}) = p_t(\sigma_t^{(k)}, x_{t+1}^{(k)})$ is the probability density of the realized event in terms of the forecast distribution and $\sigma_t^{(k)} = \sigma_t^{(k)}(\alpha_0^{(k)}, \alpha_1^{(k)}, \beta_1^{(k)})$ is the variance of the forecast distribution in accordance with the GARCH(1, 1) expression 16. For stationarity of the GARCH(1, 1) process, the optimization is done with constraint $\alpha_0^{(k)} + \alpha_1^{(k)} + \beta_1^{(k)} < 1$. (We note here that the J. P. Morgan RiskMetrics process equation violates this stationarity condition). The solution space was searched using the genetic algorithms method – but convergence was aided by strategic use of the BHHH algorithm Berndt et al. (1974) at the end of every generationDacorogna et al. (1997).

With Table 1 and the GARCH(1, 1) recursion formula for the variance $\sigma_t^{(k)}$ of the Gaussian distribution $p_t^{(k)}$, we construct the 1000 *prediction-realization pairs* over the period $t \in [1, 1001]$ that is the starting point for the evaluation of this model in the *univariate* context.

Multivariate context: Here – once again – we proceed via the construction of the covariance matrix Σ_t but in this case we introduce a different methodology. Instead of making the straightforward bivariate generalization of the GARCH(1, 1), we proceed by fitting the 45 sum series,

$$x_t^{(jk)} = x_t^{(j)} + x_t^{(k)}, \tag{18}$$

(note new notation to reference sum series) to the GARCH(1, 1). The fit is done in the same in-sample period and using the same mix of the genetics algorithm and BHHH approach used in the *univariate* context. The fit parameters for the 45 series are presented in Table 2.

The $\sigma_t^{(jk)}$ are then determined for the sum series using the same recursion expression and build up criteria used in the *univariate* context. The off-diagonal elements of Σ_t are then computed as an implied covariance:

$$s_t^{(jk)} = s_t^{(kj)} = \frac{\sigma_t^{(jk)} - \sigma_t^{(j)} - \sigma_t^{(k)}}{2}, \quad i \neq j. \tag{19}$$

The following additional specification of diagonal elements:

$$s_t^{(kk)} = \sigma_t^{(k)\,2}, \tag{20}$$

where $\sigma_t^{(k)}$ is determined in the *univariate* context, concludes the full prescription on how Σ_t is generated in this model.

Finally, using standard multivariate probability theory the forecast distribution $p_t^{(P)}$ for $x_{t+1}^{(P)}$ is constructed from Σ_t in the same manner as expressed at the end of section 3.2 (expression 13). The 1000 *prediction-realization pairs*, $(p_t^{(P)}, x_{t+1}^{(P)})$, over the period $t \in [1, 1001]$ is the starting point for the evaluation of this model in the *multivariate* context.

	Garch(1,1)				Tail emphasized Garch(1,1)			
Series	$\alpha_0 \times 10^5$	α_1	β_1	LL	$\alpha_0 \times 10^5$	α_1	β_1	LL
CHF	0.1887	0.04185	0.9279	3.4410	0.6404	0.01669	0.9361	3.1113
DEM	0.1806	0.06171	0.9053	3.5187	0.9698	0.02959	0.8880	3.1732
FRF	0.1488	0.06580	0.9050	3.5659	0.9808	0.05182	0.8646	3.2155
GBP	0.1912	0 07187	0.8925	3.5427	0.9695	0.02959	0 8864	3.1809
ITL	0.1601	0.08730	0.8836	3.5648	0 9808	0.05182	0.8658	3.1927
JPY	0.1074	0.05991	0.9132	3.6761	0.9748	0.05153	0.8431	3.3196
NLG	0.1803	0.06020	0.9061	3.5275	0.9715	0.05133	0.8728	3.1813
SEK	0.0069	0.04783	0.9558	3.6714	0.9626	0.02959	0.8670	3.2376
XAG	1 2128	0.06719	0.8441	3.0388	0.8269	0.00749	0.9653	2.6915
XAU	0.0212	0.02793	0.9680	3.4656	0.4518	0.00840	0.9582	3.0685

Table 1: Fit parameters $(\alpha_0, \alpha_1, \beta_1)$ for variance forecasts and the mean log-likelihood (LL) over the in-sample for GARCH(1,1) (using 1000 points) and tail emphasized GARCH(1,1) processes (using 500 points). See sections 3.4 and 3.5 for additional details. The series are as defined in section 2.1.

3.5 Tail emphasized GARCH(1,1)

The implementation of this model differs from the GARCH(1,1) implementation only in the manner of optimization of the process over the 10 series $x_t^{(k)}$ (needed in the *univariate* context) and the additional 45 sum series $x_t^{(jk)}$ (needed in the *multivariate context*). Instead of maximizing the average of all 1000 $\mathcal{L}_t^{(k)}$ or $\mathcal{L}_t^{(jk)}$ (for the sum series) in the in-sample – we maximize only the average of the most adverse 500. In describing this method as *tail emphasized* GARCH(1,1), we note that the tail referred to in this context is not the tail of the distributions of $x_t^{(k)}$ and $x_t^{(jk)}$, but the tail of the distributions of $\mathcal{L}_t^{(k)}$ and $\mathcal{L}_t^{(jk)}$. We however do expect considerable overlap of events ($t \in [-1000, 0]$) in the tails of the 2 distributions.

Since the conditional distributions $p_t^{(k,jk)}$ (notation (k, jk) indicating reference to the series k and the sum series jk) is Gaussian with variance $\sigma_t^{(k,jk)\,2}$ we can expand $\mathcal{L}_t^{(k,jk)}$, $t \in [-1000, -1]$, more concretely as:

$$\mathcal{L}_t^{(k,jk)} = -\frac{1}{2}\ln(2\pi) - \ln(\sigma_t^{(k,jk)}) - \frac{x_{t+1}^{(k,jk)\,2}}{2\sigma_t^{(k,jk)\,2}}. \tag{21}$$

	Garch(1,1)				Tail emphasized Garch(1,1)			
Series	$\alpha_0 \times 10^5$	α_1	β_1	LL	$\alpha_0 \times 10^5$	α_1	β_1	LL
DEM+CHF	0.6877	0.05098	0.9184	2.8053	0.9193	0.01906	0.9675	2.4652
FRF+CHF	0.6354	0.05278	0.9178	2.8283	0.9927	0.01411	0.9675	2.4890
FRF+DEM	0.6543	0.06359	0.9053	2.8509	0.9926	0.01434	0.9665	2.4990
GBP+CHF	0.5459	0.05642	0.9178	2.8513	0.9925	0.01406	0.9667	2.5069
GBP+DEM	0.5400	0.06301	0.9102	2.8814	0.9913	0.01360	0.9657	2.5257
GBP+FRF	0.4976	0.06509	0.9095	2.9031	0.9884	0.01360	0.9647	2.5450
GBP+ITL	0.5374	0.07466	0.8984	2.9057	0.9901	0.01360	0.9647	2.5344
GBP+JPY	0.5536	0.05651	0.9056	3.0168	0.8425	0.00992	0.9644	2.6733
GBP+NLG	0.5330	0.06163	0.9116	2.8857	0.9914	0.01360	0.9654	2 5303
GBP+SEK	0.2070	0.05930	0.9320	2.9429	0.9432	0.01360	0.9649	2.5612
ITL+CHF	0.6234	0.05523	0.9157	2.8355	0.9914	0.01383	0.9673	2.4921
ITL+DEM	0.6732	0.06993	0.8979	2.8600	0.9912	0.01383	0.9665	2.5014
ITL+FRF	0.6211	0.07343	0.8960	2.8824	0.9926	0.01406	0.9652	2.5211
ITL+JPY	0.6008	0.05949	0.8991	3.0180	0.8425	0.00992	0.9642	2.6728
JPY+CHF	0.5489	0.03985	0.9261	2.9578	0.8897	0.01096	0.9648	2.6301
JPY+DEM	0.6172	0.04760	0.9114	2.9923	0.8582	0.00980	0.9649	2.6573
JPY+FRF	0.5591	0.05111	0.9099	3.0193	0.8426	0.00992	0.9639	2.6824
NLG+CHF	0.6784	0.04999	0.9196	2.8093	0.9194	0.01907	0.9673	2.4691
NLG+DEM	0.7250	0.06102	0.9055	2.8306	0.9915	0.01383	0.9678	2.4809
NLG+FRF	0.6535	0.06272	0.9058	2.8553	0.9927	0.01406	0.9665	2.5039
NLG+ITL	0.6812	0.06939	0.8977	2.8639	0.9915	0.01359	0.9664	2.5058
NLG+JPY	0.5970	0.04709	0.9128	2.9973	0.8583	0.00992	0.9646	2.6617
SEK+CHF	0.2163	0.04709	0.9438	2.8810	0.9912	0.01360	0.9660	2.5219
SEK+DEM	0.1637	0.05014	0.9437	2.9063	0.9802	0.01360	0.9654	2.5314
SEK+FRF	0.1435	0.05186	0.9429	2 9313	0.9649	0.01360	0.9646	2.5528
SEK+ITL	0.1439	0.05694	0.9385	2.9369	0.9432	0.01360	0.9652	2.5476
SEK+JPY	0.3178	0.04254	0.9349	3 0461	0.8165	0.00583	0.9658	2.7015
SEK+NLG	0.1574	0.04853	0.9454	2.9106	0.9866	0.01360	0.9650	2.5359
XAG+CHF	1.3111	0.03395	0.9057	2.7977	0.9193	0.01906	0.9680	2.4649
XAG+DEM	1.1672	0.04607	0.8969	2.8316	0.9928	0.01410	0.9671	2.4979
XAG+FRF	1.1351	0.04487	0.8978	2.8472	0.9911	0.01371	0.9666	2.5136
XAG+GBP	0.6276	0.03225	0.9375	2.8249	0.9914	0.01383	0.9678	2.4904
XAG+ITL	1.2375	0 04736	0.8904	2.8464	0.9913	0 01371	0.9667	2.5127
XAG+JPY	0.3264	0.02016	0.9616	2.8848	0.9900	0.01360	0.9648	2.5500
XAG+NLG	1.0297	0.04256	0.9068	2.8339	0.9914	0.01383	0.9672	2.5001
XAG+SEK	1.4667	0.05619	0.8673	2.8667	0.9912	0.01359	0.9660	2.5266
XAU+CHF	0.4364	0.02503	0.9462	2.9736	0.8897	0.00922	0.9650	2.6454
XAU+DEM	0.5821	0.03243	0.9261	3.0152	0 8269	0.00749	0.9655	2.6902
XAU+FRF	0.5088	0.03200	0.9298	3.0406	0.8198	0.00672	0.9647	2.7147
XAU+GBP	0.3598	0.03306	0.9406	3.0339	0.8267	0.00749	0.9649	2.6998
XAU+ITL	0.6591	0.04087	0.9104	3.0367	0.8199	0.00672	0.9651	2.7082
XAU+JPY	0.0177	0.01519	0.9824	3.1589	0.6367	0.00840	0.9655	2.8000
XAU+NLG	0.4568	0.02875	0.9383	3.0203	0.8267	0.00749	0.9652	2.6951
XAU+SEK	11.2385	0.09932	0.0000	3.0703	0.8132	0.00495	0.9642	2.7376
XAU+XAG	0.1431	0.01737	0.9776	2.6099	0 9908	0.01565	0.9756	2 2537

Table 2: Fit parameters (α_0, α_1, β_1) for variance forecasts and the mean log-likelihood (LL) over the in-sample for GARCH(1,1) (using 1000 points) and tail emphasized GARCH(1,1) processes (using 500 points). See sections 3.4 and 3.5 for additional details. The series are the sum (+) of the series defined in section 2.1. This optimization is needed in the *multivariate* context to compute the implied covariances used in these methods.

It is clear from the above expression that the dominant contribution to adverse (low) values of $\mathcal{L}_t^{(k,jk)}$ come from the third term when the model underestimates the risk and predicts a small variance compared to the market move on the next day. It is expected that by optimizing with the objective of maximizing the average of the 500 most adverse $\mathcal{L}_t^{(k,jk)}$, the model will provide better day to day consistency in forecasting performance than the traditional method which includes all events with equal emphasis.

4 Performance measures

Whether for the *univariate* case or the *multivariate* case, each model specification ends with a clear recipe for constructing 1000 *prediction-realization pairs*, $(p_t^{(k,P)}, x_{t+1}^{(k,P)})$, $t \in [1, 1000]$. These *prediction-realization pairs* are the starting point for performance measurement of the models in the context of risk management.

4.1 Exceedence ratio against confidence level

Introduction: Basically this measure is a count of the events for which the loss of assets exceeds the loss predicted by the model as a function of the confidence level c. The count is normalized by the theoretical expectation of the exceedence count – and hence we refer to this as the exceedence ratio. This measure is a relatively coarse measure since it is not sensitive in distinguishing models which have the same exceedence count but a different *degree* of exceedence.

Specification: As a function of confidence level c (in %), it is convenient to first define the exceedence limits $\chi_t^{+(k,P)}(c)$ and $-\chi_t^{-(k,P)}(c)$ so that:

$$\int_{-\infty}^{\chi_t^{+(k,P)}(c)} p_t^{(k,P)}(x)dx = \int_{-\chi_t^{-(k,P)}(c)}^{\infty} p_t^{(k,P)}(x)dx = \frac{c}{100}. \tag{22}$$

It should be clear from expression 22 that $\chi_t^{+(k,P)}(c)$ is the loss that will be exceeded by a long (+) position on the US Dollar against series k or portfolio P with probability $(1 - \frac{c}{100})$. Similarly the exceedence limit $\chi_t^{-(k,P)}(c)$ is the correspondingly probable loss for a short position on the US Dollar against series k or portfolio P. We note that for all the models except historical simulation – the conditional distribution $p_t^{(k,P)}$ is symmetric around zero and $\chi_t^{+(k,P)}(c) = \chi_t^{-(k,P)}(c)$.

Next, we define the set of loss exceedence events $\Phi^{+(k,P)}(c)$ and $\Phi^{-(k,P)}(c)$ so that:

$$\Phi^{\pm(k,P)}(c) = \{t | \pm x_t^{(k,P)} > \chi_{t-1}^{\pm(k,P)}(c), t \in [2, 1001]\}. \tag{23}$$

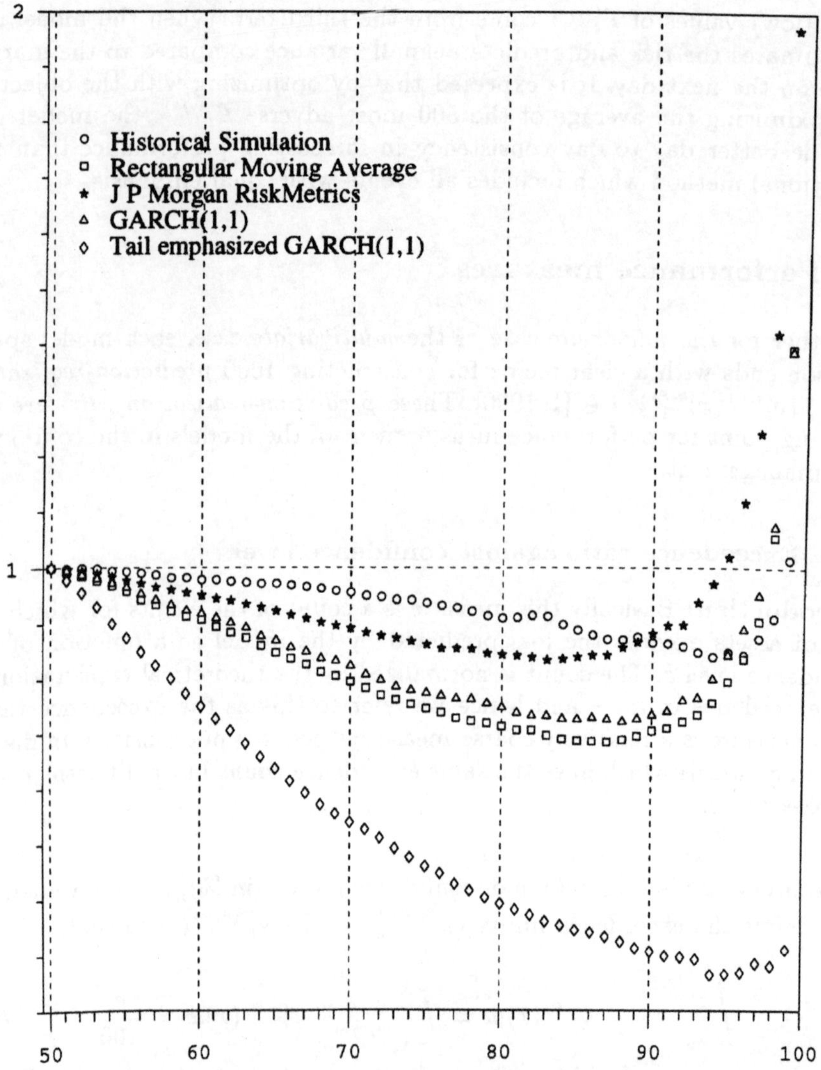

Figure 1u: Ratio of observed and expected loss exceedence count against confidence level in the *univariate* context for all candidate models. This plot shows the ratio (X axis) of the frequency of exceedences and the expected exceedence count against confidence level c (Y axis) for all the 5 models considered in this paper. Ratio values larger than unity indicate that the model underestimates risk at the confidence level for which this occurs. The value plotted for a given model is the average of the ratios computed over 20 trivial portfolios – each consisting of a *long* or *short* position of the US Dollar against the 10 series k. The curves are based on the analysis of 10000 *prediction-realization pairs* in the out-of-sample period. See section 6.3 for additional comments.

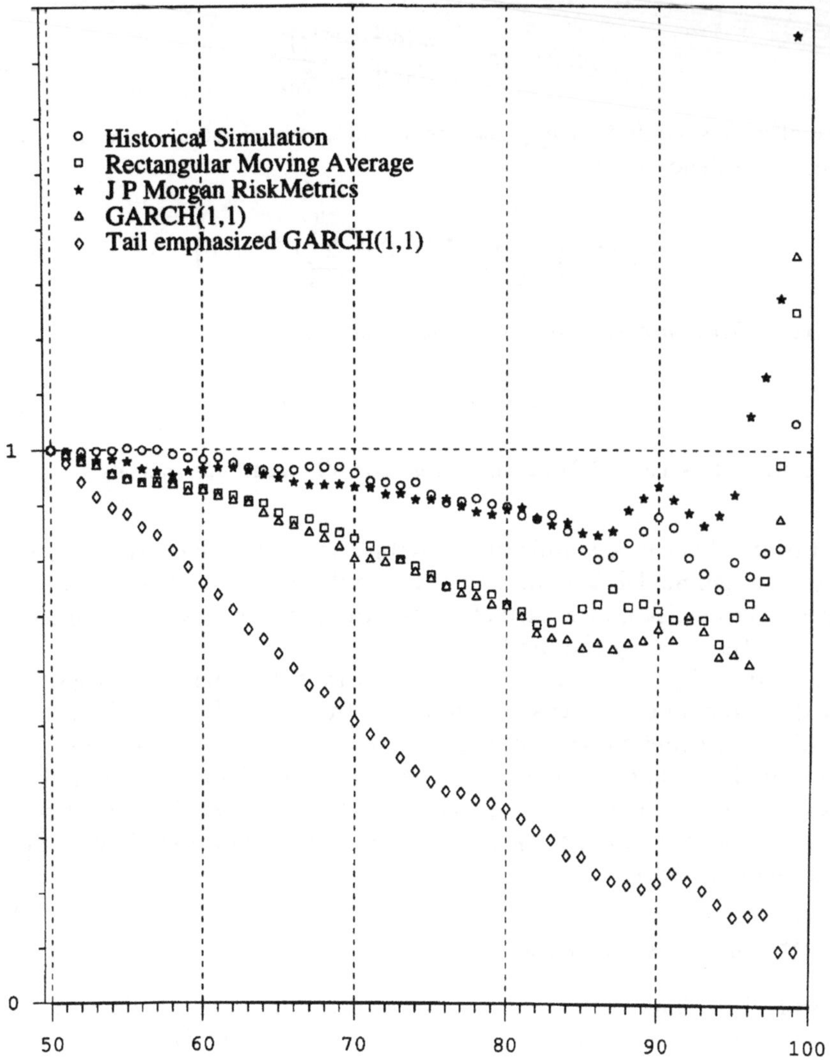

Figure 1m: Ratio of observed and expected loss exceedence count against confidence level in the *multivariate* context for all candidate models. This plot shows the ratio (X axis) of the frequency of exceedences and the expected exceedence count against confidence level c (Y axis) for all the 5 models considered in this paper. Ratio values larger than unity indicate that the model underestimates risk at the confidence level for which this occurs. The value plotted for a given model is the average of the ratios computed over 2 portfolios – each consisting of uniformly *long* or uniformly *short* positions of the US Dollar against the 10 series k. The curves are based on the analysis of 1000 *prediction-realization pairs* in the out-of-sample period. See section 6.3 for additional comments.

Using sets $\Phi^{\pm(k,P)}(c)$ we can finally specify the exceedence ratio measure

$$r^{\pm(k,P)}(c) = \frac{\#(\Phi^{\pm(k,P)}(c))}{1000(1 - \frac{c}{100})}. \qquad (24)$$

Finally, as specified by expressions 1 and 2, we construct the *univariate* average exceedence ratio

$$r_u(c) = \frac{1}{20}\left(\sum_{k=1}^{10} r^{+(k)}(c) + \sum_{k=1}^{10} r^{-(k)}(c)\right) \qquad (25)$$

and the *multivariate* average exceedence ratio

$$r_m(c) = \frac{1}{2}\left(r^{+(P)}(c) + r^{-(P)}(c)\right) \qquad (26)$$

plotted in figures 1u and 1m respectively for all candidate models.

Comment: This is a simple and absolute measure which shows general success of a risk model as a function of the confidence level. The region above unity represents risk under-estimation and the region below unity represents risk over-estimation by the model.

Comparing figures 1u and 1m for the same model – can give an insight into the performance of the covariance forecast by the model. If the performance in the *multivariate* context drops drastically compared to the *univariate* context, this may indicate that the model may be salvaged by refining the methodology for covariance forecasting in conjunction with the model. (This is of course generally true when comparing any u figure with a corresponding m figure for the same model.)

4.2 BIS colour frequency

Introduction: This measure is inspired by the BIS Green, Yellow and Red zone designations associated with a risk model defined as follows:

Green zone: A risk model is stated to be in this zone on day t if there are 4 or fewer events of losses beyond the 99% confidence level in the last 250 days – which is precisely the set $X_t^{(k,P)}$ defined in expression 5.

Yellow zone: A risk model is stated to be in this zone on day t if there are between 5 and 8 events of losses beyond the 99% confidence level in set $X_t^{(k,P)}$.

Red zone: A risk model is stated to be in this zone on day t if there are 9 or more events of losses beyond the 99% confidence level in set $X_t^{(k,P)}$.

The BIS uses the above zones as the basis for assessing the reliability of VaR models.

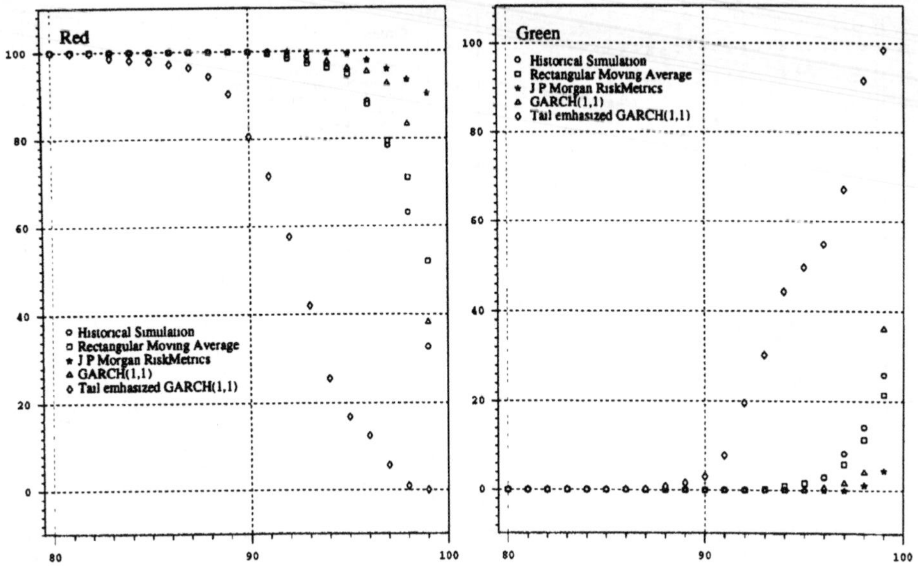

Figure 2u: BIS *red* and *green* zone frequency (Y axis) against confidence level (X axis) in the *univariate* context for all candidate models. The frequency is expressed as a fraction of the 751 days in the 1000 day out-of-sample period for which the the BIS *red* and *green* designations can be assigned to a model (249 days are needed for build up of this measure). The varying confidence level generalizes the BIS definition which is fixed at the 99% level. It is expected that the better model will approach the 100% green and 0% red frequencies faster (at lower confidence levels). The frequency plotted for a given model is the mean over 2 portfolios – each consisting of uniformly *long* or uniformly *short* positions of the US Dollar against the 10 series k. The curves are based on the analysis of 10000 *prediction-realization pairs* in the out-of-sample period. See section 6.4 for additional comments.

To construct our measure we first allow the prescribed confidence (99%) to become a variable c and then determine the frequency of the *red* and *green* zone designations associated with the model in the out-of-sample period as a function of c. Since our out-of-sample has 1000 *prediction-realization pairs*, we have 4 non-overlapping periods of 250 days but 751 distinct but overlapping moving samples, $X_t^{(k,P)}$, $t \in [251, 1001]$, in which to determine the frequency of the BIS colour designations. (We omit the *yellow* frequency from the presentation since the frequency of this middle designation leads to no clarity in the comparison of models.)

Specification: Re-using expression 22 defining the exceedence level $\chi_t^{\pm(k,P)}$ and expression 5 defining the set $X_t^{(k,P)}$ we extend definition 23 so that

$$\Phi_t^{\pm(k,P)}(c) = \{t \mid \pm x_t^{(k,P)} > \chi_{t-1}^{\pm(k,P)}(c), t \in X_t^{(k,P)}\}. \tag{27}$$

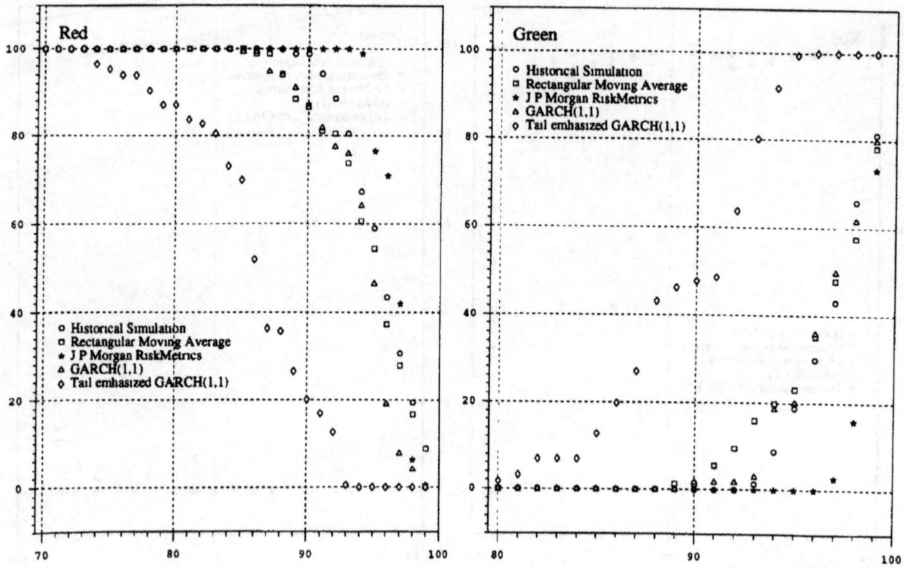

Figure 2m: BIS *red* and *green* zone frequency (Y axis) against confidence level (X axis) in the *multivariate* context for all candidate models. The frequency is expressed as a fraction of the 751 days in the 1000 day out-of-sample period for which the the BIS *red* and *green* designations can be assigned to a model (249 days are needed for build up of this measure). The varying confidence level generalizes the BIS definition which is fixed at the 99% level. It is expected that the better model will approach the 100% green and 0% red frequencies faster (at lower confidence levels). The frequency plotted for a given model is the mean over 2 portfolios – each consisting of uniformly *long* or uniformly *short* positions of the US Dollar against the 10 series k. The curves are based on the analysis of 1000 *prediction-realization pairs* in the out-of-sample period. See section 6.4 for additional comments.

Note that here we have defined sets Φ against a universal set of a moving sample of events rather than the entire out-of-sample set of events as expressed in 23. Next, using $\Phi_t^{\pm(k,P)}(c)$ defined as above, we can construct *booleans* $R_t^{\pm(k,P)}(c)$ and $G_t^{\pm(k,P)}(c)$ $(\in \{0,1\})$ such that:

$$(G_t^{\pm(k,P)}(c), R_t^{\pm(k,P)}(c)) = \begin{cases} (1,0) \text{ if } \#(\Phi_t^{\pm(k,P)}(c)) \leq 4 \\ (0,0) \text{ if } 5 \leq \#(\Phi_t^{\pm(k,P)}(c)) \leq 8 \\ (0,1) \text{ if } \#(\Phi_t^{\pm(k,P)}(c)) \geq 9 \end{cases} \tag{28}$$

Clearly, through expression 28, one can identify that $G_t^{\pm(k,P)}(99) = 1$ and $R_t^{\pm(k,P)}(99) = 1$ are the BIS defined colour designation for a risk model with respect to the assets on day t. As usual, the portfolio represented by the superscripts covers the cases of the long (+) or short (−) position in the US Dollar with respect to the series k or portfolio P. Using the *booleans* defined by expression 28 we can construct the colour frequency measures

$$(G,R)^{\pm(k,P)}(c) = \frac{1}{751} \sum_{t\in[251,1001]} (G,R)_t^{\pm(k,P)}(c) \tag{29}$$

where the notation (G, R) transparently refers to 2 expressions – one each for $G^{\pm(k,P)}(c)$ and $R^{\pm(k,P)}(c)$ respectively. $(G,R)^{\pm(k,P)}(c)$ is of course the fraction of events in the out-of-sample period when the *green* and *red* designation is applicable. Note again that we have extended the BIS definition for this measure and allowed confidence level c to be a variable.

Finally, as specified by expressions 1 and 2, we construct the *univariate* average BIS colour frequency

$$(G,R)_u(c) = \frac{1}{20}\left(\sum_{k=1}^{10}(G,R)^{+(k)}(c) + \sum_{k=1}^{10}(G,R)^{-(k)}(c)\right) \tag{30}$$

and the *multivariate* average BIS colour frequency

$$(G,R)_m(c) = \frac{1}{2}\left((G,R)^{+(P)}(c) + (G,R)^{-(P)}(c)\right) \tag{31}$$

plotted in figures 2u and 2m respectively for all candidate models.

Comment: When c is low, it is naturally expected that the models will be predominantly 100% in the *red.* As c increases, *red* frequency drops and the *green* frequency increases. The *yellow* frequency – not plotted – peaks somewhere between 50% and 100% – and is different for different models. The X-axis of the plots are restricted to high c to highlight the portions which are most interesting for comparing the models.

This absolute measure does not carry much more information than the exceedence ratio presented in the last section. It also inherits the *digital* property of the exceedence ratio – in that it is not sensitive to the degree of exceedence. The main motivation of this measure is only to present exceedence information in the more popular framework of the BIS colour designations.

4.3 Serial exceedence ratio against confidence level

Introduction: Neither the exceedence ratio of section 4.1 nor the BIS colour frequency of the last section measure the clustering of confidence level exceedences. We now present the *serial exceedence ratio* which tests the propensity of the model for consecutive prediction failures.

Specification: In order to construct the measure we define the set of consecutive loss exceedence events $\Psi^{\pm(k,P)}(c)$ so that:

$$\Psi^{\pm(k,P)}(c) = \{t|\{t,t-1\} \subset \Phi^{\pm(k,P)}(c), t \in [3,1001]\}. \tag{32}$$

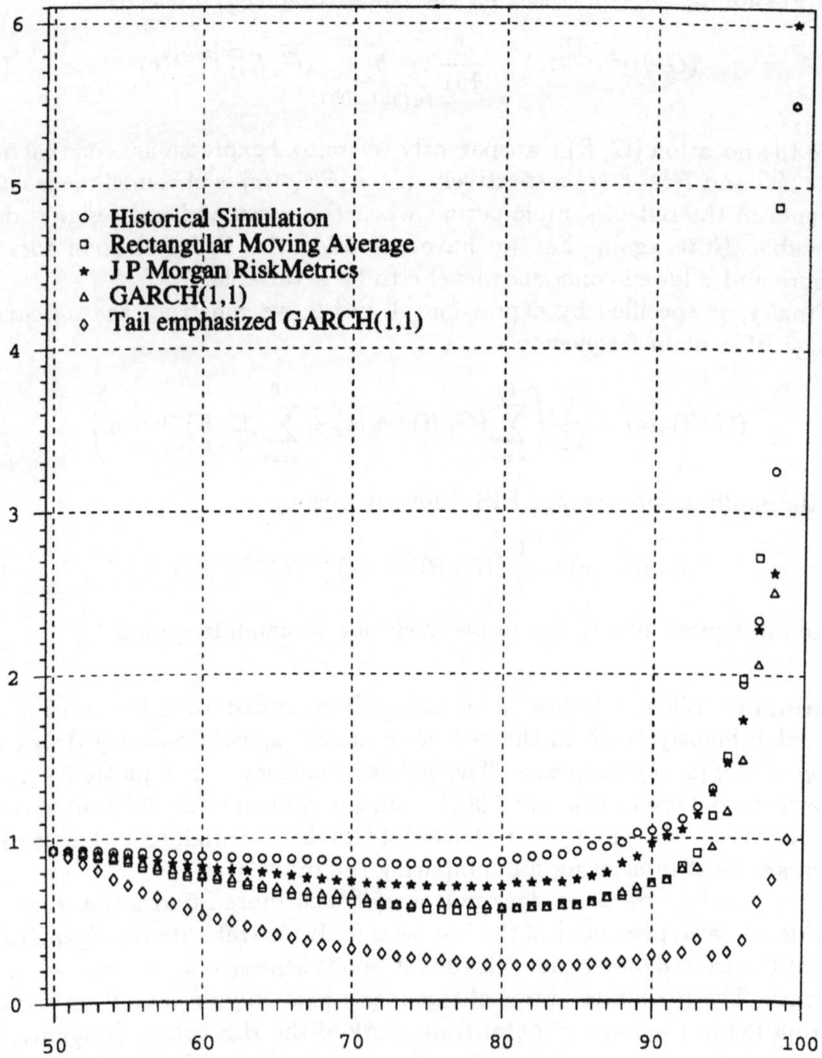

Figure 3u: Ratio of observed and expected consecutive loss exceedence counts against confidence level in the *univariate* context for all candidate models. This plot shows the ratio (X axis) of the frequency of consecutive exceedence events and the expected frequency of such events against the confidence level c (Y axis) for which the events are defined – for all the 5 models considered in this paper. Ratio values larger than unity indicate that the model is prone to greater than expected consecutive prediction failures at the confidence level for which this occurs. The value plotted for a given model is the average of the ratios computed over 20 trivial portfolios – each consisting of a *long* or *short* position of the US Dollar against the 10 series k. The curves are based on the analysis of 10000 *prediction-realization pairs* in the out-of-sample period. See section 6.5 for additional comments.

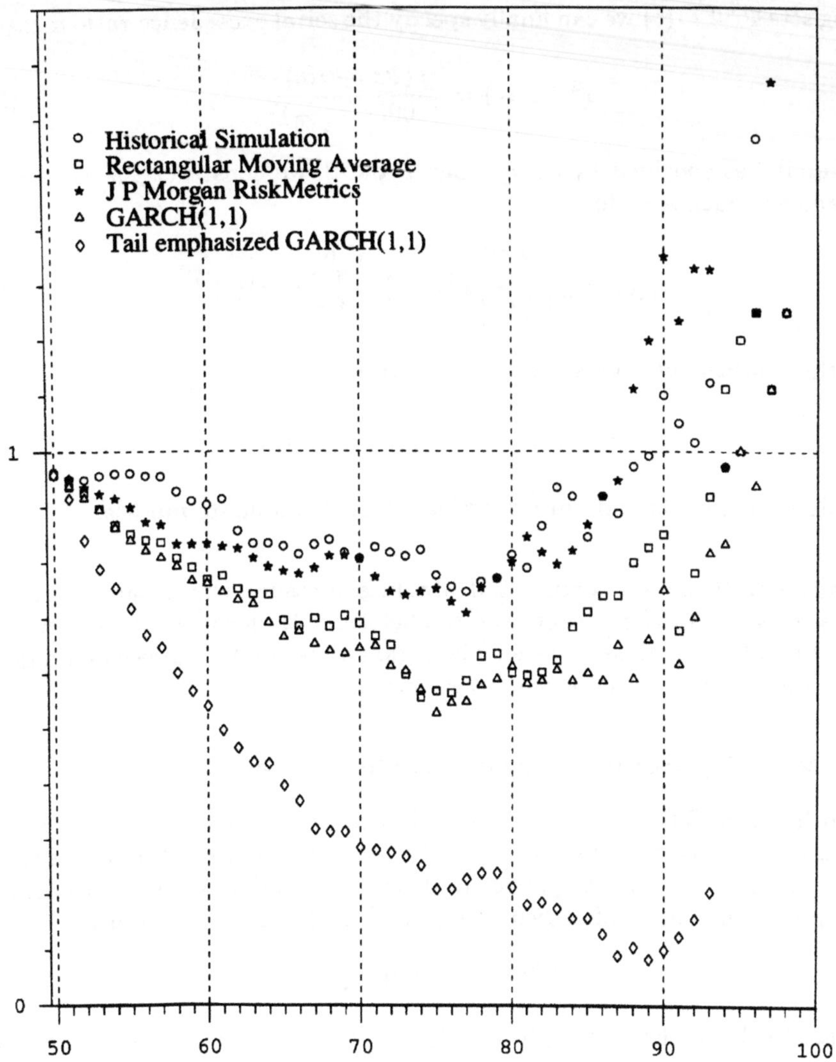

Figure 3m: Ratio of observed and expected consecutive loss exceedence counts against confidence level in the *multivariate* context for all candidate models. This plot shows the ratio (X axis) of the frequency of consecutive exceedence events and the expected frequency of such events against the confidence level c (Y axis) for which the events are defined – for all the 5 models considered in this paper. Ratio values larger than unity indicate that the model is prone greater than expected to consecutive prediction failures at the confidence level for which this occurs. The value plotted for a given model is the average of the ratios computed over 2 portfolios – each consisting of uniformly *long* or uniformly *short* positions of the US Dollar against the 10 series k. The curves are based on the analysis of 1000 *prediction-realization pairs* in the out-of-sample period. See section 6.5 for additional comments.

Using sets $\Psi^{\pm(k,P)}(c)$ we can finally specify the serial exceedence ratio measure

$$s^{\pm(k,P)}(c) = \frac{\#(\Psi^{\pm(k,P)}(c))}{1000(1-\frac{c}{100})^2} \tag{33}$$

Finally, as specified by expressions 1 and 2, we construct the *univariate* average exceedence ratio

$$s_u(c) = \frac{1}{20}\left(\sum_{k=1}^{10} s^{+(k)}(c) + \sum_{k=1}^{10} s^{-(k)}(c)\right) \tag{34}$$

and the *multivariate* average exceedence ratio

$$s_m(c) = \frac{1}{2}\left(s^{+(P)}(c) + s^{-(P)}(c)\right) \tag{35}$$

plotted in figures 3u and 3m respectively for all candidate models.

Comment: This is a simple and absolute measure which shows general success of a risk model in capturing the heteroskedasticity of the market – as a function of the confidence level. The region above unity represents a higher than expected rate of successive exceedences.

4.4 Mean log-likelihood against confidence level

Introduction: The main drawback of the exceedence ratio as a performance measure is that it is only sensitive to the frequency and not the *degree* with which the loss exceeds that predicted at a certain confidence level. A natural measure for the *degree* of exceedence is the log-likelihood contribution

$$\mathcal{L}_t^{(k,P)} = \ln(p_t^{(k,P)}(x_{t+1}^{(k,P)})) \tag{36}$$

which is the logarithm of the probability density of the realized event in terms of the forecast probability distribution $p_t^{(k,P)}$. While we have used the mean log-likelihood as a criteria for optimization in the in-sample, we now extend it to measure performance out-of-sample. Although the expression 21 refers only to a Gaussian conditional distribution – it does qualitatively show through its third term that it is appropriate to regard the log-likelihood contribution $\mathcal{L}_t^{(k,P)}$ of a single event as a comparative (between models) measure of the *degree* of exceedence.

Specification: Using expressions 23 and 36 we can directly construct

$$\ell^{\pm(k,P)}(c) = \frac{\sum_{t\in\Phi^{\pm(k,P)}(c)} \mathcal{L}_{t-1}^{(k,P)}.}{\#(\Phi^{\pm(k,P)}(c))} \tag{37}$$

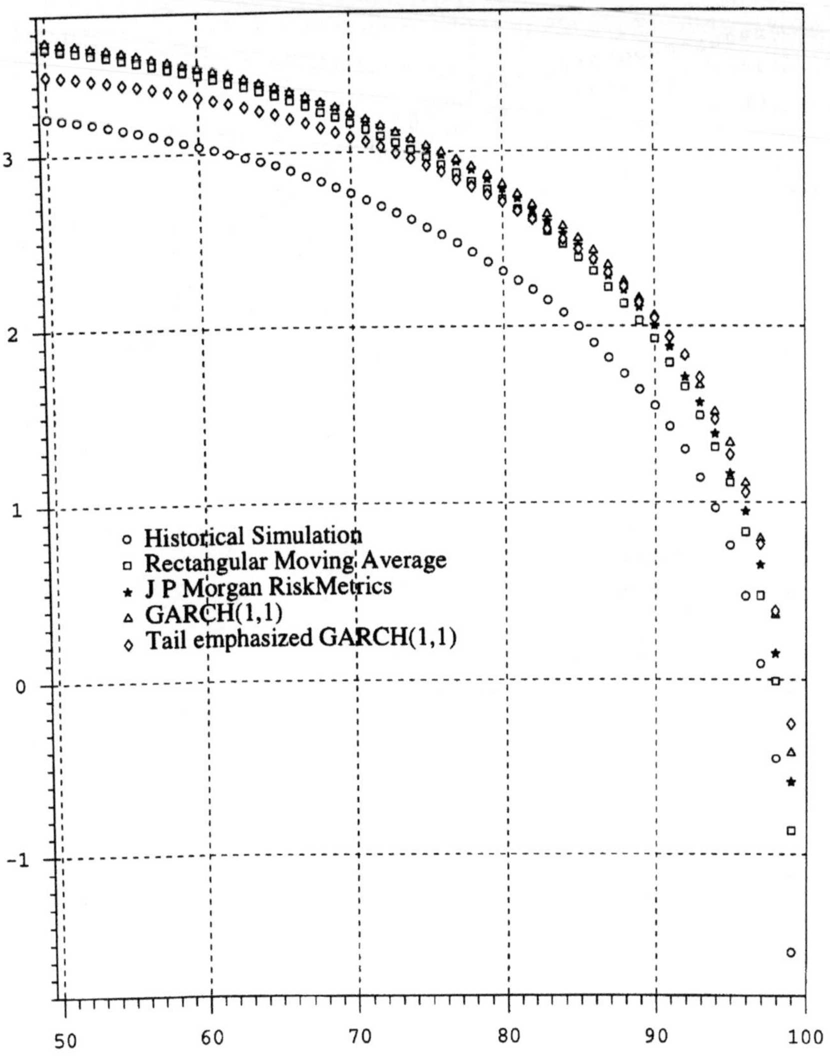

Figure 4u: Mean log-likelihood of exceedence events against confidence level in the *univariate* context for all candidate models. This plot shows the mean log-likelihood (X axis) of all exceedence events against the confidence level c (Y axis) for which the events are defined – for all the 5 models considered in this paper. Higher mean log-likelihood values at a given confidence level indicate better models and lower degrees of exceedence at that confidence level. The value plotted for a given model is the average of the mean log-likelihoods of exceedence events computed over 20 trivial portfolios – each consisting of a *long* or *short* position of the US Dollar against the 10 series k. The curves are based on the analysis of 10000 *prediction-realization pairs* in the out-of-sample period. See section 6.6 for additional comments.

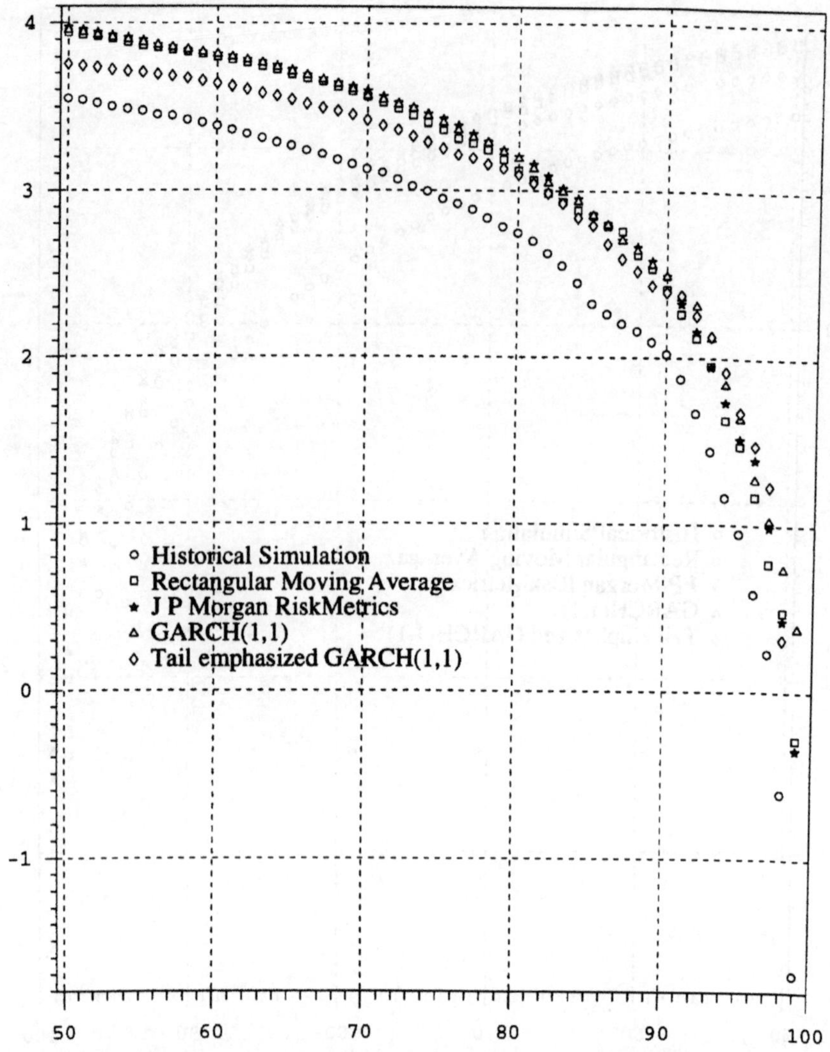

Figure 4m: Mean log-likelihood of exceedence events against confidence level in the *multivariate* context for all candidate models. This plot shows the mean log-likelihood (X axis) of all exceedence events against the confidence level c (Y axis) for which the events are defined – for all the 5 models considered in this paper. Higher mean log-likelihood values at a given confidence level indicate better models and lower degrees of exceedence at that confidence level. The value plotted for a given model is the average of the mean log-likelihoods of exceedence events computed over 2 portfolios – each consisting of uniformly *long* or uniformly *short* positions of the US Dollar against the 10 series k. The curves are based on the analysis of 1000 *prediction-realization pairs* in the out-of-sample See section 6.6 for additional comments.

which is the mean log-likelihood of all loss exceedences over confidence level c for a long (+) or short (−) position in the US Dollar with respect to the series k or portfolio P.

Finally, as specified by expressions 1 and 2, we construct the *univariate* average mean log-likelihood

$$\ell_u(c) = \frac{1}{20}\left(\sum_{k=1}^{10} \ell^{+(k)}(c) + \sum_{k=1}^{10} \ell^{-(k)}(c)\right) \tag{38}$$

and the *multivariate* average mean log-likelihood

$$\ell_m(c) = \frac{1}{2}\left(\ell^{+(P)}(c) + \ell^{-(P)}(c)\right) \tag{39}$$

plotted in figures 4u and 4m respectively for all candidate models. Plots are always for the confidence range of 50% and higher, since this defines the division between loss making and profitable events. Due to *symmetrization* (consideration of both *long* and *short*, positions) the value of $\ell_u(c = 50)$ is based on all 10000 out-of-sample events $x_t^{(k)}$ $(t \in [2, 1001])$ and the value $\ell_m(c = 50)$ is based on all 1000 out-of-sample events $x_t^{(P)}$ $t \in [2, 1001]$ $(t \in [2, 1001])$.

Comment: This relative measure, $\ell_{u,m}$, is maximized when the distribution of events in $\Phi^{\pm(k,P)}(c)$ are predicted with the correct probability by the model. Of course this applies strictly only when $\#(\Phi^{\pm(k,P)}(c))$ is large. In figures 4u and 4m we have plotted data in steps of 1% starting from 50% up to 99% – which is for increasing confidence levels but decreasing population of contributing events. It is precisely because the robustness of the statistic is in question for very high levels of confidence, that we have taken the approach of visualizing the measure as a function of the confidence level. When one model shows consistently superior performance than another in the entire high confidence level range – can we be sure that it is a superior model in the context of risk management.

The figures 4u and 4m inform us about the relative performance of the models for events deemed improbable by the model itself. As already discussed in section 2.5 the sets $\Phi^{\pm(k,P)}(c)$ for different models do not refer to the same subset of $t \in [2, 1001]$. The motivation of the next measure is to evaluate how the model performs explicitly for large price moves and to allow comparison of models over exactly the same data.

4.5 Mean log-likelihood against volatility percentile

Introduction: Instead of computing the mean log-likelihood over all events above a certain confidence level – we now compute the mean log-likelihood over all events wherein the predicted movement exceeds a certain absolute

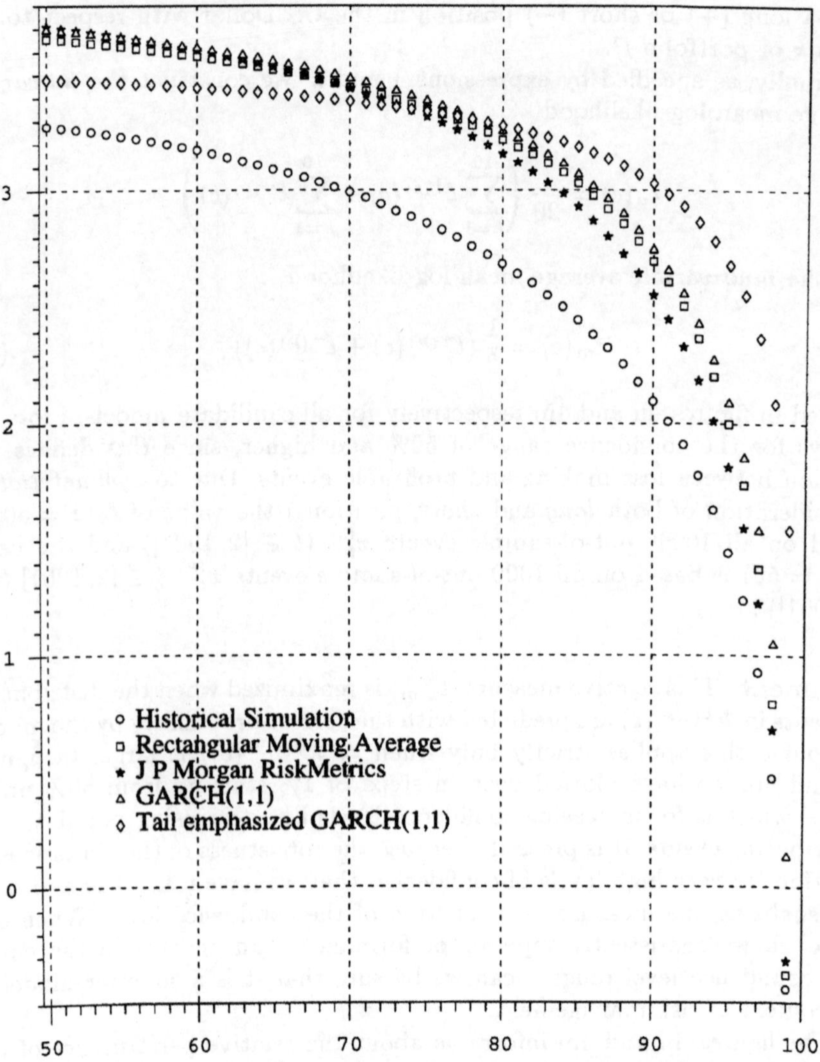

Figure 5u: Mean log-likelihood of loss making events against exceedence percentile of events in the *univariate* context for all candidate models. This plot shows the mean log-likelihood (X axis) of all loss inducing price changes above a certain percentile level in magnitude. High mean log-likelihood values at large percentile levels indicate models which have better captured the dynamics of large financial movements. The value plotted for a given model is the average of the mean log-likelihoods of events beyond a certain percentile level computed over 20 trivial portfolios - each consisting of a *long* or *short* position of the US Dollar against the 10 series k. The curves are based on the analysis of 10000 *prediction-realization pairs* in the out-of-sample period. See section 6.7 for additional comments.

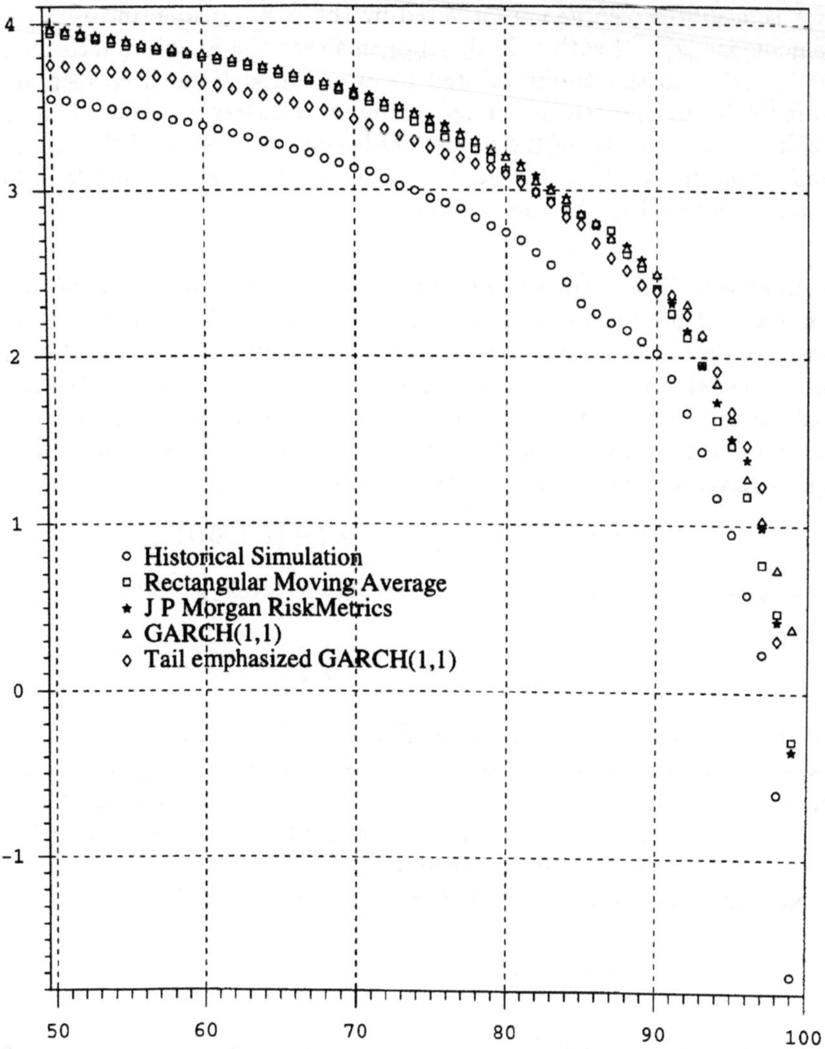

Figure 5m: Mean log-likelihood of loss making events against exceedence percentile of events in the *multivariate* context for all candidate models. This plot shows the mean log-likelihood (X axis) of all loss inducing price changes above a certain percentile level in magnitude. High mean log-likelihood values at large percentile levels indicate models which have better captured the dynamics of large financial movements. The value plotted for a given model is the average of the mean log-likelihoods of events beyond a certain percentile level computed over 2 portfolios – each consisting of uniformly *long* or uniformly *short* positions of the US Dollar against the 10 series k. The curves are based on the analysis of 1000 *prediction-realization pairs* in the out-of-sample See section 6.7 for additional comments.

size – parametrized by the percentile level of the event size. (The nomenclature *volatility percentile* is motivated by the popular identification of the movement size $|x_t^{(k,P)}|$ with volatility.) In this case the X-axis is independent of the model – so the points plotted for every model which correspond to the same volatility percentile – refer to the same extreme events. This measure tests the conformity of the model predictions to the tail of the empirical *unconditional* distribution instead of the tail of the forecast *conditional* distribution (presented in the last section).

Specification: The empirical unconditional distribution of out-of-sample events may not be symmetric around zero – especially because of long term increases or decreases in one or more series k. We must however still be able to put losses with respect to the *long* and *short* positions in the US Dollar against series k and portfolio P on equal footing when assigning percentile levels to large movements on both extremes. To impose symmetry in the percentile assignments we first construct the event sets:

$$X^{\pm(k,P)}(\chi) = \{t| \pm x_t^{(k,P)} > \chi, t \in [2, 1001]\}. \tag{40}$$

Using sets 40 we can implicitly define exceedence levels $\chi^{\pm(k,P)}(\varphi)$

$$\frac{\#(X^{\pm(k,P)}(\chi^{\pm(k,P)}(\varphi)) - \frac{1}{2}}{\#(X^{\pm(k,P)}(0))} = \frac{1}{2} + \frac{1}{2}\left(\frac{\varphi}{100}\right), \quad \varphi > 50, \tag{41}$$

where the exceedence percentile is φ. The right hand side of expression 41 imposes a mapping so that the loss making data (relative to any portfolio and position thereof) will always cover a 50% range of percentile values. (Cases of $x_t^{(k,P)} = 0$ may be ignored or assigned arbitrarily to either of sets $X^{\pm(k,P)}(0)$, without substantially affecting the results.)

Next, using expressions 23 and 36 we can directly construct

$$\ell^{\pm(k,P)}(\varphi) = \frac{\sum_{t \in X^{\pm(k,P)}(\varphi)} \mathcal{L}_{t-1}^{(k,P)}.}{\#(X^{\pm(k,P)}(\varphi))} \tag{42}$$

which is the mean log-likelihood of all loss exceedences over volatility percentile φ for a long (+) or short (−) position in the US Dollar with respect to the series k or portfolio P.

Finally, as specified by expressions 1 and 2, we construct the *univariate* average mean log-likelihood

$$\ell_u(\varphi) = \frac{1}{20}\left(\sum_{k=1}^{10} \ell^{+(k)}(\varphi) + \sum_{k=1}^{10} \ell^{-(k)}(\varphi)\right) \tag{43}$$

and the *multivariate* average mean log-likelihood

$$\ell_m(\varphi) = \frac{1}{2}\left(\ell^{+(P)}(\varphi) + \ell^{-(P)}(\varphi)\right) \tag{44}$$

plotted in figures 5u and 5m respectively for all candidate models. The construction of expression 41 guarantees that what is true naturally for the 50% confidence level (based on the forecast distribution) is also true for the 50% percentile level (of the empirical out-of-sample distribution) – that this level marks the division of profitable and loss making events in the out-of-sample. As a consequence of this and *symmetrization* (consideration of both *long* and *short*, positions) the value of $\ell_u(\varphi = 50)$ is based on all 10000 out-of-sample events $x_t^{(k)}$ $(t \in [2, 1001])$ and the value $\ell_m(\varphi = 50)$ is based on all 1000 out-of-sample events $x_t^{(P)}$ $(t \in [2, 1001])$ – as was the case for $\ell_u(c = 50)$ and $\ell_m(c = 50)$ specified in the last section.

Comment: This relative measure clearly tests the success of models when large movements are encountered. Figures 5u and 5m show that the *tail emphasized* GARCH(1, 1) stands out in this respect. In figures 5u and 5m the mean log-likelihood value at the 50% percentile level is the mean log-likelihood over all out-of-sample events. As we move to higher percentile levels, small movements are ignored. From this it is clear that the *tail emphasized* GARCH(1, 1) slightly sacrifices performance in predicting small movements for better performance when movements are large.

5 Salient features of methodology

5.1 Modeling

Implied covariances: In the GARCH based models 4 and 5 we construct covariance forecasts for series i and j by looking at the variance forecasts for the series i and j as well as the forecast for the sum series of i and j (expression 19). While applying this procedure we are well aware that formally the sum of two GARCH processes is not a GARCH process. This fact however does not intrinsically pose any caveat to the methodology – since the GARCH parameters for the three series (i, j and the sum thereof) are separately optimized in the in sample for all three – and no implicit constructions are used extending the parameters for the series i and j to their sum series.

What can however be of serious concern is that we have no guarantee that Σ_t has positive eigen-values. If Σ_t does have one or more negative eigen-values - this implies that there exists a portfolio with a forecast variance which is negative. Unless the dynamics of the market changes drastically between the in-sample and the out-of-sample, it is extremely unlikely that the method of implied covariances will lead to this pathological condition. We can confirm that in our analysis, with the fixed portfolios, we did not run into this situation – but this does not imply that Σ_t never had negative eigen-values. In section 7 we again discuss this issues of pathology in Σ_t.

Tail emphasis: In model 5 we introduce the idea of tail emphasized fitting. As explained in section 3.5 – in this method, for every candidate parameter set the mean log-likelihood is computed only over the adverse half of the total number of contributing events in the in sample. To be precise, the log-likelihood contributions of all events in the in-sample are sorted and the mean is computed over the lowest half of the set. The consequent fitness landscape is considerably fractured since small changes in parameters can lead to the inclusion and exclusion of one or more events in the computation of the mean. Without the use of genetic algorithms Dacorogna et al. (1997) – for example by using only the BHHH algorithm – it would be quite impractical to find the optimal solution for such a fitness landscape.

It is important to note that adverse log-likelihood values are not necessarily contributed by the events of largest magnitude in the in-sample data. The tail referred to in the current context – is the tail in terms of bad predictions by a model – the tail of log-likelihood contributions. The tail emphasized optimization allows the model to find a parameter set which alleviates the problem with its worst predictions by compromising on its better predictions. It makes the model more homogeneous in the quality of its predictions and as a by product of this effect this model also shows better performance for large movements.

In-sample and out-of-sample: We again emphasize that all optimization – wherever applicable – was done in an in-sample period $t \in [-1249, 0]$ and all performance evaluation was done in the out-of-sample period $t \in [1, 1001]$.

HARCH and EMA-HARCH: Although HARCH Müller et al. (1997) and EMA-HARCH Dacorogna et al. (1997) processes were constructed in order to capture the information from high frequency data, we have also studied these models (in addition to the 5 presented in detail) in the present context with daily data. The results we obtained with the 4 parameter processes HARCH(3) and EMA-HARCH(3) and the 5 parameter processes HARCH(4) and EMA-HARCH(4) (all implemented with tail emphasized optimization) showed marginal improvement when compared with tail emphasized GARCH(1, 1) (model 5). We have however not emphasized these models in the present study because we did not see the kind of improvement per degree of freedom to warrant their use with daily data. The comparative improvement of these models over tail emphasized GARCH(1, 1) when run with high frequency data – which they were designed for – is currently under investigation to be presented in a separate paper.

5.2 Performance measures

Measuring performance over extreme subsets: One of the main problems of evaluating models in the context of risk management is that the performance is gauged by looking at the few adverse events beyond the 99% confidence level or associated with the few strongest adverse price movements in relation to some portfolio. Even with a back testing history of 1000 days (as used in this paper) we expect only 10 events beyond the 99% confidence level. Statistics and performance measures based solely on these few events cannot help clearly distinguish good models from bad. Our approach of plotting model performance against confidence level or volatility percentile is motivated by exactly this problem. As discussed in section 2.5, the confidence level or the volatility percentile (plotted on the X axis of all figures) parametrize a class of increasingly extreme partial sets of events by differing definitions. By probing model performance as a *function* of increasingly extreme (but at the same time increasingly depopulated) event classes – we can get a better feel for the behaviour in the limit of extreme behaviour for that class type. We propose that the consistent superiority of one model over another over an entire range of confidence levels or volatility percentiles – approaching the 99% level (say from 75% to 99%) – increases confidence in the model. In addition by looking at the extrapolated approach to the 99% level from the less depopulated classes – we can better appreciate the error in the empirically determined measure at this level.

Separation of variance and covariance forecasting: Another important issue we cover is the separate analysis of the model with regard to variance and covariance forecasting. As already discussed in section 2.5 the results plotted in the *univariate* context determine the quality of the model for variance forecasting. The measures plotted in the *multivariate* context determine the quality of the model for variance and covariance forecasting. If it has been determined that the model has good performance in the *univariate* context but bad performance in the *multivariate* context – this must mean that some modifications are necessary only in the covariance forecasting aspect of the model. Aside from the latter important information – the analysis in the *univariate* context is based on far more data and helps give more robust estimates of variance forecasting performance by a model even at the 99% level where exceedence data is sparse. (Note how the figures m are noisier than figures u.)

Symmetrization: As part of our performance measurement methodology we propose that measures must always be presented as a mean over opposing portfolios. As discussed in section 2.2 this procedure gets rid of any bias in the measure arising out of long term increases or decreases in the series k.

Increasing robustness – averaging over more portfolios: For completeness we mention here that we have evaluated the models in the *multivariate context* as an average over 2 portfolios with opposing position vectors

– equally weighted and *long* on the US Dollar against all series k and equally weighted and *short* on the US Dollar against all series k. For a more complete analysis it is viable to extend the methodology by averaging over a larger set of position vectors (such as for example on a sphere of dimension k with some appropriate radius), always ensuring that the symmetrically opposed portfolio is also included.

6 Discussion and results

In this section we present a number of possibly disjoint but important comments with the purpose of achieving two objectives. One is to discuss the results of how the various models performed in comparison with each other. The second is to discuss the details of what the various performance measures convey, how they complement each other and what their drawbacks are. For convenience we shall refer to the models by the numbers associated with them in section 2.4 and to the performance measures by the numbers associated with them in section 2.5. The numbers are also consistent with the order in which we presented the models and the measures.

6.1 Measures sensitive to number of exceedence

Performance measures 1, 2 and 3 are *digital* measures sensitive to the concept of exceedence as defined by the event set Φ (expression 23) and are entirely based on the cardinality of these or similar (Ψ for measure 3) sets as a function of confidence level c. As already remarked (in the detailed specification) – these measures are oblivious to the *degree* of exceedence – and hence lacking in this respect. It is quite possible that two models may have the same number of exceedences – even at the same points in time – but one model barely overshoots the exceedence limit $\chi(c)$ whereas another goes well beyond. Clearly measures 1, 2 and 3 will be unable to discriminate between two such models.

6.2 Measures sensitive to degree of exceedence

To capture the degree of exceedence – we have presented performance measures 4 and 5. The use of the mean log-likelihood of exceedent or large events (measures 4 and 5 respectively) in the out of sample is a sophisticated measure. For the conditional Gaussian distributions used in models 2 through 5, the probability density is strictly monotonically decreasing on both sides of the zero mean – and we are thus guaranteed that as desired $\ln(p_t(x_{t+1}))$ for each exceedent event becomes increasingly negative as the degree of exceedence of x_{t+1} increases. In the case of historical simulation however (model 1), this is not strictly true – since p_t can be and is most likely multi-modal. If according to the model – a large event is more likely than a smaller event of lower size (as in the case of a multi-modal distribution) – and if the out-of-sample data confirms this distribution – then this is a truth which is captured by the model and the model should not be penalized for such an occurrence. On the other hand, if the conditional distribution is multi-modal, but the realizations do not fit this distribution – then the model should be penalized. This is exactly what is achieved by the mean log-likelihood measures 4 and 5 – by expressing exceedence as a function of the probability density of the event rather than simply looking at the size of the exceedence. In general, for large populations, the mean log-likelihood maximizes when the predicted and realized distributions of an event class match up.

6.3 Measure 1

It is interesting to note that when viewed in terms of measure 1, historical simulation does appears to be a reasonable model which neither significantly overestimates or underestimates risk over the full range of confidence levels. In fact, on the basis of this measure, one could perhaps reject model 5 (tail emphasized GARCH(1, 1)) as being too conservative – in the sense of always over-estimating risk. In general the message from figures 1u and 1m is that models 1 and 3 have similar behaviours at low confidence level but in the regime beyond the 95% level – tables are turned and – model 1 becomes far superior than all models except model 5. We also find that there is not much too choose between models 2 and 4.

6.4 Measure 2

Measure 2 is interesting only because of its relationship to BIS recommendations. In terms of this measure, if we looked only at the 99% level – which is the legally defined relevant level – the best model is clearly model 5 – based on figure 2m for the *green* zone frequency. This is however not unexpected in view of the appearance that model 5 is extremely conservative in terms of figures 1u and 1m. While model 5 is the only one which is 100% in the *green*, figure 2m shows that none of the models – barring model 2 – are ever significantly in the *red* zone.

6.5 Measure 3

In spite of the fact that measure 3 is what we refer to as a *digital* measure (like measures 1 and 2) – it brings brings out an important point which is missed by measures 1 and 2. This is the clustering property of exceedences. This measure determines if the frequency of consecutive exceedences is higher than expected at a given confidence level. It is expected that models 1 and 2 which make no attempt to capture the autocorrelation of volatility should behave the worst in terms of this measure. It is therefore surprising to find that model 3 fares worse than model 1 with respect to this measure. (The consecutive loss exceedence ratio of model 2 in the *univariate* context at the 99% confidence level is 11.01 – not plotted in figure 3u for clarity of presentation.)

We have already pointed out in section 5.2 the serious caveat with regard to low number of exceedence events when focusing on the 99% confidence levels. We again emphasize that while it is legitimate that any regulatory authority should be looking at these high levels for the evaluation of risk on a day to day basis once a model is accepted – the criteria of accepting this model should be based on performance of the model over an entire range of extreme levels starting from say the 75% level. Figure 3m provides an example of this problem. In figure 3m we have only plotted data when there is at least one exceedence event beyond a certain confidence level (contributed by at least one of the two portfolios over which the measurement is averaged). Thus at the 99% level the absence of any entry in figure 3m indicates that none of the models admit a single event of consecutive exceedences over a 1000 day out-of-sample period. It is only when viewing the approach to the 99% level that we can pick out the relative performance of the models and the consistent superiority of model 5.

6.6 Measure 4

Measure 4 looks at the mean log-likelihood of all exceedent events beyond the confidence level plotted on the X axis. We present figures 4u and 4m only to illustrate the difference in relation to measure 5 which we strongly recommend as a solid performance measure. While the mean log-likelihood against confidence level does help in establishing the *degree* of exceedence – it fails as a valid comparative measure between models. This is because the exceedent events referred to by a specific choice on the X axis do not correspond to the same set of events for different models. Differing sets of exceedent events at the same confidence level imply differing theoretical maximums for the mean log-likelihood of exceedence – making direct comparison fallacious. (In addition we note that figure 4m has no expression of mean log-likelihood for model 5 at the 99% level because both contributing portfolios do not contribute an exceedent event. This again highlights the problem of passing legislation based entirely on evaluations at the 99% level.)

6.7 Measure 5

What is most important in the context of risk management is *not* whether a large number of run-of-the-mill movements were predicted well at the expense of a few large movements – but exactly the contrary. In fact, we can well afford a general over-estimate or under-estimate of risk with regard to small movements in exchange for better performance in the tail. Measure 5 presents the mean log-likelihood of all movements beyond a certain percentile level rather than all movements exceeding a certain confidence level. As discussed in section 5.1 the results of model 5 in figures 5u and 5m clearly show that tail emphasized optimization used in model 5 achieves decreased sensitivity of forecasting performance to the size of the event as desired. In addition, figures 5u and 5m are the only figures presented where the X and Y axis are not both model dependent. The X axis depends only on the data and the corresponding mean log-likelihoods refer to the same set of events – making comparison of models meaningful.

6.8 Model 1

In terms of the measures presented it turns out that historical simulation is not as bad as one would expect for such a simplistic model – which makes no attempt to capture the dynamics of the market and which is based entirely on the series $x^{(P)}$ without referring to the characteristics of the individual risk factors $x^{(k)}$. It has reasonable results in terms of the *digital* measures but has very bad quality in terms of the *degree* of exceedence as reported in figures 4 and 5. We believe that by introducing smoothing in the conditional distribution determined from the 250 day empirical distribution – with the constraint that the probability density must monotonically decrease away from the center – the performance of this model can be considerably improved. If in addition the smoothed 250 day empirical distribution based on X_t (expression 5) could be modulated so as to confirm to the variance of a GARCH(1, 1) process (optimized with tail emphasis) then the resultant model may be extremely powerful. The caveat of such a scheme however is the complexity introduced in the portfolio function if this methodology is employed in the usual way – where the stochastic process predicts each series $x^{(k)}$ rather than directly predicting the portfolio series $x^{(P)}$. For now, we leave this proposal of such a hybrid approach – mixing historical simulation and stochastic methods – as food for thought for future investigations.

6.9 Model 2

Rectangular simulation is the simplest of the models employing a covariance matrix at the basis of the VaR forecast and completely satisfies the BIS framework. The individual analysis of each series and series pair for the determination of the covariance matrix is the most important element distinguishing this and other models from historical simulation (model 1). This model (like historical simulation) makes no attempt to capture the dynamics of the financial markets – particularly with regard to the observed autocorrelation of volatility.

6.10 Model 3

We have already pointed out that the J. P. Morgan RiskMetrics process is a non-stationary GARCH(1, 1) process. The generally poor performance of this process which is fully multivariate in the sense of being based on the construction of a covariance matrix and which does also attempt to capture the autocorrelation of volatility is rather surprising. The main drawback of this process is *possibly* the en-masse assignment of the same parameter set for all financial series. In defense of the en-masse RiskMetrics setting of $\beta_1 = 0.94$ we would like to point out that the parameters for tail emphasized GARCH(1,1) in table 1 may very well be approximated by an en-masse setting of $\beta_1 \approx 0.96$ but with non-zero and diverse minimum variance α_0. It would seem that adding a minimum variance ($\alpha_0 \neq 0$) to the RiskMetrics variance prediction might go a long way in improving its performance. We believe that the difference in values for β_1 may arise simply out of our portfolio choice which does not have the variations in instrument class which are attempted to be captured by the RiskMetrics settings.

6.11 Models 4 and 5

Table 1 shows the comparison between the parameters fit for GARCH(1,1) and tail emphasized GARCH(1,1). The main observation is that α_0 for tail emphasised GARCH(1,1) – which represents a minimal variance – is considerably larger than regular GARCH(1,1). Also, by and large, β_1 is comparatively smaller for the individual series but comparatively larger for the sum series for tail emphasised GARCH(1,1). This opposing movement of β_1 on the individual and sum series as a consequence of tail emphasis may point to possible volatility conditional effects on covariance to be discussed in paper Dave et al. (1997).

While increasing the size of the minimal variance α_0 may appear as a panacea for bad performance from the regulatory perspective – it should be noted that the mean log-likelihood will begin to deteriorate rapidly – if α_0 is increased further – causing the model to predominantly over-estimate risk. Thus, in our view, the mean log-likelihood function achieves the right balance for penalizing risk underestimation and risk overestimation. In the beginning of this section we had commented that figures 1u and 1m make model 5 appear too conservative. Through figure 5u and 5m we can appreciate however that model 5 is reasonable and particularly so in the context of large movements. This highlights the deficiency of *digital* measures like exceedence count as the sole criteria for comparative evaluation of model performance in the context of risk management.

7 Problems of pathological VaR estimates

In this section we comment on various sources of pathological estimates of VaR and finally speculate on a framework for defining and handling pathologically underestimated risk.

7.1 Rank defects

In typical implementation of VaR methodologies it is not uncommon that the number of risk factors involved, d, is far greater than the number of data, n, in the history used for forecasting the covariance matrix Σ_t (of dimension d). In particular, if $n < d$ and Σ_t is constructed with the rectangular moving average model (model 2) of section 3.2 or the RiskMetrics model (model 3) of section 3.3 it is guaranteed that the covariance matrix will have a rank defect. The existence of a rank defect implies that Σ_t admits the existence of a portfolio (or a set of portfolios) $\lambda = \lambda(\Sigma_t)$ such that $\lambda^T \Sigma_t \lambda = 0$ (or $\mathrm{VaR}(\lambda, \Sigma_t, c) = 0$ for all c, where c is the confidence level used – usually 99%).

In the rectangular moving average model every $x_{t-i}^{(j)}$ entering the computation 12 of $s^{(jk)}$ for any k does so with the same weight – unity. In the J. P. Morgan RiskMetrics case the computation 15 expresses $s^{(jk)}$ in recursive form - but when expanded in terms of the data history it is stll true that every participation of $x_{t-i}^{(j)}$ in the construction of the covariance matrix occurs with the same weight or coefficient (although not unity).

In general - any covariance matrix construction which can be expressed in the form:

$$\Sigma_t = W_t^T W_t \quad \text{where}$$

$$W_t^T = \begin{pmatrix} w_t^{(1)} x_t^{(1)} & w_{t-1}^{(1)} x_{t-1}^{(1)} & \cdots & w_{t-n+1}^{(1)} x_{t-n+1}^{(1)} \\ w_t^{(2)} x_t^{(2)} & w_{t-1}^{(2)} x_{t-1}^{(2)} & \cdots & w_{t-n+1}^{(2)} x_{t-n+1}^{(2)} \\ \vdots & \vdots & \vdots & \vdots \\ \vdots & \vdots & \vdots & \vdots \\ w_t^{(d)} x_t^{(d)} & w_{t-1}^{(d)} x_{t-1}^{(d)} & \cdots & w_{t-n+1}^{(d)} x_{t-n+1}^{(d)} \end{pmatrix} \tag{45}$$

(W^T is a matrix of the weighted data history) will suffer from a rank defect if $n < d$. (Theoretically, the rank of Σ_t is the the rank of W_t – which is the minimum of d and n. If $n < d$, then the rank n of Σ_t is smaller than its dimension d and hence Σ_t must have a rank defect.)

7.2 Near singularity

We have illustrated above that the rectangular moving average model satisfying the BIS parameters (model 2) as well as the J. P. Morgan RiskMetrics model (model 3) necessarily have rank defects if the number of risk factors d exceeds the number of historic data n from which the correlation matrix is constructed. This *overt* existence of the rank defect however points to a deeper problem. Even when the methodology is more complex – such as when we are using model 4 or 5 with implied covariances – while it is unlikely that Σ_t is exactly singular (and hence admits a rank defect), it can very well be *near* singular. This means that while there will not exist a portfolio λ (aside from the trivial null portfolio $\lambda_i = 0$) such that $\mathrm{VaR}(\lambda, \Sigma_t, c) = 0$ (for all c)

there will still exist portfolios for which risk is pathologically underestimated (as defined in section 7.6).

7.3 Stochastic errors

Even if we have very good models (and even if $n > d$), simply due to inevitable stochastic errors in the estimation of Σ_t it is possible for the covariance matrix to become near singular and admit portfolios for which risk is pathologically underestimated. More often however, stochastic errors will lead to spurious minima in the VaR – which can also lead to a systematic underestimation of risk as explained in the next subsection.

7.4 Bias due to portfolio optimization

In our entire analysis of models and measures in the earlier sections, we worked with fixed, pre-defined portfolios. During this analysis, spurious minima could have occasionally appeared in the vicinity of the fixed portfolios leading to risk underestimation but spurious maxima at other times would have largely balanced this effect in the overall performance evaluation of the model.

In typical scenarios of the application of VaR methodology, however, the actual portfolio is determined with the help of the *estimated* covariance matrix Σ_t. This means that, on a frequent basis, investors are likely to take advantage of rank defects, near singularities and spurious minima due to stochastic errors – and tend to select portfolios in that portion of the portfolio space which necessarily underestimates risk.

7.5 Negative variance

We have already mentioned in section 5.1 the possibility for Σ_t having negative eigen-values and hence admitting the existence of portfolios for which the forecast variance of assets is negative. This is clearly a pathological situation that requires to be addressed. Below we present a preliminary framework for addressing all of the above problems which also serves to partially address the problem of negative variance.

7.6 Treatment of pathological VaR estimates

All the problems listed above deal with the existence of unreliable regions of portfolio space in which risk is underestimated due to estimation errors in the covariance matrix. Our proposal to handle this problem is to determine $\kappa(\lambda, \Sigma_t, c)$ as the minimum acceptable risk for a portfolio at confidence level c. The reported risk (at level c) is then the larger of $\mathrm{VaR}(\lambda, \Sigma_t, c)$ and $\kappa(\lambda, \Sigma_t, c)$. As a corollary, we assert that $\mathrm{VaR}(\lambda, \Sigma_t, c)$ is defined as pathologically underestimated if $\mathrm{VaR}(\lambda, \Sigma_t, c) < \kappa(\lambda, \Sigma_t, c)$.

To build this machinery we first introduce the function:

$$\nu(\lambda) = \sum_{i=1}^{d} |\lambda_i| \tag{46}$$

and refer to this as the *collateral function* – since this function approximates the collateral needed to support the investment. (The picture here is consistent with the view that trading is conducted without leverage and via a broker who allows one to take long or short positions on paper – as long as the value of the collateral function is deposited with the broker. The true collateral may be different due to several reasons – but modeling the true collateral is tangential to our objective.) Our purpose is simply to define a meaningful norm, $||\lambda|| = \nu(\lambda)$, against which the size of $\mathrm{VaR}(\lambda, \Sigma_t, c)$ may be compared, and which satisfies the condition that the ratio

$$\frac{\mathrm{VaR}(\lambda, \Sigma_t, c)}{\nu(\lambda)} \tag{47}$$

is invariant under scalar multiplication of λ (true because both the numerator and the denominator satisfy the property, $f(\alpha\lambda) = \alpha f(\lambda)$).

The scale invariance of the ratio 47 has the important repercussion that distinct values must correspond to distinct *directions* in the portfolio space of dimension d (from the origin). Since only *direction* matters, for further analysis it is convenient to define the surface S of normalized portfolios, as the locus of all portfolios for which $\nu = 1$. In this notation, the components of $\lambda^{(s)} = \frac{\lambda}{||\lambda||} \in \mathcal{S}$ may be denoted:

$$\lambda_i^s = \frac{\lambda_i}{\sum_{i=1}^{d} |\lambda_i|}. \tag{48}$$

(For example, if $d = 3$, then $\mathcal{S}$ is a regular octahedron joining the points on each axis at positive and negative unity.)

We still need two more mathematical objects to define the methodology for determining $\kappa(\lambda, \Sigma_t, c)$. One is the function $\mathcal{P}$ defined by:

$$\mathcal{P}(\lambda^{(1)}, \lambda^{(2)}) = \frac{1}{d} \sum_{i=1}^{d} \log \left(\frac{\lambda_i^{(1)}}{\lambda_i^{(2)}} \right)^2 \tag{49}$$

which measures the strength of a perturbation that transforms $\lambda^{(1)}$ to $\lambda^{(2)}$. The other is the more complicated object, λ_o^s, determined as a function of λ^s. This object has different definition depending on whether $\mathrm{VaR}(\lambda^s, \Sigma_t, c)$ is negative or positive.

- If $\mathrm{VaR}(\lambda^s, \Sigma_t, c) > 0$ then λ_o^s is the local minimum for the VaR on surface $\mathcal{S}$ which can be reached from λ^s via a continuous path of decreasing VaR. If the local minima is negative – λ_o^s is the first point on the steepest descent where VaR becomes zero.

- If $\mathrm{VaR}(\lambda^s, \Sigma_t, c) < 0$ then λ^s_o is the point on the surface $\mathcal{S}$ such that $\mathrm{VaR}(\lambda^s_o, \Sigma_t, c) = 0$ which requires the smallest perturbation $\mathcal{P}(\lambda^s, \lambda^s_o)$.
- If $\mathrm{VaR}(\lambda^s, \Sigma_t, c) = 0$ then $\lambda^s_o = \lambda^s$.

We are now in a position to describe a procedure for reporting VaR which largely handles the underestimation problems when reporting risk:

1. From λ, determine $\nu(\lambda)$ and $\lambda^{(s)} = \frac{\lambda}{\|\lambda\|} \in \mathcal{S}$.
2. From $\lambda^{(s)}$, determine λ^s_o.
3. From λ^s_o follow the path of steepest ascent (of VaR) – constrained to the surface $\mathcal{S}$ – until $\tilde{\lambda}^s$ such that $\mathcal{P}(\lambda^s_o, \tilde{\lambda}^s) = p$, where p is a pre-established (but ad-hoc) perturbation strength.
4. From $\tilde{\lambda}^s$, define:
$$\kappa(\lambda, \Sigma_t, c) = \nu(\lambda)\mathrm{VaR}(\tilde{\lambda}^s, \Sigma_t, c) \tag{50}$$
5. The reported VaR at confidence level c is then the larger of $\mathrm{Var}(\lambda, c)$ and $\kappa(\lambda, \Sigma_t, c)$.

Admittedly, our skeletal presentation above, is not an exhaustive consideration of all possible topological complexities that can arise – particularly when Σ_t admits negative eigen-values. Our motivation, in this section, is only to introduce the concept of minimum acceptable risk for a portfolio λ with respect to perturbation level p – as a useful concept which can go a long way in handling the various problems discussed in this section.

8 Conclusions

- Model judgement is dramatically improved by visualization of model performance over an adequately large range of high confidence levels.
- A sophisticated measurement strategy should include the following in addition to the conventional measures based on exceedence counts:
  - Measures (or variants thereof) sensitive to degree of exceedence (such as measures 4 and 5).
  - Measures discriminating model behaviour for large and small movements (such as measure 5).
  - Measures which assess model behaviour with respect to market dynamics (such as measure 3).
- The comparative performance of models in the *univariate* and *multivariate* contexts provide a systematic approach to model building which allows discrimination between the quality of the diagonal and off-diagonal elements of the covariance matrix.
- The method of tail-emphasized optimization brings striking improvement in model performance – particularly for the large movements to which VaR methodologies are most vulnerable.

- The usage of *implied covariances* in models 4 and 5 to estimate Σ_t has shown remarkable stability and performance.
- The issues of underestimated VaR due to rank defects or near-singularity of Σ_t, and stochastic errors in estimation of Σ_t – particularly in conjunction with frequent portfolio adjustments – are non trivial.
- Underestimated VaR may be improved by analyzing VaR as a function of portfolio space in the neighbourhood of the candidate portfolio.

Acknowledgements

The authors acknowledge useful discussions with Giuseppe Ballocchi, the software support provided by Olivier V. Pictet and the careful review of the manuscript by Michel M. Dacorogna.

References

C. Alexander. History debunked. *RISK*, 1994

C. Alexander. After the event: A review of the volatility and correlation forecasts in the third edition of RiskMetrics. *Risk*, 1995.

C. Alexander. *Handbook of Risk Management and Analysis.* Wiley, 1996.

C. Alexander and A. M. Chidumba. Multivariate orthogonal factor garch. *Preprint*, 1995.

P. Artzner, F. Delbaen, J. M. Eber and D. Heath. A characterization of measures of risk. H Risk 1997.

Basle Euro-currency standing committee of the central banks of ten countries, Bank of International Settlements Public disclosure of market and credit risks by financial intermediaries. *Fisher report*, 1994.

Basle Committee on Banking Supervision. Amendment to the Capital Accord To Incorporate Market Risks, 1996.

R. A. Beckström and A. R. Campbell, editors. *An Introduction to VaR.* Cats Software, Palo Alto, CA, 1995.

T. S. Beder. VaR: Seductive but dangerous. *Financial Analysts Journal*, 12–24, 1995.

E.K. Berndt, B.H. Hall, R.E. Hall, and J.A. Hausman. Estimation and inference in nonlinear structural models. *Annals of Economic and Social Measurement*, 3:653–665, 1974.

Tim Bollerslev, Ray Y. Chou, and Kenneth F. Kroner. ARCH modeling in finance. *Journal of Econometrics*, 52:5–59, January 1992.

W. Bühler, O. Korn and A. Schmidt. Ermittlung von Eigenkapitalanforderungen mit internen Modellen. *Preprint, University of Mannheim, Germany*, 1997.

M. Chen T. S. Y. Ho and F. Eng. VaR analytics: Portfolio structure, key rate convexities and var betas. *Journal of Portfolio Management*, 89–98, Fall 1996.

C. Crnkovic and J. Drachman. A universal tool to discriminate between risk management techniques. *Risk*, 1996.

Michel M. Dacorogna, Ulrich A. Müller, Richard B. Olsen, and Olivier V. Pictet Internal document MMD.1997-01-08, Olsen & Associates, Seefeldstrasse 233, 8008 Zürich, Switzerland, January 08 1997.

Rakhal D. Davé, Gerhard Stahl, Giuseppe Ballocchi, Olivier V. Pictet and Richard B. Olsen. Volatility conditional correlation between financial markets. *Paper presented at 6th International Workshop on Parallel Applications in Statistics and Economics, PASE-97, Mariánské Lázně, November 9-12, 1997*, 1997.

F. Drudi, A. Generale and G. Majoni. Sensitivity of VaR measures to different risk models. *Preprint, Banca d'Italia*, 1996.

E. Eberlein, U. Keller and K. Prause. New insights into smile – mispricing and value at risk: the hyperbolic model. *Preprint, Freiburger Zentrum für Datenanalyse und Modellbildung*, 1997.

E. Ghysels, A. Harvey and E. Renault. Stochastic volatility. *CORE discussion paper No. 9569*, 1995.

Group of thirty. Derivatives: Practices and principles. New York, 1993.

D. Hendricks. Evaluation of value-at-risk models using historical data. *Risk measurement and systemic risk: Proceedings of a joint central bank research conference, Washington D. C, November 16-17, 1995*, 1995.

P. Jackson, D. Maude and W. Perraudin. Value-at-risk techniques: An empirical study. *Risk measurement and systemic risk: Proceedings of a joint central bank research conference, Washington D. C, November 16-17, 1995.*

F. Jamshidian and Y. Zhu. Scenario simulation: Theory and methodology. *Journal of Finance and Stochastics*, 1:43–68, 1997.

P. Jorion. Value at risk. *IRWIN, Chicago*, 1997.

P. Kupiec. Techniques for verifying the accuracy of risk management models. *Risk measurement and systemic risk: Proceedings of a joint central bank research conference, Washington D. C, November 16-17, 1995.*

José A. Lopez. Regulatory evaluation of value-at-risk models. *Research Paper # 9710, Federal Reserve Bank of New York*, 1997.

J. M. Mahoney. Empirical based versus model-based approaches to value-at-risk: an examination of foreign exchange and global equity portfolios. *Risk measurement and systemic risk: Proceedings of a joint central bank research conference, Washington D. C, November 16-17, 1995.*

C. Marshall and M. Siegel. Value at risk: Implementing a risk measurement standard. *Preprint, MIT Finance Research Center*, 1996.

J. P. Morgan (Risk management advisory) and Reuters Ltd. (International marketing). RiskMetrics – technical document, fourth edition. 1996.

U. A. Müller, M. M. Dacorogna, R. B. Olsen, O. V. Pictet, M. Schwarz, and C. Morgenegg Statistical study of foreign exchange rates, empirical evidence of a price change scaling law, and intraday analysis. *Journal of Banking and Finance*, 1:1189–1208, 12, 1990.

Ulrich A. Müller, Michel M. Dacorogna, Rakhal D. Davé, Richard B. Olsen, Olivier V. Pictet, and Jakob E. von Weizsäcker. Volatilities of different time resolutions – analyzing the dynamics of market components. *Forthcoming in the Journal of Empirical Finance*, 1997.

Confidence Intervals for the Value-at-Risk

Stefan Huschens

TU Dresden, Fakultät Wirtschaftswissenschaften, 01062 Dresden, Germany

Abstract. Exact and asymptotic confidence intervals for the Value-at-Risk (VaR) are derived in a parametric context with linear portfolio structure and multinormal distributed returns.[1]

Keywords. Value-at-Risk, VaR, confidence interval

1 Introduction

The determination of Value-at-Risk (VaR) is from a statistical point of view not a simple computation but a statistical point estimation of an unkown parameter of the loss distribution. Point estimates of VaR should be completed by interval estimates because sampling errors are not negligable for realistic volatilities and applicable sample sizes.

The admission of in-house models for the banking sector to calculate capital requirements for market risk[2] imposes the use of backtesting procedures[3] based on the comparison of daily losses with daily estimated VaR. The national banking supervision fixes a bank specific multiplier depending on the results of backtesting. Capital requirements are computed with this multiplier from VaR. Suggested backtesting procedures[4] base on the number of so called exceptions which result when daily losses exceed estimated VaR. This backtesting approach treats daily losses as random variables wheras the sampling error of the VaR estimates are ignored. An exception might result not only from a very high daily loss but also from a loss coinciding with an underestimated VaR caused by the sampling error. This motivates one to analyze the sampling error and to complete point estimates with interval estimates. If exceptions should signal only model errors but not sampling errors, a useful approach is to define an upper confidence bound for VaR. Daily losses above this bound signal exceptions which are with high probability not caused by sampling errors.

---

[1] A german version is published in the *Proceedings of the Machine Learning and Data Mining Days* (December 16/17, 1996, Daimler-Benz AG, Forschung und Technik, Ulm) Physica-Verlag, Heidelberg, 1997. G. Nakhaeizadeh (Hrsg.), Data Mining, Theoretische Aspekte und Anwendungen, Physica-Verlag, Heidelberg, 1998.

[2] Basle (1996a).

[3] Basle (1996b).

[4] E.g. Basle (1996b), Kupiec (1995).

In this article confidence intervals for VaR are derived in a standard model with linear portfolio structure and multinormal distributed returns. This allows the use of parametric estimation methods.[5]

Let us start with a portfolio of K assets with quantities $b_1, b_2, \ldots, b_K$ held constant in time. The price of asset j ($j = 1, \ldots, K$) at time t ($t = 0, 1, \ldots, n, n+1$) is denoted by $p_{t,j}$. The daily loss from today ($t = n$) to tomorrow ($t = n+1$) is the difference of portfolio values,

$$L_{n+1} = \sum_{j=1}^{K} b_j p_{n,j} - \sum_{j=1}^{K} b_j p_{n+1,j} = \sum_{j=1}^{K} b_j (p_{n,j} - p_{n+1,j})$$
$$= -\sum_{j=1}^{K} w_j X_{n+1,j} = -\mathbf{w}'\mathbf{X}_{n+1},$$

where

$$w_j := b_j p_{n,j}$$

denotes the market value of the j-th position at time $t = n$,

$$X_{t,j} := \frac{p_{t,j} - p_{t-1,j}}{p_{t-1,j}} \qquad (t = 1, \ldots, n, n+1) \tag{1}$$

denotes daily returns of asset j, and

$$\mathbf{w} := (w_1, \ldots, w_K)', \qquad \mathbf{X}_t := (X_{t,1}, \ldots, X_{t,K})'.$$

The computation of VaR bases on the probability distribution of L_{n+1} conditionally on prices known at time $t = n$.

Assumption 1: $\mathbf{X}_t$ ($t = 1, \ldots, n, n+1$) are independent and identically multivariate normal distributed,

$$\mathbf{X}_t \sim \mathcal{N}_K(\boldsymbol{\mu}, \boldsymbol{\Sigma}). \tag{2}$$

Assumption 1 implies that daily losses follow a normal distribution,

$$L_{n+1} = -\mathbf{w}'\mathbf{X}_{n+1} \sim \mathcal{N}(\mu, \sigma^2) = \mathcal{N}(-\mathbf{w}'\boldsymbol{\mu}, \mathbf{w}'\boldsymbol{\Sigma}\mathbf{w}). \tag{3}$$

The VaR for a given prediction probability p (e. g. $p = 99\%$) can be defined in general as p-quantile of the loss distribution,[6]

$$\nu_p := \mu + z_p\sigma = -\mathbf{w}'\boldsymbol{\mu} + z_p\sqrt{\mathbf{w}'\boldsymbol{\Sigma}\mathbf{w}}, \tag{4}$$

[5] Confidence intervals based on nonparametric quantile estimation are described by Ridder (1996) and Jorion (1996).

[6] This definition supposes $\mu + z_p\sigma > 0$. To comprise the less interesting case $\mu + z_p\sigma \leq 0$ too, where expected profit is so high that a loss is possible only with probabilty less than or equal $1 - p$, VaR has to be defined as $\max\{0, \nu_p\}$.

where z_p denotes the p-quantile of the standard normal (e. g. $z_{0.99} = 2.326$). The daily loss L_{n+1} exceeds ν_p with probability $1 - p$,

$$\mathbf{P}(L_{n+1} > \nu_p) = 1 - p. \tag{5}$$

There are two ways to use n independent K-dimensional observations $\mathbf{X_t}$ $(t = 1, \ldots, n)$ to estimate ν_p.

Firstly, fictitious losses

$$L_t = -\mathbf{w}'\mathbf{X}_t, \quad t = 1, \ldots, n \tag{6}$$

are computed from present market values $\mathbf{w}$ and historical price changes $\mathbf{X_t}$, so that the parameters of the loss distribution can be estimated from these losses.

Secondly, the $\mathbf{X}_t$ are used to estimate the parameters of (2) and with (4) one gets plug-in-estimators for $\boldsymbol{\mu}$, σ and ν_p. The first way should be used, if the distribution of L_{n+1} and its properties are of interest for a fixed given portfolio. The second way is more useful if the VaR should be computed for alternative portfolio weights $\mathbf{w}$ with the same estimated covariance matrix. Using standard estimators both ways lead to the same result.[7]

The second section of this article deals with exact and asymptotic confidence intervals for the case of known μ and estimated σ. In the third section exact and asymptotic confidence intervals are derived for the case where μ and σ are estimated. The fourth section compares these intervals.

2 Confidence intervals with known means

The mean μ is known, if we assume that the vector $\boldsymbol{\mu}$ is zero in all components.[8] The case $\mu = 0$ also arises, if VaR is defined not as p-quantile of the loss distribution, but as difference between the p-quantile and the mean of the loss distribution.[9]

If $\boldsymbol{\mu} \neq \mathbf{0}$ a trend in prices can be considered. For example, it could be appropriate, to suppose for some portfolio a growth with the risk free interest rate. With $\boldsymbol{\mu}$ the parameter $\mu = -\mathbf{w}'\boldsymbol{\mu}$ of the loss distribution is known too.

---

[7] Compare (8) and (14) to (16).

[8] Morgan/Reuters (1996) assume in general, that daily returns have zero expectation. Cp. p. 18, 51, 59, 74, 81, 90, 92 and 132. In literature it is proposed (cp. Morgan/Reuters (1996), p. 92 and Jorion (1997), p. 79), to estimate σ and VaR in setting $\mu = 0$ if μ is expected to be near to zero, because then the induced bias of variance estimation can be ignored. This recommandation seems to be problematically because of the fact that the tradeoff between the bias, if μ is not estimated, and the additional sampling error, if μ is estimated, depends on the unknown proportionality between μ and σ.

[9] E. g. in Jorion (1996).

If μ is known, an appropriate estimator for ν_p is

$$\hat{\nu}_p = \mu + z_p \hat{\sigma} \tag{7}$$

with

$$\hat{\sigma} = \sqrt{\frac{1}{n}\sum_{t=1}^{n}(L_t - \mu)^2} = \sqrt{\mathbf{w}'\hat{\boldsymbol{\Sigma}}\mathbf{w}} \tag{8}$$

and

$$\hat{\boldsymbol{\Sigma}} = \frac{1}{n}\sum_{t=1}^{n}(\mathbf{X}_t - \boldsymbol{\mu})(\mathbf{X}_t - \boldsymbol{\mu})'. \tag{9}$$

Equation (8) shows, that $\hat{\sigma}$ can be defined by fictitious losses L_t or by estimating the covariance matrix $\boldsymbol{\Sigma}$ first where $\boldsymbol{\mu}$ is assumed to be known.

2.1 Exact confidence intervals

The exact distribution of $\hat{\nu}_p$ under assumption 1 is given by

$$n\left(\frac{\hat{\nu}_p - \mu}{\nu_p - \mu}\right)^2 = \frac{n\hat{\sigma}^2}{\sigma^2} \sim \chi^2(n). \tag{10}$$

If $\chi^2_{n,\alpha}$ denotes α-quantile of the chi-square distribution with n degrees of freedom,

$$\mathbf{P}\left(\nu_p < \mu + c_{n,\alpha}(\hat{\nu}_p - \mu)\right) = \alpha$$

for all $0 < \alpha < 1$ follows immediately from (10) with

$$c_{n,\alpha} := \sqrt{\frac{n}{\chi^2_{n,1-\alpha}}}.$$

An *upper confidence bound with confidence level* $1-\alpha$, i. e.

$$\mathbf{P}(\nu_p \leq U) = 1 - \alpha,$$

is given for all $0 < \alpha < 1$ by

$$U := \mu + c_{n,1-\alpha}(\hat{\nu}_p - \mu).$$

A *lower confidence bound with confidence level* $1-\alpha$, i. e.

$$\mathbf{P}(\nu_p \geq L) = 1 - \alpha,$$

is given for all $0 < \alpha < 1$ by

$$L := \mu + c_{n,\alpha}(\hat{\nu}_p - \mu).$$

For $0 < \alpha_\star$, $0 < \alpha^\star$ with $\alpha_\star + \alpha^\star < 1$ a *confidence interval with confidence level* $1 - \alpha_\star - \alpha^\star$, i. e.

$$\mathbf{P}(\nu_p \in \mathcal{I}_1) = 1 - \alpha_\star - \alpha^\star,$$

is given by

$$\mathcal{I}_1 := [\mu + c_{n,\alpha_\star}(\hat{\nu}_p - \mu), \mu + c_{n,1-\alpha^\star}(\hat{\nu}_p - \mu)].$$

2.2 Asymptotic confidence intervals

Assumption 1 implies that $Q_t := (L_t - \mu)^2$ are independent random variables with $\mathrm{E}(Q_t) = \sigma^2$ and $\mathrm{E}(Q_t^2) = 3\sigma^4$. From the central limit theorem of statistics[10] follows

$$\sqrt{n}\left(\hat{\sigma}^2 - \sigma^2\right) \overset{Law}{\to} \mathcal{N}\left(0, 2\sigma^4\right).$$

The so called delta method of asymptotic statistics[11] allows to determine the asymptotic normal distribution for the estimators $\hat{\sigma}$ and $\hat{\nu}_p$. One gets[12]

$$\sqrt{n}\left(\hat{\sigma} - \sigma\right) \overset{Law}{\to} \mathcal{N}\left(0, \frac{\sigma^2}{2}\right).$$

From (7) follows

$$\hat{\nu}_p - \nu_p = z_p(\hat{\sigma} - \sigma)$$

and further

$$\sqrt{n}\left(\hat{\nu}_p - \nu_p\right) \overset{Law}{\to} \mathcal{N}\left(0, \frac{(\nu_p - \mu)^2}{2}\right). \tag{11}$$

(11) leads to

$$\lim_{n\to\infty} \mathbf{P}\left(\hat{\nu}_p + \frac{z_{\alpha_\star}}{\sqrt{2n}}(\nu_p - \mu) \le \nu_p \le \hat{\nu}_p + \frac{z_{1-\alpha^\star}}{\sqrt{2n}}(\nu_p - \mu)\right) = 1 - \alpha_\star - \alpha^\star. \tag{12}$$

Equation (12) serves as starting point for the construction of asymptotic confidence intervals.

1. (12)implies

$$\lim_{n\to\infty} \mathbf{P}\left(\mu + d_{n,\alpha_\star}(\hat{\nu}_p - \mu) \le \nu_p \le \mu + d_{n,1-\alpha^\star}(\hat{\nu}_p - \mu)\right) = 1 - \alpha_\star - \alpha^\star$$

with

$$d_{n,\alpha} := \frac{1}{1 - \frac{z_\alpha}{\sqrt{2n}}}.$$

This gives the confidence interval

$$\mathcal{I}_2 := \left[\mu + d_{n,\alpha_\star}(\hat{\nu}_p - \mu), \mu + d_{n,1-\alpha^\star}(\hat{\nu}_p - \mu)\right]$$

with asymptotic confidence level $1 - \alpha_\star - \alpha^\star$.

2. Asymptotic confidence intervals also can be gained replacing in (12) the unknown difference $\nu_p - \mu$ with the consistent estimator $\hat{\nu}_p - \mu$. From (12) and Slutsky's lemma follows[13]

$$\lim_{n\to\infty} \mathbf{P}\left(\hat{\nu}_p + \frac{z_{\alpha_\star}}{\sqrt{2n}}(\hat{\nu}_p - \mu) \le \nu_p \le \hat{\nu}_p + \frac{z_{1-\alpha^\star}}{\sqrt{2n}}(\hat{\nu}_p - \mu)\right) = 1 - \alpha_\star - \alpha^\star.$$

[10] Cp. e. g. Theorem 1.3 in Lehmann (1983), p. 336.

[11] See Appendix.

[12] E. g. with Theorem 1.5 in Lehmann (1983), p. 337.

[13] Cp. Witting/Müller-Funk (1995), p. 76.

This implies that

$$\begin{aligned}\mathcal{I}_3 &:= \left[\hat{\nu}_p + \frac{z_{\alpha_\star}}{\sqrt{2n}}(\hat{\nu}_p - \mu), \hat{\nu}_p + \frac{z_{1-\alpha^\star}}{\sqrt{2n}}(\hat{\nu}_p - \mu)\right] \\ &= \left[\mu + d'_{n,\alpha_\star}(\hat{\nu}_p - \mu), \mu + d'_{n,1-\alpha^\star}(\hat{\nu}_p - \mu)\right]\end{aligned}$$

with

$$d'_{n,\alpha} := 1 + \frac{z_\alpha}{\sqrt{2n}}$$

is a confidence interval with asymptotic confidence level $1 - \alpha_\star - \alpha^\star$.

3. Another approach to reach a confidence interval from (11) is with the help of a *variance stabilizing transformation* and the delta method again to construct an asymptotic normal distributed statistic in such a way that the asymptotic normal distribution does not depend on unknown parameters. Starting with

$$g'(\nu_p) = \frac{\sqrt{2}}{\nu_p - \mu}$$

we find

$$g(\nu_p) = \sqrt{2}\ln(\nu_p - \mu)$$

as an appropriate variance stabilizing transformation. We get

$$\sqrt{n}\,(g(\hat{\nu}_p) - g(\nu_p)) \overset{Law}{\to} \mathcal{N}(0,1). \tag{13}$$

$$\begin{aligned}&\mathbf{P}\left(g(\hat{\nu}_p) + \frac{z_{\alpha_\star}}{\sqrt{n}} \le g(\nu_p) \le g(\hat{\nu}_p) + \frac{z_{1-\alpha^\star}}{\sqrt{n}}\right) \\ &= \mathbf{P}\left(\ln(\hat{\nu}_p - \mu) + \frac{z_{\alpha_\star}}{\sqrt{2n}} \le \ln(\nu_p - \mu) \le \ln(\hat{\nu}_p - \mu) + \frac{z_{1-\alpha^\star}}{\sqrt{2n}}\right)\end{aligned}$$

gives

$$\lim_{n\to\infty} \mathbf{P}\left(\mu + d''_{n,\alpha_\star}(\hat{\nu}_p - \mu) \le \nu_p \le \mu + d''_{n,1-\alpha^\star}(\hat{\nu}_p - \mu)\right) = 1 - \alpha_\star - \alpha^\star$$

with

$$d''_{n,\alpha} := e^{\frac{z_\alpha}{\sqrt{2n}}}$$

and the confidence interval

$$\mathcal{I}_4 := \left[\mu + d''_{n,\alpha_\star}(\hat{\nu}_p - \mu), \mu + d''_{n,1-\alpha^\star}(\hat{\nu}_p - \mu)\right]$$

with asymptotic confidence level $1 - \alpha_\star - \alpha^\star$.

4. Furthermore, the fact that the chi-square distribution approximates the $\mathcal{N}(n, 2n)$-distribution with increasing number of degrees of freedom can be used to determine the asymptotic intervals

$$\mathcal{I}_5 := \left[\mu + c'_{n,\alpha_\star}(\hat{\nu}_p - \mu), \mu + c'_{n,1-\alpha^\star}(\hat{\nu}_p - \mu)\right]$$

with

$$c'_{n,\alpha} := \sqrt{\frac{n}{z_{1-\alpha}\sqrt{2n}+n}} = \sqrt{\frac{\sqrt{n}}{z_{1-\alpha}\sqrt{2}+\sqrt{n}}}$$

from the exact intervals $\mathcal{I}_1$.

$c'_{n,\alpha}$ follows from $c_{n,\alpha}$ with the normal approximation

$$\lim_{n\to\infty} \frac{\chi^2_{n,1-\alpha}-n}{\sqrt{2n}} = z_{1-\alpha}$$

for the chi-square distribution.

3 Confidence intervals with estimated means

If μ has to be estimated, then

$$\hat{\nu}_p = \hat{\mu} + z_p\hat{\sigma} \tag{14}$$

with

$$\hat{\mu} = \frac{1}{n}\sum_{t=1}^{n} L_t, \quad \hat{\sigma} = \sqrt{\frac{1}{n}\sum_{t=1}^{n}(L_t-\hat{\mu})^2} \tag{15}$$

is an appropriate estimator[14] for VaR. The same estimator $\hat{\nu}_p$ results from

$$\hat{\mu} = -\mathbf{w}'\hat{\boldsymbol{\mu}}, \qquad \hat{\sigma} = \sqrt{\mathbf{w}'\hat{\boldsymbol{\Sigma}}\mathbf{w}}$$

with

$$\hat{\boldsymbol{\mu}} := (\hat{\mu}_1, \ldots, \hat{\mu}_K)', \quad \hat{\mu}_j := \bar{X}_j = \frac{1}{n}\sum_{t=1}^{n} X_{t,j}$$

and

$$\hat{\boldsymbol{\Sigma}} = \frac{1}{n}\sum_{t=1}^{n}(\mathbf{X}_t - \hat{\boldsymbol{\mu}})(\mathbf{X}_t - \hat{\boldsymbol{\mu}})'. \tag{16}$$

---

[14] Instead of the maximum-likelihood estimators other estimators for σ can be used e. g. derived from an unbiased estimator for σ^2 dividing the sum of squares by $n-1$. The differences are small. Furthermore, an unbiased estimator T for σ^2 does not imply unbiasedness for the estimators $\sqrt{T}$ for σ and $\hat{\mu} + z_p\sqrt{T}$ for ν_p.

Another approach is to use an unbiased estimator $\tilde{\sigma} = K_n\hat{\sigma}$ for σ whith a correction factor K_n only depending on n. See e. g. Lehmann (1983), p. 85. The unbiasedness of $\hat{\mu}$ and $\tilde{\sigma}$ implies unbiasedness of the modified estimator $\tilde{\nu}_p := \hat{\mu} + z_p\tilde{\sigma}$ for ν_p.

The biased estimator for σ given in (8) may by modified analogously. In the same way an unbiased estimator for (7) can be gained. Of course, the correction factors are derived from the assumption of normal distributed observations. In general, σ^2 can be estimated unbiasedly, whitout the assumption of a special parametric distribution family, but not σ.

3.1 Exact confidence intervals

The independent random variables

$$U := \frac{\mu - \hat{\mu}}{\sigma}\sqrt{n} \sim \mathcal{N}(0,1), \qquad V := \frac{n\hat{\sigma}^2}{\sigma^2} \sim \chi^2(n-1) \tag{17}$$

serve as starting point for the construction of exact confidence intervals.

An appropriate pivotal quantity with a distribution depending only on the known parameters n and p, but not on μ and σ, is

$$W := \frac{\nu_p - \hat{\nu}_p}{z_p \hat{\sigma}} = \frac{1}{z_p}\frac{\mu - \hat{\mu}}{\sigma}\frac{\sigma}{\hat{\sigma}} + \left(\frac{\sigma}{\hat{\sigma}} - 1\right) = \frac{1}{z_p}\frac{U}{\sqrt{V}} + \left(\sqrt{\frac{n}{V}} - 1\right).$$

The last equation allows one to determine the quantiles $w_{\alpha_\star}$ and $w_{1-\alpha^\star}$ of the distribution of W by Monte-Carlo simulation or other numerical methods. From

$$\mathbf{P}\left(w_{\alpha_\star} \le W \le w_{1-\alpha^\star}\right) = 1 - \alpha_\star - \alpha^\star$$

results the exact confidence interval

$$\begin{aligned}\mathcal{I}_6 &:= [\hat{\nu}_p + w_{\alpha_\star}(\hat{\nu}_p - \hat{\mu}), \hat{\nu}_p + w_{1-\alpha^\star}(\hat{\nu}_p - \hat{\mu})] \\ &= [\hat{\mu} + (w_{\alpha_\star} + 1)(\hat{\nu}_p - \hat{\mu}), \hat{\mu} + (w_{1-\alpha^\star} + 1)(\hat{\nu}_p - \hat{\mu})]\end{aligned}$$

for ν_p with confidence level $1 - \alpha_\star - \alpha^\star$.

In addition, there is the connection

$$\sqrt{n-1}\, z_p(W+1) \sim t(n-1, z_p\sqrt{n})$$

to the noncentral t-distribution.[15] Here $t(m,\delta)$ denotes a noncentral t-distribution with m degrees of freedom and with noncentrality parameter δ. $t_\alpha(m,\delta)$ denotes the α-quantile.

The α-quantile of the distribution of W, w_α is connected with the α-quantile of a noncentral t-distribution by equation

$$w_\alpha + 1 = \frac{t_\alpha(n-1, z_p\sqrt{n})}{z_p\sqrt{n-1}} =: C_{n,\alpha,p}.$$

Thus, the confidence interval $\mathcal{I}_6$ can also be written as

$$\mathcal{I}_6 = [\hat{\mu} + C_{n,\alpha_\star,p}(\hat{\nu}_p - \hat{\mu}), \hat{\mu} + C_{n,1-\alpha^\star,p}(\hat{\nu}_p - \hat{\mu})].$$

[15] Cp. e. g. Johnson et al (1995), 508-544.

To determine these intervals we can use either tables of statistical tolerance intervals[16] or statistical standard software to compute the needed quantiles of the noncentral t-distribution.[17]

3.2 Asymptotic confidence intervals

It is well-known that $\hat{\mu}$ and $\hat{\sigma}$ are stochastic independent random variables with limit distributions [18]

$$\sqrt{n}\,(\hat{\mu}-\mu) \stackrel{Law}{\to} \mathcal{N}\left(0,\sigma^2\right)$$

and

$$\sqrt{n}\,\left(\hat{\sigma}^2-\sigma^2\right) \stackrel{Law}{\to} \mathcal{N}\left(0,2\sigma^4\right).$$

Using the delta method (see Appendix) it easily follows

$$\sqrt{n}\,(\hat{\sigma}-\sigma) \stackrel{Law}{\to} \mathcal{N}\left(0,\frac{\sigma^2}{2}\right).$$

From the stochastic independence of $\hat{\mu}$ and $\hat{\sigma}$ results the limit distribution

$$\sqrt{n}\,(\hat{\nu}_p-\nu_p) \stackrel{Law}{\to} \mathcal{N}\left(0,\frac{2+z_p^2}{2}\sigma^2\right).$$

However, this form is not suitable to give a confidence interval directly because σ^2 is unknown. If σ is replaced by the consistent estimator $\hat{\sigma}$ we get[19]

$$\sqrt{n}\left(\frac{\hat{\nu}_p-\nu_p}{\hat{\sigma}}\right) \stackrel{Law}{\to} \mathcal{N}\left(0,\frac{2+z_p^2}{2}\right).$$

As a result, the confidence interval with asymptotic level $1-\alpha_\star-\alpha^\star$ is

$$\begin{aligned}\mathcal{I}_7 &:= \left[\hat{\nu}_p+\frac{z_{\alpha_\star}}{\sqrt{2n}}\sqrt{2+z_p^2}\,\hat{\sigma},\,\hat{\nu}_p+\frac{z_{1-\alpha^\star}}{\sqrt{2n}}\sqrt{2+z_p^2}\,\hat{\sigma}\right]\\ &= \left[\hat{\mu}+D_{n,\alpha_\star,p}(\hat{\nu}_p-\hat{\mu}),\,\hat{\mu}+D_{n,1-\alpha^\star,p}(\hat{\nu}_p-\hat{\mu})\right]\end{aligned}$$

---

16 For instance, a one-sided upper 95%-tolerance bound, which includes 99% of the population, is formally the same as an upper 95%-confidence bound for the 99%-quantile this population. Cp. e. g. Hahn/Meeker (1991).
Between $C_{n,\alpha,p}$ and the values $g'_{(1-\alpha;p,n)}$ tabulated in Hahn/Meeker (1991) for $n=2(1)30,35,40,50,60,120$ and 480, $p=0.6(0.1)0.9,0.95$ and 0.99, and for $\alpha=0.01,0.025,0.05,0.1,0.2,0.8,0.9,0.95$ and 0.99 exists the connection

$$C_{n,\alpha,p}z_p\sqrt{n-1}=g'_{(1-\alpha;p,n)}\sqrt{n}.$$

More detailed tables may be found in Odeh/Owen (1980).

17 E. g. using procedure *tinv* in SAS or or *cdftnc* in GAUSS.

18 Cp. Witting/Müller-Funk (1995), p. 95f.

19 Cp. Slutsky's Lemma e. g. in Witting/Müller-Funk (1995), p. 76.

with

$$D_{n,\alpha,p} := 1 + \frac{z_\alpha}{\sqrt{2n}} \frac{\sqrt{2+z_p^2}}{z_p}.$$

4 Comparison of the confidence intervals

The intervals given above are compared in the following for the case $\mu = \hat{\mu} = 0$. This implies that $z_p \hat{\sigma} = \hat{\nu}_p$ holds also for the intervals $\mathcal{I}_6$ and $\mathcal{I}_7$, where μ is estimated. $\hat{\mu} \neq 0$ has two effects: a change of interval length and a shift of the interval by $\hat{\mu}$.

In case $\mu = \hat{\mu} = 0$ the intervals $\mathcal{I}_j$ $(j = 1, \ldots, 7)$ are

$$\mathcal{I}_1 = [c_{n,\alpha_\star}\hat{\nu}_p, c_{n,1-\alpha^\star}\hat{\nu}_p],$$

$$\mathcal{I}_2 = [d_{n,\alpha_\star}\hat{\nu}_p, d_{n,1-\alpha^\star}\hat{\nu}_p],$$

$$\mathcal{I}_3 = [d'_{n,\alpha_\star}\hat{\nu}_p, d'_{n,1-\alpha^\star}\hat{\nu}_p],$$

$$\mathcal{I}_4 = [d''_{n,\alpha_\star}\hat{\nu}_p, d''_{n,1-\alpha^\star}\hat{\nu}_p],$$

$$\mathcal{I}_5 = [c'_{n,\alpha_\star}\hat{\nu}_p, c'_{n,1-\alpha^\star}\hat{\nu}_p],$$

$$\mathcal{I}_6 = [C_{n,\alpha_\star,p}\hat{\nu}_p, C_{n,1-\alpha^\star,p}\hat{\nu}_p],$$

$$\mathcal{I}_7 = [D_{n,\alpha_\star,p}\hat{\nu}_p, D_{n,1-\alpha^\star,p}\hat{\nu}_p].$$

In the following table confidence intervals for the symmetric case $\alpha_\star = \alpha^\star = \alpha/2$ are given. An interval $[a\hat{\nu}_p, b\hat{\nu}_p]$ is given in the more compact form $[a,b]\hat{\nu}_p$.

Table: Confidence intervals for VaR

	$p = 99\%, \alpha_\star = \alpha^\star = \alpha/2$			
	$\alpha = 1\%$		$\alpha = 0.1\%$	
	$n = 50$	$n = 250$	$n = 50$	$n = 250$
$\mathcal{I}_1$	$[.793, 1.337]\hat{\nu}_p$	$[.896, 1.129]\hat{\nu}_p$	$[.747, 1.460]\hat{\nu}_p$	$[.870, 1.169]\hat{\nu}_p$
$\mathcal{I}_2$	$[.795, 1.347]\hat{\nu}_p$	$[.897, 1.130]\hat{\nu}_p$	$[.752, 1.490]\hat{\nu}_p$	$[.872, 1.173]\hat{\nu}_p$
$\mathcal{I}_3$	$[.742, 1.258]\hat{\nu}_p$	$[.885, 1.115]\hat{\nu}_p$	$[.671, 1.329]\hat{\nu}_p$	$[.853, 1.147]\hat{\nu}_p$
$\mathcal{I}_4$	$[.773, 1.294]\hat{\nu}_p$	$[.891, 1.122]\hat{\nu}_p$	$[.720, 1.390]\hat{\nu}_p$	$[.863, 1.159]\hat{\nu}_p$
$\mathcal{I}_5$	$[.812, 1.436]\hat{\nu}_p$	$[.902, 1.140]\hat{\nu}_p$	$[.777, 1.710]\hat{\nu}_p$	$[.879, 1.190]\hat{\nu}_p$
$\mathcal{I}_6$	$[.757, 1.402]\hat{\nu}_p$	$[.879, 1.152]\hat{\nu}_p$	$[.700, 1.544]\hat{\nu}_p$	$[.847, 1.198]\hat{\nu}_p$
$\mathcal{I}_7$	$[.699, 1.301]\hat{\nu}_p$	$[.865, 1.135]\hat{\nu}_p$	$[.615, 1.385]\hat{\nu}_p$	$[.828, 1.172]\hat{\nu}_p$

The interval $\mathcal{I}_2$ reaches the best approximation to the exact interval $\mathcal{I}_1$ compared with the other approximate intervals $\mathcal{I}_3$ to $\mathcal{I}_5$. Compared with the exact interval $\mathcal{I}_1$ the interval $\mathcal{I}_3$ shows a great shift of both interval ends to zero. The interval $\mathcal{I}_4$ shows a shift to zero too, while $\mathcal{I}_5$ is shifted in the opposite direction.

The intervals $\mathcal{I}_6$ and $\mathcal{I}_7$, which also take into consideration the sampling error of the mean estimation, are longer than $\mathcal{I}_1$ to $\mathcal{I}_5$.

With the same parameters, the interval $\mathcal{I}_7$ is by the factor

$$Q_p := \frac{\sqrt{2 + z_p^2}}{z_p} > 1$$

longer than $\mathcal{I}_3$. This factor is $Q_{0.99} = 1.17$ for $p = 0.99$. When $p = 0.95$ this factor increases to $Q_{0.95} = 1.32$, so that the additional estimation error caused by estimating μ is relativly less important for large p. The proportionality between the two exact confidence intervals $\mathcal{I}_6$ and $\mathcal{I}_1$ is of the same order as Q_p.

All given confidence intervals do not depend on K, the number of portfolio positions.[20] This seems to be strange at the first sight, because it is well-known that the estimators (9) and (16) follow a Wishart-distribution with parameters depending on K. But the distribution of $\hat{\nu}_p$ depends on the distribution of the quadratic form $\mathbf{w}'\hat{\Sigma}\mathbf{w}$ which is chi-square distribution with a parameter depending only on n.

Appendix: delta method

The delta method of asymptotic statistics bases on the following Theorem:[21]

Let $(\hat{\theta}_n)$ be a sequence of estimators for the parameter θ and let (r_n) be a sequence of positive numbers with $r_n \to \infty$. Let the function h be differentiable at θ with derivate $h'(\theta) \neq 0$. Then holds:

$$r_n(\hat{\theta}_n - \theta) \overset{Law}{\to} \mathcal{N}(0,1) \quad \Longrightarrow \quad r_n \frac{h(\hat{\theta}_n) - h(\theta)}{|h'(\theta)|} \overset{Law}{\to} \mathcal{N}(0,1).$$

Example:

$$\sqrt{n}\frac{\hat{\sigma}^2 - \sigma^2}{\sqrt{2}\sigma^2} \overset{Law}{\to} \mathcal{N}(0,1) \quad \Longrightarrow \quad \sqrt{2n}\frac{\hat{\sigma} - \sigma}{\sigma} \overset{Law}{\to} \mathcal{N}(0,1).$$

20 The number of portfolio positions plays a role e. g. if a confidence interval for a single element of the covariance matrix is determined. This is beyond the scope of this article.

21 Cp. e. g. Theorem 5.78 in Witting/Müller-Funk (1995), p. 73 and Theorem 1.5 in Lehmann (1983), p. 337. Multivariate versions are given in Witting/Müller-Funk (1995), p. 107 and Lehmann (1983), p. 344.

References

Basle Committee on Banking Supervision (1996a): Supplement to the Capital Accord to Incorporate Market Risks. Basle.

Basle Committee on Banking Supervision (1996b): Supervisory Framework for the Use of "Backtesting" in Conjunction with the Internal Models Approach to Market Risk Capital Requirements. Basle.

Hahn, G. J., Meeker, W. Q. (1991): Statistical Intervals: A Guide for Practitioners. Wiley, New York.

Johnson, N. L., Kotz, S., Balakrishnan, N. (1995): Continuous Univariate Distributions. Vol 2. 2nd Ed. Wiley, New York.

Jorion, P. (1996): Risk^2: Measuring the risk in value at risk. Financial Analysts Journal, November/December, 47-56.

Jorion, P. (1997): Value at Risk: The New Benchmark for Controlling Market Risk. Irwin, Chicago.

Kupiec, P. H. (1995): Techniques for verifying the accuracy of risk measurement models. Journal of Derivatives 3(2), 73-84.

Lehmann, E. L. (1983): Theory of Point Estimation. Wiley, New York.

Odeh, R. E., Owen, D. B. (1980): Tables for Normal Tolerance Limits, Sampling Plans, and Screening. Dekker, New York.

Morgan Guaranty Trust Company/Reuters Ltd (1996): RiskMetrics – Technical Document. 4th ed.

Ridder, T. (1996): Quantilschätzer zur Analyse des Value at Risk. Preprint. SGZ-Bank, Frankfurt a. M./Karlsruhe.

Witting, H., Müller-Funk, U. (1995): Mathematische Statistik II; Asymptotische Statistik: Parametrische Modelle und nichtparametrische Funktionale. Teubner, Stuttgart.

Regulatory Framework for the Risk Management of German Credit Institutions

Hermann Schulte-Mattler

Dr. Hermann Schulte-Mattler is Professor of Finance at Fachhochschule Dortmund, Emil-Figge-Strasse 44, D-44047 Dortmund

Abstract. The banking industry and the national and international financial markets have undergone radical structural changes over the past two decades. In order to prevent undesirable developments, the supervisory authorities constantly adapt their regulatory framework to the changing circumstances. Until the late 80s, countries adopted a specifically national approach to the question of banking regulation. The picture has completely changed. Internationally uniform supervisory rules particularly for financial price risks have been established over the past few years. International harmonization of banking rules has been accompanied by increasing complexity of regulation. Thus, it is worthwhile to give a brief overview of the development and basic structure of the German Principle I, which sets the capital requirements for credit and market risk and thus the regulatory framework for German credit institutions.[1]

Keywords. Backtesting, Banking supervision, Capital Adequacy Directive, Credit and counterparty risk, Internal risk measurement models, Market risks, Principle I, Sixth Amendment to the German Banking Act, Value at Risk.

1 Introduction: Supervisory Triple-B

The banking industry and the national and international financial markets have undergone radical structural changes over the past two decades. This development can be summed up in three key words: globalization, securitization and futurization. Round-the-clock trading has led to an increasing integration of financial markets - that is, to *globalization* - combined with low transaction costs and increased market transparency. The ever-closer links between national and international financial markets, which ultimately means growing mutual interdependence, have undoubtedly been encouraged and accelerated by rapid technological progress.

This structural change has gone hand in hand with the decline of the classical forms of bank credit and an increase in *securitization* - that is, securities-backed financing. The almost inevitable result has been intensified competition for borrowers of unquestionable credit standing and thus decreasing

[1] This paper was presented at the 6. Ökonometrie-Workshop, Risikometrie, Ökonometrie und neuronale Netze, at the University of Karlsruhe, March 19-21, 1997. I thank Edward Graham and Andre O. Weidlich for valuable comments.

margins. Banks and other financial institutions have been constantly supplying their good customers with ever-newer and cheaper products.

Along with this trend, there has been a growing need on the part of market participants to protect themselves against the risk of price changes - particularly exchange rate and interest rate fluctuations - produced by increasingly volatile and dynamic financial markets. This explains to a great extent the expansion of futures markets, although speculation has undoubtedly also been a prime motive. In response to the change in market requirements, banks have developed as part of the *futurization* process - that is, in the growing pressure to move towards active risk management - a whole number of new derivative instruments.

In order to prevent undesirable developments, the supervisory authorities constantly adapt their regulatory framework to the changing circumstances. All supervisory rules limit in a more or less restrictive way the volume of business and the risk exposure of a financial institution. Until the late 80s, countries adopted a specifically national approach to the question of banking regulation. The picture has completely changed in the 90s. Today's approach to shaping supervisory regulations for the banking industry is internationalized, mainly influenced by the work of the *Basle Committee on Banking Supervision* (the Committee), which usually meets at the Bank for International Settlements. This committee was established by the central-bank Governors of the Group of Ten countries in 1975.[2] Findings of the Basle Committee are published as proposals for internationally operating banks. Although these findings are not legally binding, they are transformed into national law or regulations either directly or through legislative acts of the *European Union* (Brussels). As far as the Federal Republic of Germany is concerned the Ministry of Finance (*Ministerium der Finanzen*) in Bonn and the Federal Banking Supervisory Office (*Bundesaufsichtsamt für das Kreditwesen*) in Berlin are responsible for transferring international capital rules into the German Banking Act (Kreditwesengesetz) and Principle I (Grundsatz I).[3]

All legislation and the regulatory framework governing German financial institutions are therefore influenced by the *supervisory Triple-B*: Basle - Brussels - Bonn/Berlin. The starting point for new regulations is the proposals of the Basle Committee. After a certain time lag, many of these proposals find their way into European banking directives and are thereafter transformed into national legislation. Conflicts between national supervisory authorities and banks occur merely in those cases where proposals of the Basle Com-

[2] Today the Basle Committee consists of senior representatives of bank supervisory authorities and central banks from Belgium, Canada, France, Germany, Italy, Japan, Luxembourg, the Netherlands, Sweden, Switzerland, the United Kingdom and the United States.

[3] See for details Schulte-Mattler and Traber (1997).

mittee or European regulations are translated into more restrictive national legislation.

The centerpiece of all legislation concerning the supervision of the banking industry is basically *capital requirements*, which impose strict rules upon financial institutions on how to capture and measure typical risks. The basic feature of such requirements is that the amount of a bank's business which potentially threatens its capital is limited to a certain amount of the same. Therefore, a crucial factor in the process of shaping supervisory rules is a precise description of how to define capital and quantify risk.

Due to the changes in banking regulation which accompany the harmonization of international banking legislation, risk measurement techniques have become increasingly complex. Depending on the degree of complexity of such techniques, which are used to quantify a potential risk of losses arising from movements of certain risk factors, banking rules can be divided into *three generations*. The higher the generation, the more complex the mathematical and statistical algorithms required by the supervisory authorities to estimate the potential risk will be.

This paper aims to illustrate the above-mentioned general development by way of Principle I, a banking regulation of major importance to German banks. We start in Section 2 with the current capital requirements for credit and counterparty risk. In Section 3, we look at the new capital cushion for price risks, particularly those arising from the trading activities, to which banks are exposed. Alternative use of proprietary in-house models for measuring price risk instead of a standardized measurement framework will be presented in Section 4. In addition, we will cover the backtesting procedure with which supervisory authorities plan to make ex-post comparisons between model estimates and actual performance. Section 5 then concludes our paper.[4]

2 Credit and Counterparty Risk

The *first-generation* rules incorporate capital requirements for credit-risk-related on- and off-balance-sheet business including OTC-derivatives. The starting point for international harmonization was the agreement concluded in July 1988 by the Basle Committee. The so-called *Basle Accord* represents a first fundamental international consensus on the capital requirements for credit and counterparty risk. It contains a definition of capital, a risk weighting for business to be taken into account and a standard ratio of capital to risk-weighted assets.[5]

The *Basle Accord* determined to a large extent the content of the Own Funds Directive and the Solvency Ratio Directive adopted in Brussels as part

[4] Especially regarding rules and regulations for OTC derivatives see Schulte-Mattler (1997).

[5] See Basle Committee on Banking Supervision (1988)

of the harmonization of European banking regulation. Contrary to the Basle Accord, an agreement between participating supervisory authorities for the banking industry, all directives of the European Union are to be formally transposed into national law and regulations. The national rules may be stricter but never less restrictive than the European directives.

At the end of December 1993, the Basle standards and the EU directives had been implemented in Germany and are contained in the so-called *Fourth Amendment to the German Banking Act* and in today's Principle I. German banking regulation in this context is highly peculiar. Whereas the bank's capital is defined in § 10 of the German Banking Act, the assessment of the capital to risk assets ratio, meaning capital adequacy, is subject to Principle I.

Principle I quantifies the capital requirement for potential risk of losses arising in respect of the credit-risk, including the counterparty risk of OTC-derivatives, in both the trading and the banking book. The capital in this context is the combined amount of the so-called tier 1 capital (shareholders' equity and retained earnings) and the tier 2 capital (supplementary capital); however, the latter is only recognized up to the total amount of the tier 1 capital.

Like the Basle Accord and the Solvency Ratio Directive, Principle I provides for an *8 percent ratio* of total capital to credit risk-weighted assets, with a distinction made between tier 1 and tier 2 capital. In addition to the 8 percent standard, a 4 percent standard for the tier 1 capital also always applies. Formally, this can be stated as:

$$Capital = Tier_1 + Tier_2 \leq 0.08 * (Assets * Risk\ Weights) \quad \text{and}$$
$$Tier_1 \geq 0.04 * (Assets * Risk\ Weights)$$

The main reason why we call the 8 percent figure a *magic number* is that for some reason international supervisory authorities like to quantify not only credit risks but almost all typical bank risks with this number.[6] However, no empirical studies have been published to verify the adequacy of this number for various risk types. This means that the authorities use 8 percent in the belief that this figure correctly and adequately assesses the potential losses due to various risk factors. It can be assumed that on average, estimating the potential losses at 8 percent exaggerates the risk. Otherwise, it is inexplicable that supervisory authorities usually come to an agreement during international negotiation at just this number.

3 Market Risks

The *Second Banking Directive*, effective as of January 1st, 1993, finally established the single European market for the financial industry. With this

[6] Like equity position risk, foreign exchange risk, large exposure position risk and settlement risk.

directive, all banks in Europe were in possession of the European passport enabling them to carry out operations in all member states of the European Union. Any bank wishing to conduct business in another country no longer needs to apply for a separate business license for their branch in the host country; instead, a valid business license issued for the corporate headquarters in the home country will be honored by the authorities in the host country. In this context, business operations are supervised solely by the authorities in the corporate country of origin.[7] However, the European passport is only valid for banks whose business purpose, according to the *First Banking Directive*, is to grant loans and accept deposits at the same time (EU bank).

Anglo-Saxon-type *investment firms*, however, are not recognized as EU banks. Therefore, they do not automatically receive the privileges of conducting business freely anywhere within the EU. Especially the United Kingdom in this context had expressed its wish to have a directive giving the European passport also to investment firms. However, those states within the EU that favor the universal banking system, headed by the Federal Republic of Germany, had in return insisted on implementing capital requirements for investment firms. Their argument for doing so was the fear that separate rules and legislation governing commercial banks and investment firms might interfere with the market and hinder competition. For both types of competitors in the financial market a *level playing field* should be created in order to avoid favoritism of one or the other banking system. The problem in this respect was to create harmonized rules for the different types of financial institutions by applying the principle of same business, same risk and same supervisory rules.

As a result, the *second-generation* capital requirements for market-risk-related business were developed against the background of increasing price volatility and stronger growth in derivative products. At the end of April 1993, the Basle Committee submitted initial proposals for capturing and limiting market risks incurred by internationally operating banks. These proposals are largely identical with the provisions of the *Capital Adequacy Directive* already adopted in March 1993. The Directive sets minimum capital requirements for interest-rate and equity-price risk arising from the *trading book*, that means securities transactions for own account, and for foreign-exchange risk attached to all banking business. Due to the limited scope of business activities of investment firms as mentioned before, investment firms and universal banks basically just compete in the area of securities trading and their

[7] This only applies to the area of banking supervision for which there is existing, EC-wide harmonization of banking legislation, i. e. not in the area of supervising liquidity of banks.

associated derivatives. Therefore, there arose the need to first create special regulations for this area of business activities.[8]

The Capital Adequacy Directive was to be transposed into national supervisory law by December 31st, 1995. However, adoption and implementation is still overdue in Germany. The market-risk capital requirements will be implemented and put into effect within the framework of the Sixth Amendment to the German Banking Act and the new version of Principle I no earlier than the end of 1997.

The proposal for the *Sixth Amendment* to the German Banking Act as well as the Capital Adequacy Directive contain two new elements of capital.[9] Institutions may cover market risks with tier 1 and tier 2 capital but may alternatively also employ short-term subordinated debt and trading profits (tier 3 and tier 4 capital) limited to 250 percent of a bank's tier 1 capital that is available to cover market risks, i.e. tier 1 capital which is not used for credit risk in the banking book.

The proposal for the *revised Principle I* adopts the standard approach of the Capital Adequacy Directive for calculating the capital requirement for market risks, which involve all risks that result from activities in the trading book as well as all transactions involving foreign currencies and commodities. For the purpose of determining the adequacy of the capital requirement in each and every one of the risk areas all accountable transactions are first identified. Usually all transactions are valued at market price. The open position is then determined as the difference between long and short positions (hedging). Finally, all the capital requirements needed to cover the different types of risk are added. This produces the total amount of capital needed to cover all potential risks which the financial institution is to keep in hand at the end of every business day.

The risk of open positions due to changes in a risk factor is quantified with a *standardized measurement framework*. In order to do so, the banking supervisory authorities explicitly use first (partial) derivatives of a price function (like delta of an option or modified duration) in order to estimate the sensitivity of an open position. If one multiplies the sensitivity of a position by an assumed figure for the volatility of a risk factor (see Table 1), set by the banking supervision authority, the result is the assumed probability of losses.

Usually this probability will be smaller than the magic 8 percent. The amount of capital needed for the risk of a certain open position is determined by multiplying the position by the probability of losses. The supervisory regulation of the second generation, therefore, basically raises the level of

[8] The trading-book is basically financial instruments held by any financial institution for trading purposes and containing a price risk (such as stocks, bonds, certificates in commodities, interest rate futures, options, interest rate and foreign currency swaps as well as money market instruments).

[9] See Bundesministerium der Finanzen (1996).

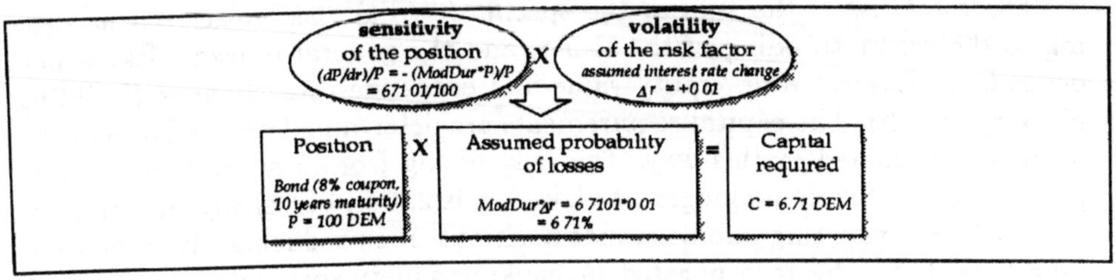

Fig. 1. The assumed probability of losses

complexity to obtain a better estimate of the potential losses of an individual position.

The *third-generation* capital requirements for market-risk-related business are included in the Basle Committee's final Amendment to the Capital Accord to incorporate Market Risks.[10] The paper is largely identical with the Basle initial proposals and therefore with the provisions of the Capital Adequacy Directive but, in addition, allows *internal risk measurement models* to calculate the capital requirements. These models usually use even more complex methods to estimate potential losses of a portfolio of financial assets over a predetermined time period. We will explain this in the following Section.

4 Internal Risk Models

To determine capital requirements in-house risk measurement models may, as an alternative to standardized procedures, also be permitted by supervisory authorities. This possibility is based on the belief that standardized methods are far less accurate in their estimate of potential risks, seen as the probability of devaluation of a bank's portfolio value, than those procedures currently used by financial institutions. An amendment of the Capital Adequacy Directive (so-called CAD II) to include the internal model alternative is in preparation and is to be ratified by the end of 1997.

For existing regulation in Germany the Basle market risk paper and the translation of the Capital Adequacy Directive into national banking regulation results in capital requirements also for price risks, which have so far merely been limited in relation to own funds in *Principle Ia* (Grundsatz Ia) of the German Banking Act. Open positions will, therefore, involve capital costs in future, i.e. not only credit risk - the risk of counterparty failure -

[10] See Basle Committee on Banking Supervision (1996a).

but also market risk will require capital backing. Banks are being given external standards also for their market risk policy as part of their global risk management.

The *total risk* is the sum of the specific (credit) risk, measured according to the credit standing of the issuer, and the general market risk, which depends on interest-rate, equity-value and exchange-rate changes (building block approach). The capital requirements are determined on the basis of the total risk measured in this way. The risk arising from a specific assets-side position will in future no longer be viewed in isolation but always in conjunction with any offsetting liabilities-side positions. Risk-reducing diversification aspects will thus be incorporated in banking supervisory rules for the first time. Banking books will be seen as portfolios of financial assets. In this way, regulators will be taking into account the possibility of avoiding or reducing risk by holding offsetting long and short positions. That means that regulation for market risks will include incentives to hedge and to net opposite positions in order to reduce capital requirements.

The market risk of a portfolio refers to the probability of financial losses due to the movement of risk factors, e. g. interest rate and exchange rate changes. Almost all of the risk models used by institutions follow the concept of *Value-at-Risk* (VaR), which quantifies the amount of expected change in the value of the portfolio.[11] The calculation of the VaR requires a probability distribution of changes in portfolio value.

Two basic assumptions have to be made on how the portfolio function V is approximated and how the risk factors P (state variables) are modeled. From the supervisory point of view, the statistic to measure the VaR is defined as a one-sided 99 percent confidence interval on portfolio losses over a holding period of 10 days:

$$Prob[\Delta\tilde{V}(\Delta t, \Delta\tilde{P}) > -VaR] = 1 - \alpha = 1 - 0.99 = 0.01$$

where $\Delta\tilde{V}(\Delta t, \Delta\tilde{P})$ is the change in the market value of the bank's portfolio, which is a function of the forecast horizon or liquidation period Δt (10 days) and the vector of the random risk variables changes $\Delta\tilde{P}$ (i.e. exchange rates). Parameter α is the confidence level. The interpretation of the VaR is as follows: Over a large number of trading days, e. g. 100 days, the value of a bank's portfolio will decline by no more than VaR on one business day (which is 1 percent of the time).[12]

In general, it is not feasible to calculate the *true distribution of the change in the market value of the portfolio* $\Delta\tilde{V}$ because of the large computational burden in revaluing an entire portfolio. Therefore, the supervisory authorities will allow local approximations for the distribution of the change in the market value of the portfolio. The easiest way to do so is to use Taylor's Theorem

[11] See the excellent papers of Jorion (1997) and Alexander (1996) as well as the RiskMetrics Technical Document by J. P. Morgan (1996).

[12] See Fallon (1996), especially for the VaR calculation in the following paragraphs.

and approximate the distribution in the neighborhood of the current market value of the portfolio $V(t_0, P_0)$ with first derivatives V_t, V_P (delta model):

$$\Delta\tilde{V}(\Delta t, \Delta\tilde{P}) \equiv V(t, \tilde{P}) - V(t_0, P_0) = V_t \Delta t + V_P \Delta\tilde{P}.$$

In more sophisticated models, higher derivatives could be added to the first derivative in order to obtain an even closer approximation of value changes of the portfolio function (e. g. delta-gamma model). This can intuitively be seen as a ***multivariate sensitivity*** of the portfolio function.

The second assumption refers to the way *changes (returns) of the risk factors* over the forecast horizon are modeled. One simple way is to assume that the Log-returns r of the underlying asset are jointly normally distributed with mean vector zero O and constant covariance matrix Σ, that means $r \sim N(O, \Sigma)$. Thus, we can characterize the distribution of the changes of the risk factors $\Delta\tilde{P}$ in the following way:

$$\Delta\tilde{P}_t \sim N(O, P_t' \Sigma P_t).$$

On the basis of this assumption, one can derive an estimate of the *multivariate volatility* of the risk factors which could change the portfolio value. Although the models in practice are more complex, this simplified description makes the essential point clear. First, one has to estimate the sensitivity of the portfolio function and, second, the volatility of the risk factors. Combining the two approximations gives the normal distribution of portfolio value changes

$$\Delta\tilde{V}(\Delta t, \Delta\tilde{P}) \sim N(V_t \Delta t, V_P' \Sigma V_P)$$

with mean $V_t \Delta t$ and variance $V_P' \Sigma V_P$. Using this characterization and the above-mentioned VaR definition, we can standardize $\Delta\tilde{V}$ to a unit normal by deducing the mean and dividing by the standard deviation:

$$\begin{aligned} Prob[\frac{\Delta\tilde{V}(\Delta t, \Delta\tilde{P}) - V_t \Delta t}{\sqrt{V_P' \Sigma V_P}} &> \frac{-VaR - V_t \Delta t}{\sqrt{V_P' \Sigma V_P}}] \\ = Prob[\tilde{Z} &> \frac{-VaR - V_t \Delta t}{\sqrt{V_P' \Sigma V_P}}] = 1 - \alpha. \end{aligned}$$

Solving this equation for the VaR, we obtain:

$$\begin{aligned} Z(\alpha = 0.99) &= \frac{-VaR - V_t \Delta t}{\sqrt{V_P' \Sigma V_P}} \\ VaR &= -V_t \Delta t - Z(\alpha = 0.01)\sqrt{V_P' \Sigma V_P} \\ &= -V_t \Delta t - 2.33\sqrt{V_P' \Sigma V_P}. \end{aligned}$$

The parameter $Z(\alpha)$ is the α-percentile of the standard unit normal distribution. To sum up, this kind of model estimates the potential losses (-VaR) by combining first derivatives (theta and delta) of the portfolio pricing function and the multivariate normality of changes in risk factors. The distribution of the first-order approximation of the portfolio value changes is characterized as univariate normal.

The required capital for market risks will be the greater amount of the following two: VaR of the previous day or average VaR of the last sixty trading days, multiplied by an additional *scaling factor*. This scaling factor should not be lower than three. The main purpose of this scaling factor is to take into consideration that while calculating the VaR normal market conditions are assumed. Irregular but considerable changes in price and value, market crashes, corner situations and the like are not taken into consideration during this process. All of these so-called stress situations, however, have to be considered by financial institutions within their internal risk models in order to assure qualitative compliance, and the amount of risk is to be quantified but neither directly or indirectly covered with capital.

It was a conditio sine qua non for the Basle Committee in their decision to permit in-house models that in order to ensure complete compliance with supervisory regulations concerning capital coverage of market risks all methods in use must be identical with those methods and risk models used by financial institutions during the process of daily risk controlling (with the exemption of slight variations in quantitative parameters). Precisely for this reason, all of the following *qualitative standards* must be fulfilled:

- All models in use have to be completely integrated into the system of in-house risk controlling mechanisms and are to be the basis for daily decisions to be made by that financial institution as far as all positions are concerned. In addition, these models are to give the framework within which the limit system for use in daily trade is created. This enables the financial institution to determine a near real-time determination of market risk exposure. This requires a particular corporate organizational structure.
- All levels within the hierarchy of the bank responsible for trading activities are to be informed about the results of the model in use and all of those levels have to base their entire decisions solely on these results.
- Routinely stressed market situations have to be simulated in order to ensure the level of accuracy and reality as well as the adequacy of all parameters in use within the internal models.
- A risk controlling department, which is to be completely independent from the trading department, is to be enrolled in an ongoing process of constantly checking all models in use as to their level of accuracy and reality as far as results of calculations and processes in use are concerned.
- Additionally, all models are to be made subject to internal and external controlling and auditing in order to ensure precision-enhancing revisions without time lag.

Not taking the advantages and disadvantages of different in-house models into consideration, the Basle Committee devised a tool for conducting ex-post tests to check on the accuracy of risk models in use by banks (*back-testing*).[13] Thus, a bank must compare its VaR estimate with the following day's outcome. Regulators agreed on a method based on the proportion of exceptions, i. e. occasions where the observed change in the portfolio value is greater than estimated.

The *backtesting procedure* is based on a binomial test. Depending on the number of exceptions, a virtual traffic light grades the models. The division into the three colors - green, yellow and red - is the result of the cumulative, single probability for the occurrence of a certain number of exceptions k of observed portfolio value changes x in a sample of size n trading days, given a coverage level of α:

$$Prob(x \leq k) = \sum_{x \leq k} \alpha^x * (1-\alpha)^{n-x} * \binom{n}{x}.$$

If the number of exceptions (change in value of portfolio greater than expected) within the last 250 trading days is smaller or equals 4, the traffic light is green - the model will be approved by the supervisory authority without addition to the scaling factor ($Prob(x \leq 4) = 89.22\, percent$). The traffic light will be yellow up to 9 exceptions ($Prob(x \leq 9) = 99.97\, percent$), meaning the financial institution may continue using its risk model while applying an addition to the scaling factor it uses of up to 0.85. If the total number of exceptions within the given period of time is greater than 9, the traffic light is red. In this particular case, the addition to the scaling factor will be 1 and the supervisory authority may prohibit the further use of this particular internal risk model.

The power to authorize or to prohibit the use of certain internal risk models is a fundamentally new approach especially for the German authorities and a big step towards more qualitatively-oriented banking supervision. The Federal Banking Supervisory Office will be in charge of approval of internal risk models within the framework of the regulations of Principle I.

5 Summary

Banking supervision is a somewhat delicate matter. Even though its general objective, namely protection of banks, should be given unrestricted support, supervision always implies intervention in the free running of the market. As is the case with any regulatory undertaking, it is important to maintain a balance between the requirements of the market and the justified interests of supervisory law. Bankers are called upon to handle the risks associated with their business in a responsible manner and credit the principles of prudent

[13] See Basle Committee on Banking Supervision (1996b).

banking with the importance they deserve. External control can, at best, sharpen awareness of this need.

With regard to the changes in legislation concerning credit and market risks, it can be stated that:

- The basic regulatory framework at international and European level has been established. Rules and regulations regarding the risk of losses have been implemented and put into effect. Supervisory regulations for the financial industry have been ratified and were supposed to be transposed into national rules and regulations within Europe no later than December 31st, 1995. However, in Germany this will not happen before the end of 1997. Banks have to face the fact that there will be capital requirements for price risks. This will result in higher capital costs. However, certain incentives will be included in Principle I, enabling the financial industry to lower these capital costs by using netting and hedging techniques.
- Supervisory regulations for the financial industry are a global matter and therefore the entire bank, including subsidiaries, has to cover its risks with capital on a consolidated basis. Due to the increasing mathematical and statistical complexity in the process of estimating potential losses and the flood of data and information accompanying this development, a must will be the existence and use of powerful high-end electronic data processing devices (data warehouses).
- There is no doubt that the use of internal risk models for supervisory purposes will lead to more efficient risk determination and assessment. Whether supervisory authorities will be able to control every detail of their qualitative and quantitative standards remains somewhat doubtful. As mentioned before, for the German supervisory authorities these internal risk models mean a new approach aimed at quality-oriented supervision for banks. At any rate, there will be no way for the banking supervisory authorities to avoid having to check all kinds of risk models used by the financial industry and granting them the official seal of approval.

The Basle proposals make clear how important harmonization of supervisory rules at international, European and, finally, national level is. Only if it can be ensured that all suppliers of services doing the same business and facing the same risks are also treated in the same way will a level playing field be possible.

References

Alexander, C. (1996), Ed., The Handbook of Risk Management and Analysis, New York.

Basle Committee on Banking Supervision (1988), Capital Accord, Basle, July 1988.

Basle Committee on Banking Supervision (1996a), Amendment to the Capital Accord to incorporate Market Risks, January 1996.

Basle Committee on Banking Supervision (1996b), Supervisory Framework for the use of "Backtesting" in conjunction with the Internal Models Approach to Market Risk Capital Requirements, January 1996.

Bundesministerium der Finanzen (1996), Sechstes Gesetz zur Änderung des Gesetzes über das Kreditwesen, Draft, 29. November 1996.

Fallon, William (1996), Calculating Value-at-Risk, Working Paper 96-49, Wharton Financial Institutions Center, The Wharton School of the University of Pennsylvania.

J. P. Morgan (1996), RiskMetrics Technical Document, Fourth Edition.

Jorion, Ph. (1997), Value At Risk: The New Benchmark for Controlling Market Risk, Chicago.

Schulte-Mattler, H. (1990), Neuregelung des Grundsatzes I gemäß §§10, 10a Kreditwesengesetz, Wertpapier-Mitteilungen, p. 2061-2067.

Schulte-Mattler, H., Traber, U. (1997), Marktrisiko und Eigenkapital, Gabler Press, Second Edition, Wiesbaden.

Measuring and Managing Credit Portfolio Risk

Thomas C. Wilson[1]

McKinsey & Company, Alpenstrasse 3, 8065 Zürich, Switzerland; phone: (41) 1 874 7414; fax: (41) 1 874 77 77

Abstract. This paper describes a model for calculating the expected losses as well as economic capital required to support an arbitrary portfolio of credit exposures. It does so by explicitly modelling both the marginal and absolute conditional loss distributions for any arbitrary portfolio of credit exposures; these portfolio loss distributions can be made conditional on the current state of the economy given the counterparty's country, industry and rating. The conditioning relationships between the probability of a credit event (e.g. credit rating migrations or defaults) and the current state of the economic cycle are based on empirical regularities observed in historical data. This model differs from other credit portfolio models in several important aspects:

- First, it models the **actual, discrete** *loss distribution, dependent upon the number and size of credits, as opposed to using a normal distribution or mean-variance approximations*; this allows the model to explicitly tabulate a '*large exposure premium*' in terms of risk adjusted capital for less diversified portfolios.
- Second, the **losses** *(or gains)* are measured on a *marked-to-market* basis for credit exposures which *cannot be liquidated* (e.g. most loans or OTC trading exposure lines) as well as those which *can be liquidated* prior to the maximum maturity of the exposure; these loss distributions can therefore be tabulated for any time horizon, including one which coincides with an organisations planning and budgeting process
- Third, the tabulated loss distributions are **conditional** *on the current state of the economy* rather than being based on the *unconditional or 20 year averages which do not reflect the portfolio's true current risk.*
- Finally, a multi-factor model of systematic default risk, as opposed to a single factor model based on asset volatilities and CAPM or public rating histories, is explicitly estimated based on empirically observed 'regional- and sectoral-betas', allowing the model to mimic the actual default correlations between industries and regions at the transaction as well as portfolio level.[2]

---

[1] Thomas C. Wilson is a Partner in the Zürich office of McKinsey & Company and specialises in serving financial institutions in the areas of trading and credit risk management strategies, organisation and risk measurement.
Please direct any comments to Thomas C. Wilson, McKinsey & Company

[2] McKinsey & Company is making available, free of charge, a software implementation of this approach. For further information, please contact Thomas C. Wilson on (1)212 446 7000 or at tom_wilson@McKinsey.com

1 Introduction

Increasingly, the management at many financial institutions is focusing its attention on measuring and managing the risk from its credit exposures at a *portfolio* level, in addition to the *transaction* level. This change in perspective has occurred for a number of reasons, the most notable being:

- First, the recognition that the traditional, binary classification of credits into 'good' credits and 'bad' credits is not sufficient - a pre-condition to managing credit risk at the portfolio level is the recognition that all credits can become potentially 'bad' over time given a particular economic scenario, implying that managing the portfolio's composition through the cycle may help the organisation to better survive rough economic weather.
- Second, the narrowing profitability of traditional credit products, implying little room for error both in terms of the selection and pricing of individual transactions as well as for portfolio decisions where diversification and timing effects increasingly mean the difference between profit and loss.
- Third, the increased opportunities to proactively manage the institution's exposure after it has been originated, including the increased liquidity in the secondary loan market, the increased importance of syndicated lending, the availability of credit derivatives and third-party guarantees, etc. These evolving markets allow management to proactively manage credit exposures as opposed to passively holding the exposures as originated.

In order to take advantage of credit portfolio management, however, management must be in a position to answer several important questions. More specifically, management must be able to such questions as:

- What is the **risk** of a given portfolio? E.g.
  - What are the portfolio's *expected losses*?
  - What amount of *credit risk capital* is required to support the portfolio?
  - How do *'large', undiversified exposures* affect the portfolio's risk?
  - What is the *marginal* and *absolute risk contribution* of an individual position or sub-portfolio of credits?
- How do *different macro-economic scenarios*, at both the *regional* and *industry* sector level, effect the portfolio's risk profile?
- What is the *impact of changing the portfolio mix*? E.g.
  - What is the marginal and absolute risk contribution of a new credit risk position?
  - What is the diversification benefit from altering the credit portfolio mix along the country dimension? Along the industry dimension? Along the rating dimension?
  - How might *risk-based pricing* at the individual contract and portfolio level be influenced by the level of expected losses and credit risk capital?

This paper describes a new and intuitive method for tabulating, via Monte Carlo methods, the conditional loss distribution arising from correlated credit events for any arbitrary portfolio of counterparty exposures, down to the individual contract level. Various parameters of these distributions can then be easily tabulated including expected losses, the maximum absolute and marginal loss within a given confidence, etc., which can then be used not only for portfolio risk management and controlling purposes, but also for risk adjusted pricing and profitability measurement.

It should be stated at the outset that the focus of this paper is **not** on estimating the **potential exposure** of derivatives portfolios, as these issues have been well treated elsewhere (see, for example, Lawrence (95) and Rowe (95), both in Risk Publication's *Derivative Credit Risk: Advances in Measurement and Management*). Rather, the emphasis is on providing an efficient and intuitive method of **tabulating the actual credit loss distribution, conditional on the current state of the economy, from correlated credit defaults and credit quality migrations for credit exposures which can and cannot be liquidated**. The approach developed here differs from others in several important respects:

- First, it models the **actual, discrete** *loss distribution, dependent upon the number and size of credits, as opposed to using a normal distribution or mean-variance approximations*; this is important because, in the extreme with one large exposure, the portfolio's loss distribution is discrete and binomial as opposed to continuous and unimodal, it is highly skewed as opposed to symmetric and, finally, its shape changes dramatically as other positions are added. Because of this, the variance of loss is like a 'rubber ruler': it can be used to give a sense of the uncertainty of loss, but its actual interpretation in terms of dollars at risk depends on the degree to which the ruler has been 'stretched' by diversification or large exposure effects. In contrast, the model developed here explicitly tabulates the actual, discrete loss distribution for any given portfolio, thereby also allowing the management to explicitly tabulate a '**large exposure premium**' in terms of risk adjusted capital for less diversified portfolios; this large exposure premium might be thought of as representing the diversifiable or idiosyncratic risk of the portfolio.
- Second, the **losses** *(or gains)* are measured on a *marked-to-market, discounted* basis[3] for credit exposures which *cannot be liquidated* (e.g. most loans or OTC trading exposure lines) as well as those which *can be liquidated* prior to the maximum maturity of the exposure. This implies

[3] Gollinger and Morgan (93) use the standard deviation of a portfolio's ZETA score, calculated based on the historical correlation between industry ZETA scores over time, as a measure of risk; one drawback with this approach is that it is difficult to interpret the variance of a ZETA score as an absolute, dollar-based risk measure whereas expected losses and maximum possible dollar losses have an intuitive management appeal as risk measures.

that the model can be used to cover the credit exposures arising from liquid, secondary market positions as well as from illiquid, commercial loan positions; in addition, it also allows management to tabulate the actual loss distribution for holding period horizons which might differ from the 'liquidation' period for the commitment but which might better coincide with their capital allocation planning and budgeting period.

- Third, the tabulated loss distributions are based on credit migration and default probabilities which are **conditional** *on the current state of the economy* rather than being based on the *unconditional or 20 year averages which do not reflect the portfolio's true current risk.* This allows us to capture the systematic, or non-diversifiable, risk of the portfolio in addition to the idiosyncratic or diversifiable risk; more importantly, it allows us to capture the cyclical default effects which determine the lion's share of the risk for diversified portfolios.
- Finally, specific country and industry influences are explicitly recognised based on empirically observed 'regional- and sectoral-betas', allowing the model to mimic the actual default correlations between industries and regions at the transaction as well as portfolio level. This is important because, as noted by Gollinger and Morgan (93), 'Generally, industry credit quality moves with changes in the economy. For example, during a recession, credit quality of heavy manufacturers (primary metals, fabricated metals, industrial machinery, and stone, clay and glass products) tends to deteriorate at the same time and about the same rate, indicating a strong inter-industry relationship or high positive covariance.'

The model itself consists of two important components, each component being discussed in greater detail in later sections.

- The first is a *multi-factor model of systematic default risk.* This model is used to jointly simulate the *conditional, correlated, average default* and credit migration probabilities for each individual *country/industry/rating segment*; these average segment default probabilities are made conditional on the current state of the economy and incorporate industry sensitivities (e.g. 'high beta' industries such as construction reacting more to cyclical changes) based on aggregate historical relationships. This is in contrast to many single factor models based on asset volatilities or historical, public rating histories.
- The second is a method for tabulating the discrete loss distribution for any portfolio of credit exposures, both liquid and non-liquid, constant and non-constant, diversified and non-diversified. This is achieved by convoluting the *conditional* loss distributions of the individual positions with the default correlations between different counterparties being determined by the systematic risk driving the correlated average default rates.

Before outlining the model, however, we begin by developing a high level credit risk measurement framework. In the companion article, we develop and estimate a multi-factor credit risk measurement model which captures some of the most important empirical regularities currently challenging risk managers.

2 Credit risk measurement framework

For many financial institutions, *credit risk* is roughly defined as *the potential loss in mark-to-market value which might be incurred due to the occurrence of a credit event.* For the remainder of this article, a *credit event* is defined as *a change in the counterparty's current or future ability to perform under its contractual obligations*, manifested either as a change in credit quality (e.g. upgrading or downgrading of the counterparty's rating) or, in the extreme, an actual default. Including the marked-to-market impact of rating migrations as well as defaults in our definition of credit events is natural and necessary if the model is to be applied to credit exposures which can be liquidated in secondary markets, such as Eurobond and Brady bond positions, some classes of loans for which secondary markets exist, etc.

In addition, it also allows us to explicitly recognise the fact that a rating downgrade can lead to a real economic loss although no accounting loss has been recognised for the more traditional commercial lending products which are not marked-to-market as a matter of course. In order to illustrate the difference, consider a 10 year loan with a triple-A counterparty and suppose that, after one year, the counterparty becomes downgraded to single-A and that average credit spreads have remained unchanged for triple-A and single-A counterparties. Although no accounting loss has occurred, the holder of this commitment has clearly suffered due to the downgrading: the counterparty's future ability to fully meet its contractual obligation, as manifested by its rating, has been diminished, implying both a higher level of expected losses as well as a higher volatility of losses over the remaining life of the obligation.[4]

2.1 Loss distributions, expected losses and credit risk capital

Measuring credit risk at the transaction or portfolio level is complicated by the fact that counterparty defaults and rating migrations cannot be predicted with perfect foresight and are not perfectly correlated, implying that management is actually faced with a*distribution of potential losses* rather than one, single potential loss. In order to make the definition of credit risk more precise in the context of these loss distributions, the financial industry is converging on risk measures which summarise management relevant aspects of

[4] The actual loss in value can be calculated by recognising that, although the holder of the commitment had locked in triple-A spreads for the remaining 9 years of the 10 year commitment, it now has a 9 year single-A commitment which should carry a higher spread if priced at current market rates.

the entire loss distribution. In this regard, there are two distributional statistics which are increasingly relevant for measuring credit risk: *expected losses* and *a critical value of the loss distribution often defined as the portfolio's credit risk capital (CRC)*. Each of these measures serves a distinct and useful role in supporting management decision making and control (see Figure 1).

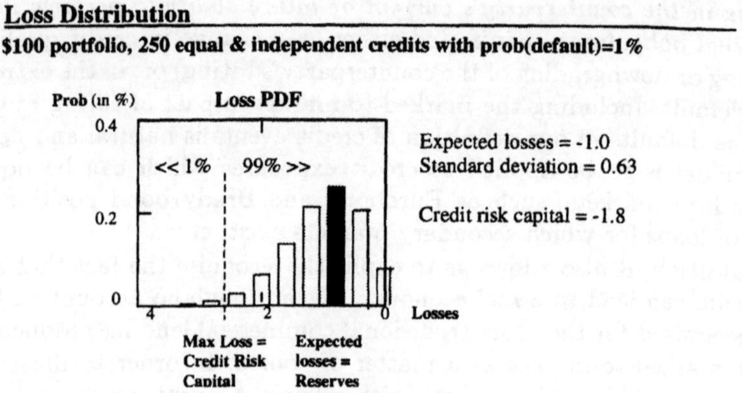

Fig. 1. Loss distribution

Expected losses, illustrated as the mean of the distribution in the figure, often serve as the basis for management's reserve policies: the higher the expected losses, the higher the reserves which need to be set aside to cover them. As such, expected losses are also an important component for determining whether the pricing of the credit risky position is adequate: normally, each transaction or netted credit portfolio should be priced with sufficient margin to cover its contribution to the portfolio's expected credit losses.

Credit risk capital, *defined as the maximum loss within a known confidence interval (e.g. 99%) over an orderly liquidation period*, illustrated in the figure, is often interpreted as the additional *economic capital* which needs to be held against a given portfolio, above and beyond the level of credit reserves, in order to cover its *unexpected* credit losses. Since it would be uneconomic to hold capital against *all* potential losses (as this would imply that equity is held against 100% of all credit exposures), some level of capital must be chosen to support the portfolio of transactions in most, but not all, cases. Just as with expected losses, CRC also plays an important role in determining whether the credit risk of a particular transaction is appropriately priced: typically, each transaction or treaty must be priced with sufficient margin to cover not only its expected losses, but also the cost of its marginal risk capital contribution.

One primary difference between the two risk measures is that, as opposed to expected losses, the marginal CRC contribution of an individual transaction is not equal to its absolute or stand-alone CRC contribution due to diversification effects; in fact, a portfolio's credit CRC will always be less than[5] the sum of the individual transaction's stand-alone CRCs. As such, only the absolute CRC measured at the portfolio level is relevant for the determination of capital adequacy; similarly, only the marginal CRC contribution of the individual transaction relative to the institution's entire portfolio is relevant for risk adjusted pricing.[6]

Figure 1 illustrates these measures with a simple example portfolio with total notional of \$100 million consisting of 250 equal sized, independent credits, each a probability of default of 1%. For this portfolio, the expected losses are equal to \$1 million while the additional maximum losses will be less than or equal to \$1.8 million with 99% confidence. Note that, as the loss distribution is not 'normal' this additional loss is not well approximated by some multiple of the portfolio's standard deviation of losses (e.g. 2.31 times the standard deviation, or \$1.45 million, in the case of a 99% one-tailed confidence interval for the normal distribution).

2.2 Risk decomposition: exposure versus default

Given that industry's convergence on expected losses and credit CRC as risk measures, tabulating an arbitrary portfolio's loss distribution is a critical task; in order to do so, management must be in a position to answer two challenging questions: first, What is the joint probability of a credit event occurring? And, second, what is the loss should such events occurs (see Figure 2)?

Credit event probabilities Most institutions answer the first question by consistently classifying, through an in-depth analysis[7], each individual counterparty into a specific rating category or rating class with a unique probabil-

[5] Or equal to, in the case of perfectly correlated credit events, a rare event in practice.

[6] Nonetheless, if used appropriately, the (scaled) absolute CRC on a transaction by transaction basis is also a useful key or index for allocating credit risk adjusted capital to a particular position. Nonetheless, if used appropriately, the (scaled) absolute CRC on a transaction by transaction basis is also a useful key or index for allocating credit risk adjusted capital to a particular position.

[7] In most institutions, this is typically accomplished by combining a subjective evaluation of counterparty quality, perhaps supported by cash flow and industry analysis, with some form of quantitative scoring model, e.g. the Z-score and Zeta models as well as standard and behavioural scoring models (see, for example, Altman (83) or Thomas, Crook and Edelman (92) for applications in the cor-

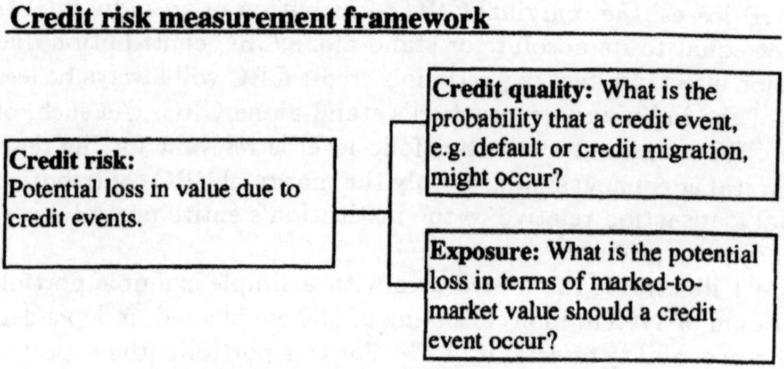

Fig. 2. Credit risk measurement framework

ity of default or rating migration implicitly or explicitly associated with each rating class. Successful completion of this process yields the marginal default or migration probabilities, with the joint distribution being determined by the correlations between rating classes. The second step is therefore to model the *joint distribution* of credit events across all of the counterparties contained in the portfolio. Roughly speaking, this is done by estimating the correlation or covariance between credit events for the different counterparties contained within the portfolio.

The most simplistic models assume that credit events between counterparties are independent or have a constant correlation, an assumption which leads to unacceptable implications as is demonstrated later. Other approaches include the estimation of implicit asset value correlations using equity prices based on the same observations made by Black Scholes which led to the KMV model (again, see Kealhofer (95b)). Using another approach, Gollinger and Morgan (93) model the joint risk of a credit portfolio by measuring the correlation of different industry segments' ZETA score over time.

porate and retail segments, respectively), or, more recently, neural networks and genetic learning algorithms. In addition to these models based on micro-data, e.g. financial statement and account information, another class of models has been developed based on the observation made by Black and Scholes (1973) that the equity and debt of a company can be viewed as a call and put option on the firm's assets, respectively. This relationship has been exploited by KMV, a firm in California, which has developed a default prediction model based on the implied volatility of a firm's asset value from the stock market as well as its current market value of debt and equity. See Kealhofer (1995a, 1995b) and McQuown (1995).

In terms of this risk measurement framework, the model developed in these articles is not a rating classification model; in fact, it takes the counterparties' current rating as given when tabulating the joint credit loss distribution for the entire portfolio of exposures. Its main contribution is that it explicitly models how individual exposures of a given rating for different industries and countries are correlated and relates these correlated outcomes to various macro-economic scenarios in the future. In fact, the primary advance of the model presented here is its parsimonious and tractable model for the joint default behaviour of an arbitrary portfolio of counterparties over time, enabling it to answer explicitly the portfolio management questions discussed earlier.

Losses in case of a credit event The second question concerns the severity of the loss should a credit event occur. The measurement of potential losses given a credit event is a trivial exercise for many standard commercial banking products: if a \$100 million, unsecured loan is in default, then the lending institution stands to lose \$100 million in value, subject to any recoveries, plus or minus some (relatively) small amount due to changing interest rates. For many other of today's banking products, however, answering this question is a non-trivial exercise.

Consider, for example, the credit exposure to a particular counterparty arising from the purchase of a stock option written by that counterparty: if the counterparty should default on its obligations, your loss is not the premium that you had paid for the option, but rather the replacement value of the option at the time that the counterparty goes into receivership. Unfortunately for credit risk managers, this value can change substantially over time as the underlying stock price changes, approaching zero if the option is very far out-of-the-money and the value of the stock less the strike price if the option is very far in-the-money. Similarly, your exposure to a particular counterparty can vary substantially for other products as well, including swaps, forwards, collateralised loans, etc., all of whose net replacement value can change substantially as interest rates, commodity, equity and real estate prices change.

It is clear from these examples that the potential severity of loss is determined by the actual structure of the individual transaction (or portfolio of transactions), the current state of financial markets which influence its value and the potential evolution of these variables in the future. Thus, your exposure to a particular counterparty or group of counterparties is itself a random variable, with the exposure level increasing or decreasing depending upon the evolution of financial markets.

It is, in general, extremely difficult to intuitively compare the potential credit exposure of an equity option with that of a loan or a commodity swap without some method to make them comparable. For this reason, the industry is converging on a standard measure of credit exposure, or a '*Loan*

Equivalent Exposure', which allows such diverse exposures to be put on a comparable basis. Following industry best practices[8], we define the Loan Equivalent Exposure as the sum of two distinct elements (see Figure 3).

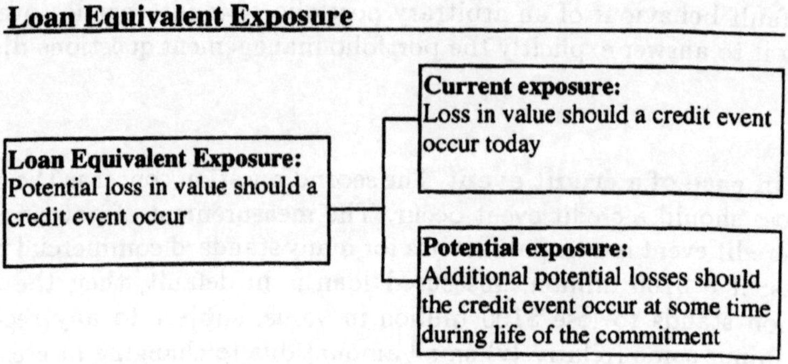

Fig. 3. Loan equivalent exposure

Current Exposure (CE), defined as the loss or replacement cost if the counterparty were to default *today*, recognising valid netting and collateral agreements.

Potential Exposure (PE): Some measure of the additional potential loss, above and beyond the current exposure, if the counterparty were to default *some time in the future*, recognising valid netting and collateral agreements. Typically, potential exposure is calculated in one of three ways:

- **BIS potential exposure add-ons**, a set of multiplicative factors which are applied against the notional of the particular transaction, determined by the remaining maturity of the transaction and the key risk factor which determines its value
- **Expected potential exposure**, which measures the expected exposure to the counterparty at any point in time over the life of the commitment, and
- **Maximum potential exposure**, which measures the maximum exposure which could occur within some given confidence interval (e.g. 99%) at any point in time over the life of the commitment.

While most market participants use the BIS method, more and more are actively experimenting with or actually implementing alternative methods

[8] See, for example, G30 (1993), DPG (1995), Rowe (95), Lawrence (95)

using Monte Carlo or Factor Push methods (see Lawrence (95), Rowe (95) and Wilson (96) for a description). In order to calculate the Loan Equivalent Exposure for an arbitrary contract using one of these methods, the institution first needs to be able to value all of its commitments at any point in time in the future and, second, to simulate those values under different market rate scenarios in order to calculate the transaction's potential value which might be put at risk in the event of a counterparty default. Since this is not the focus of this article, we refer interested readers to the references listed previously.

3 Systematic risk model

This section describes a model for an arbitrary portfolio's systematic or non-diversifiable risk: no matter how many credit exposures of whatever type (e.g. constant or non-constant exposure, liquid or illiquid, large or small), there will always remain some residual risk which cannot be diversified away. The parameters of the systematic risk model are tabulated so as to match the joint historical default and credit migration behaviour of the different country/industry/rating default indices derived from publicly available data[9]; if desired, your institution's own loss history can also be used to train the model. A prerequisite for implementation, however, is that each counterparty be rated in a manner which is consistent with the historical default and credit migration properties of publicly rated counterparties.

In terms of modelling the joint default behaviour or systematic risk of a portfolio of counterparties over time, the model developed here leverages several intuitive observations about the behaviour and impact of individual as well as aggregate credit events. In the next section, we demonstrate how to tabulate losses for an arbitrary portfolio of exposures based on these systematic risk indices.

---

[9] Several attempts have been made to model the dynamics of average default rates for a single counterparty segment using macro-economic time series data; specific examples include Altman (83), who models the dynamics of aggregate US business insolvency data using an Almon lag structure; Cumming and Saini (81), who models aggregate defaults for Britain and Japan Moody's (94), which models aggregate US speculative default data in order to forecast defaults; and Noto and Zimmermann (81), who also models US default data. Each of these econometric models could, in principle, be used to simulate aggregate defaults by sector; as of yet, no effort has been made to date either to simulate across multiple segments simultaneously or to drive the aggregate default rates down to the individual counterparty, industry and rating class level.

3.1 Observation 1: Diversification reduces idiosyncratic or diversifiable risk

The first observation is that diversification of a credit portfolio is beneficial because it reduces the portfolio's idiosyncratic or diversifiable risk, all other factors being equal. Intuitively, one would rather have a portfolio consisting of 100x $1 million exposures rather than 1x $100 million exposures. In fact, by diversifying a portfolio of equivalent risks, one can reduce the volatility of credit losses as well as the CRC which must be held against the portfolio. To support this conclusion, consider the cumulative loss distribution arising from holding a portfolio with nominal value $100 comprising n independent credits, each with a probability of default of 1%. Tabulation of this loss distribution is equivalent to tabulating the losses from a game where you must pay $1n for every black ball pulled, with replacement, from an urn and $0 for every white ball pulled from the same urn consisting of 1% black balls and 99% white balls.

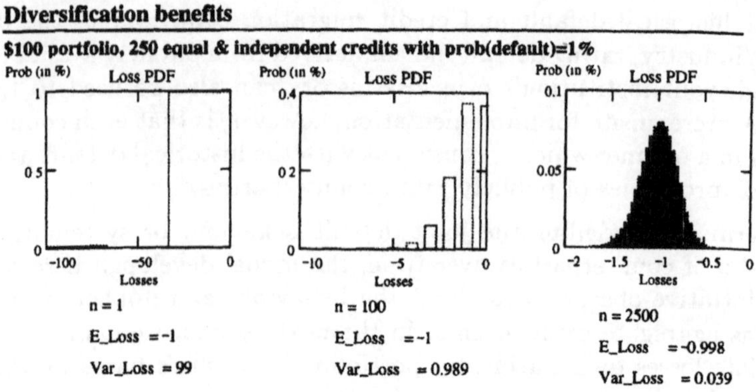

Fig. 4. Diversification benefits

As can be seen in Figure 4, the primary benefit of diversification comes from reducing the variance or uncertainty of loss. In general, diversification does not affect the mean but reduces the variance of losses significantly: for this stylised example where each of the credits are independently and identically distributed, the expected losses remain constant at 1% regardless of the number of credit exposures in the portfolio while the variance of losses decreases from 99% to 3.9% as the number of independent credits increases from 1 to 2,500.

If taken literally, the managerial implications of this simple model are that portfolios with a small number of names should bear a substantial non-diversification penalty relative to those which are more diversified. These conclusions should be taken with a grain of salt, however. For this simple pedagogical model, we have assumed that all credit events were independently and identically distributed: in terms of the distributions being identical, very often it is the case that diversification into new areas or markets is done at the cost of a decreasing quality of counterparty due to less experience and/or information; in terms of independence, even though empirical pairwise default correlations appear to be low[10], empirical data discussed in the next section seems to point to the fact that credit events in the aggregate are definitely not independent.

3.2 Observation 2: Systematic or non-diversifiable risk exists

Unfortunately, the simplistic model used in earlier figures to demonstrate the benefits of diversification has implications which are very unrealistic. Looking at the figure, it is clear that as we let the number of credits increase, the variance of loss becomes smaller and smaller, implying that losses become more certain around the 1% level. For this simple model, as the number of credits goes to infinity, the variance of loss goes to zero and the loss distribution converges to a degenerate distribution with a portfolio loss of -1 being observed *with probability 1. Thus, the independent and identically distributed assumption implies that management will never be 'surprised' by credit losses for very large, diversified portfolios.* Even more disturbing is the fact that management will rarely be surprised by large unexpected losses for portfolios containing *significantly less than* ∞ *credits*, as is illustrated in Figures 5 and 6 below.

The lines in Figure 5 give the probability that the aggregate portfolio losses will fall between a specific range of values as a function of the number of independent credits which are held in the portfolio. Thus, we can be over 99% certain that total portfolio losses will fall within the range of [0.85%,1.15%] for portfolios containing 50,000 credits, that losses will fall between [0.90%,1.10%] for portfolios containing 100,000 credits and that losses will fall between [0.95%,1.05%] for portfolios containing 250,000 credits.

The implications for the portfolio's risk adjusted capital are clear and illustrated in figure 6 below: If all counterparty credit events are independent and identically distributed, then the CRC needed to be held to cover unexpected losses will also converge to zero. This occurs because, as n**, the variance of aggregate portfolio loss approaches zero. As the figure illustrates, however, this convergence in CRC occurs at a very rapid rate, with the total CRC level converging to its asymptotic level after about 100 credits.

[10] Kealhofer (1995a) 'A typical default correlation between two investment grade counterparties (over a one year time horizon) is only around 0.01.'

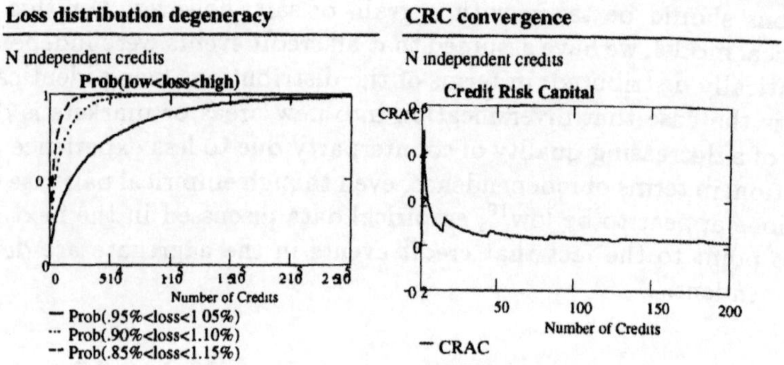

Fig. 5. and **Fig. 6.**

Note that a qualitatively and quantitatively similar effect would prevail even if the average correlation were non-zero; while this would imply that some residual or non-diversifiable risk remained, it would still be quite small as the following quote from Kealhofer (1995a) illustrates: 'The portfolio loss variance of a fully diversified portfolio approaches, in the limit, the average loss covariance in the portfolio. The later value would be in the order of 1/100 of the risk of an individual exposure, assuming that the (pairwise) correlation is in the order of 0.01.'

If we take this simplistic model literally, the implications of the earlier figures are that, with portfolios of around 1,000 credits, senior management of financial institutions should rarely, if ever, be 'surprised' by credit losses since most of the risk should, in theory, be diversified away. These implications run completely counter to common sense when we consider that a portfolio of 10,000 consumer loans or a credit card portfolio with 50,000 counterparties still has substantial capacity to generate significant surprises for top management. The reality is that financial institutions are continually surprised by credit loss variability, year after year, and not just for undiversified portfolios.

These observations are confirmed in Figure 7 below which plots the average default rates of speculative and non-speculative grades over the period 1970-1993 based on data from Moody's (94); while similar relationships hold for each of the individual rating classes, these aggregate categories were chosen for illustrative purposes. Inspection of the charts indicates that the average default rates for these (arguably well diversified portfolios) were not constant over the period, thereby leading to substantial systematic or non-diversifiable risk.

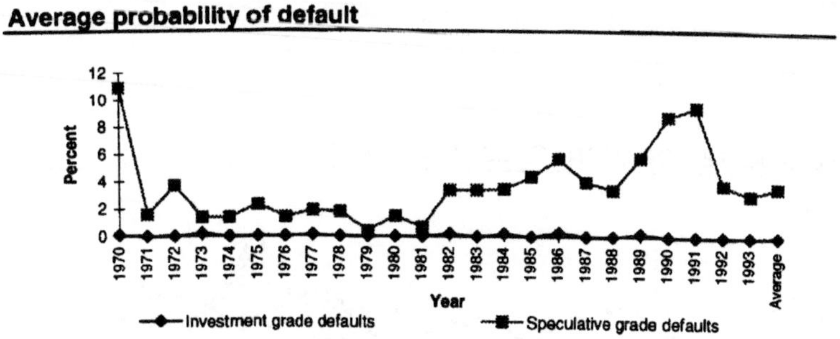

Fig. 7. Average probability of default

Figure 8 illustrates this same point by graphing a time series of (normalised[11]) average corporate default rates for several countries. The figure clearly demonstrates that *average default rates and therefore credit losses are highly volatile, even for well diversified portfolios (e.g. the entire German, French or US economy, etc.)*. The fact that average default rates change over time should not come as a surprise to those who have studied the recent experiences of Sweden, Germany, the United States, Japan as well as a host of other countries where the sudden surge in defaults in the late-80's and early-90's caught many bankers unprepared.

A second observation which can be made looking at the figure is that the *average corporate default rates for different countries seem to be highly correlated*, as is illustrated in Figure 9 which lists the correlation between average default rates for the various countries listed above as well as the standard deviation of their default rates.

Using the analogy described earlier of pulling balls out of an urn: whereas management might think that credit defaults can be represented as pulling balls out of an urn with a *known* distribution of black and white balls (e.g. 1% versus 99%), the reality is that you are drawing balls out of a very different urn every single year with a different ratio of black and white balls. The managerial implication of this observation is that, *for well-diversified portfolios, the majority of the credit loss uncertainty comes* **not** *from the default of individual names, but rather from the uncertainty of the overall default*

[11] Since the average corporate default rates by country are not directly comparable due to different rating compositions, different insolvency criteria, etc., we have normalised the default time series so that they would have the same mean as the Standard & Poor's speculative default rate between 1970-1994.

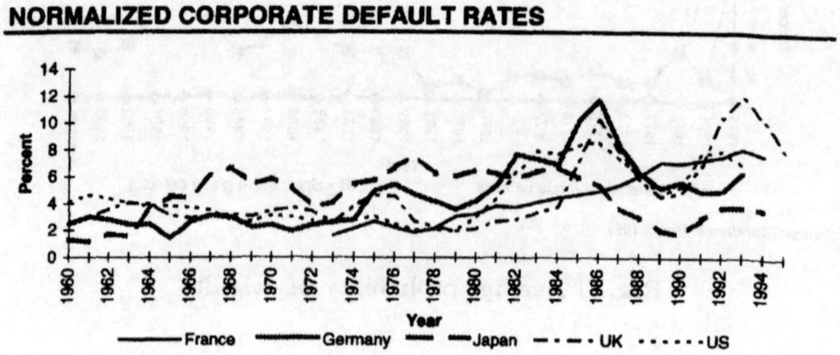

Fig. 8. Normalized corporate default rates

DEFAULT RATE CORRELATIONS & VOLATILITIES *Exhbit 9*

Correlations	FRANCE	GERMANY	JAPAN	UK	US	Standard Deviation*
FRANCE	1.00	0.32	-0.75	0.71	0.70	2.08
GERMANY		1.00	0.03	0.52	0.76	2.40
JAPAN			1.00	-0.44	-0.31	1.81
UK				1.00	0.68	2.45
US					1.00	2.08

* in percent

Fig. 9. Default rate correlations & volatilities

levels. The modelling implication is that we must fully understand not only the effect of pulling a different *number of balls from the same urn* (e.g. the concentration or idiosyncratic risk effect), but also the *distribution of urns* that we are pulling balls out of (e.g. the 'aggregate' or systematic risk effect). Looking forward, these correlated average default rates can be thought of as building the basis for a multi-factor, systematic risk model for credit events.

3.3 Observation 3: Systematic risk linked to macro-economy

Recognising that average default rates themselves are uncertain, we could consider using historical data to tabulate the distribution of these average default rates, e.g. we could answer the question: What is the probability that the *average* default rate is between [0.9%,1.0%), between [1.0%,1.1%), etc., based on historical data? We could then use these probabilities to tabulate the loss distributions for an arbitrary portfolio of credit exposures by convoluting the individual credit loss distributions on a contract by contract basis just as we did for the simple example in the earlier figures.

We would be premature in doing so, however, because we have yet to leverage another intuitive observation. If asked, many people would agree that higher defaults are associated with or even caused by weaker economies or recessions: *as unemployment rates increase and GDP growth declines, defaults and insolvencies tend to increase*. This intuition is strongly supported by the empirical facts. Figure 10 below plots the actual default rate history for Germany against the predicted default rate, where the predicted value was estimated based on statistical techniques using only macro-economic variables as explanatory variables.

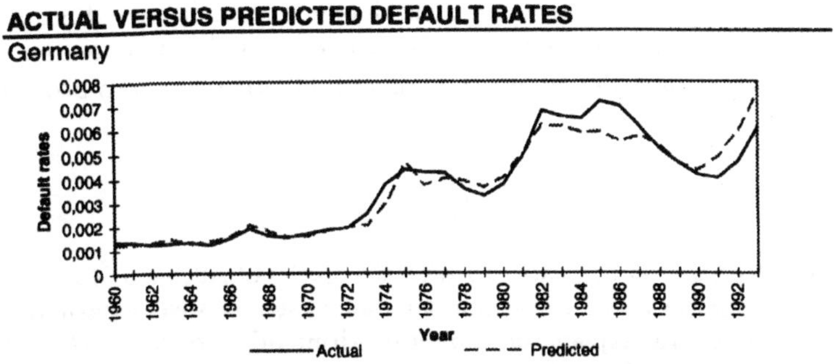

Fig. 10. Actual versus predicted default rates

As can be seen from the figure, the predicted default rates (which depend only on aggregate or macro-economic data) fit the actual default rates very well. The following table gives some impression of the model fit on a country by country basis. As is illustrated by the relatively high R-squares (a measure of the percent of total variation explained by the statistical model) found in the table below, a country's average default rate can be fairly well explained by our prediction equation which uses only macro-economic variables as explanatory variables[12] .

Default prediction model R-Square

Country	R-Square*	Explanatory variables
Belgium	96,9%	Unemployment rates
France	89,1%	GDP growth
Germany	95,7%	Long-term interest rates
Spain	95,3%	Foreign exchange rates
Switzerland	89,3%	Public disbursement
United Kingdom	65,5%	Aggregate savings rate
United States	82,6%	

* Adjusted regression R-Square

Fig. 11. Default prediction model R-square

More specifically, we model speculative default rates by estimating the parameters of a Logit function where the dependent variable is the probability of default for speculative grade counterparties and the independent variable is a segment-specific index which depends upon current macro-economic variables, e.g.

$$p_{j,t} = 1/[1 + exp(y_{j,t})] \tag{1}$$

[12] These results contrast dramatically with the results of Altmann (83), whose linear regression equations with an Almon lag had relatively weak explanatory power; the relatively poorer explanatory power of Altmann's is to some extent due to his modelling strategy where his dependent variable was changes in default rates as opposed to levels of rates. See Chapter 4 for a more in-depth discussion of other statistical models applied to aggregate default data.

where $p_{j,t}$ is the probability of default for a speculative grade counterparty in segment j at time t, $y_{j,t}$ is the segment-specific macro-economic index, the parameters of which need to be estimated. The Logit functional form was chosen over linear and exponential representations for two reasons: first, because it offered, on average, a better 'fit' as measured by R-squared and, second, because, for any value of the index y, it yields a probability between $[0,1]$. This second point is important not least because it is a desirable property for a probability to have, but also because we will be simulating these index values over a multi-year horizon over extreme economic cycles using Monte Carlo methods: without this restriction, we cannot guarantee that the resulting probability of default would be between $[0,1]$.

The macro-economic index, which measures the 'health' of the country's macro-economy, is determined by such variables as the overall unemployment rate, the gross domestic product growth rate, the rate of government disbursements, etc. More specifically, the index takes the following form:

$$y_{j,t} = \beta_{j,0} + \beta_{j,1} * X_{j,1,t} + \beta_{j,2} * X_{j,2,t} + \beta_{j,3} * X_{j,3,t} + \nu_{j,t} \tag{2}$$

where $y_{j,t}$ is the index value specific for the jth country at time t,$\beta_j = (\beta_{j,0}, \beta_{j,1}, \beta_j, 2, \beta_{j,3})$ is a set of regression coefficients to be estimated specifically for the jth country, $X_{j,t} = (X_{j,1,t}, X_{j,1,t}, X_{j,2,t}, X_{j,3,t})$ is a set of macro-economic variables at time t for the jth country, e.g. GDP growth rates, unemployment rates, etc., and $\nu_{j,t}$ is a random variable assumed to be independent and identically normally distributed, e.g.

$$\nu_{j,t} \sim \; iidN(0, \sigma_j) \, or \, \nu_t \sim iidN(0, \Sigma_\nu) \tag{3}$$

where ν is the vector of stacked index innovations and Σ_ν is the jxj covariance matrix of the index innovations. This Logit formulation can be viewed in some sense as a multi-factor model for the determination of country specific average speculative default rates: the systematic risk component is captured by the (weighted) influence of the macro-economic variables with a country specific 'surprise' captured by the error term, ν. In order to make the model country specific, we estimate individual Logit functions for each country whenever country data is available, allowing the explanatory variables to differ between countries.

Inspection of the following figure illustrates the need for a multi-factor model, as opposed to a single-factor model, for systematic risk. Performing a factor analysis of the country average default rates, a good surrogate for systematic risk by country, it emerges that the first 'factor' only captures 77.5% of the total variation in systematic default rates for 'Moody's', the US, UK, Japanese and German markets. This corresponds to the amount of systematic risk 'captured' by most single-factor models; the rest of the variation is implicitly assumed to be independent and uncorrelated. Unfortunately, the first factor only explains 23.9% of the US systematic risk index, 56.2%

for the UK and 66.8% for Germany. The figure demonstrates that the substantial correlation remaining is explained by the second and third factors, covering an additional 10.2% and 6.8% respectively of the total variation and the bulk of the risk for the US, the UK and Germany. This demonstrates that a single-factor systematic risk model like one based on asset betas or Moody's/Standard & Poor's data alone is not sufficient to capture all correlations accurately.

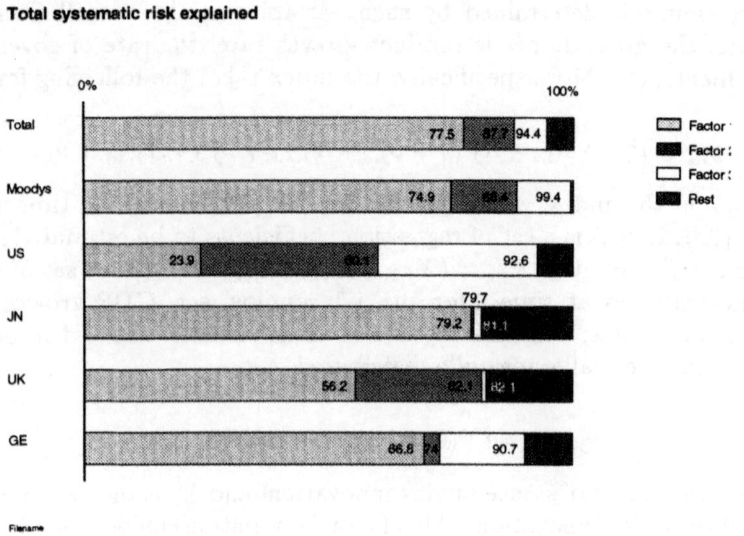

Fig. 12. Total systematic risk explained

One of the managerial implications of this observation is that, given the strong correlation between macro-economic uncertainty and credit events at the portfolio level, credit can be viewed as a play on the economy: if management's view is that the economy will improve over the next 3-5 years, then it should strive to originate longer-term commitments, possibly taking a calculated risk by accepting counterparties with lower ratings than usual; alternatively, if management expects the next 2-3 years to be difficult, then it should try to reduce the maturity of its exposure, increase the credit quality of its portfolio and try to structure contracts which allow management to exercise exit or other risk reduction options.

The implication of this observation in terms of the model is that it should be able to use current macro-economic information to help decide which of the aggregate urns is the most likely to be used when determining next period's defaults. The second modelling implication of these observations is that a multi-factor model of systematic default risk, as opposed to a single-factor model based on asset value betas or historic public default histories, is needed in order to capture the full correlation structure between customer segments.

3.4 Observation 4: Industry specific cyclical risk

The next observation is also intuitive: Different businesses behave differently during economic cycles. For example, construction and transportation firms are more likely to be adversely affected during an economic downturn than are public utilities. This observation is supported by Figure 13, which plots the industry specific default rates for Germany.

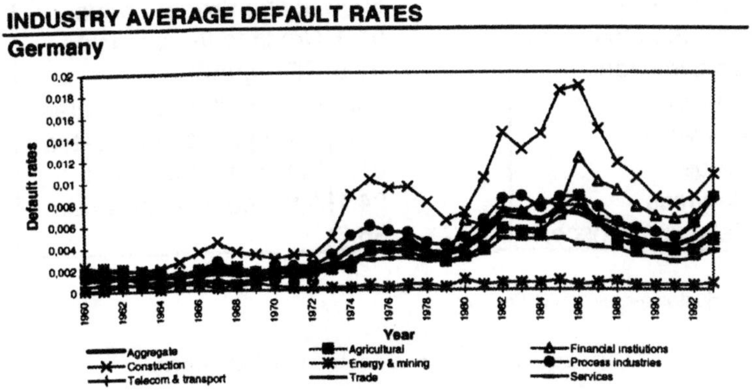

Fig. 13. Industry average default rates

As can be seen from the figure, while the default rates for different industries tend to move together in relation to the aggregate default index, some industries are more sensitive to aggregate market fluctuations than others. These industry effects have an analogy within the Capital Asset Pricing Model: different industries respond differently to aggregate market fluctuations depending upon their local market 'beta'; some industries, such as transportation with higher market betas, respond more strongly than others, such as public utilities, which have lower market betas.

However, it should not be interpreted that industry default risk is completely determined by the state of the macro-economy: It is clear that certain industries undergo dynamic pressures which put them at higher risk independently of the state of the macro-economy. For example, defaults and insolvencies increased dramatically in the early 1990's for the US financial institutions sector without a parallel extreme worsening in the economy; rather, the increased insolvencies within the financial sector were more driven by the dynamics which were unique to that sector. Nonetheless, the variation in these industry-specific default rates can also be explained to a great extent by the current state of the macro-economic environment. Figure 14 lists the Logit regression results for various German industry sectors.

Default prediction model R-Square

Country: Germany

Industry	R-Square*	Explanatory variables
Agriculture	85,4	Unemployment rates
Banking/Insurance	91,1	Credit to the economy
Construction	91,8	GDP growth
Energy	28,2	Long-term interest rates
Process	95,9	Foreign exchange rates
Trade	91,2	Public disbursement
Services	94,1	Aggregate savings rate
Telecom/transport	96,6	

* Adjusted regression R-Square

Fig. 14. Default prediction model R-square

The managerial implications of this observation is that, in times of recession, you would rather steer the portfolio away from construction and other 'high beta' industries and towards industries with lower cyclical defaults. The implications in terms of the model are that some method of capturing the potential industry, as well as country, correlation and 'beta' sensitivity effects needs to be implemented.

3.5 Observation 5: Rating migrations, as well as defaults, vary over the cycle

Up until now, we have focused on only one type of credit event, e.g. default, and then only for a specific rating category, e.g. speculative grade counterparties. In order to effectively model multi-year exposures for different credit classes, this analysis must be extended to include other rating classes as well as rating migrations (e.g. up-grades and down-grades) in addition to defaults. The reasoning is straightforward: first of all, the typical portfolio consists of exposures to counterparties with different ratings, most of which are not speculative grade; second, for multi-year exposures, although the counterparty may currently be triple-A and therefore relatively riskless, there is no guarantee that it will remain so after many years. In order to capture the migration between different rating classes, we will use some empirical relationships between rating migrations for various credit classes and the macro-economic environment. Again, the intuition is obvious: downgrades tend to happen more frequently during recessions than during recoveries and this relationship is stronger for lower rated counterparties.

The starting point is an unconditional Markov transition matrix, based on Standard & Poors' or Moody's historical data. This matrix, shown in Figure 15[13], gives the average probability of migrating from one credit class to another within a year and is estimated based on several years worth of data across many different countries and industries. For instance, the first cell in the ϕM matrix tells us that the probability of ending up with a triple-A rating after one year, given that the counterparty has started with a triple-A rating, is 87.94%; similarly, the adjacent cell tells us that the probability of ending up with a double-A rating after one year, given that the counterparty began with a triple-A rating, is 9.59%.

As mentioned before, the ϕM matrix and the implied cumulative migration probabilities are *unconditional* in the sense that they are based on many years worth of data, covering both economic recessions as well as recoveries, across many different industries and countries; further, these average default and credit migration probabilities for different rating classes are not constant but rather vary in a predictable manner over the cycle; to many, this is merely a restatement of the obvious: as the economy worsens, one would intuitively expect more migration activity for any rating class, and then typically in the downward direction. This is demonstrated in Figure 16 below, which plots rating migrations against average speculative default rates, a good proxy for the state of the economy, as we have seen.

One important observation that can be drawn from this figure is that rating migrations are not only volatile, but are also correlated with changes in

[13] The Markov transition matrices supplied by both Moodys and Standard & Poors unfortunately are not corrected for withdrawn ratings; ratings are typically withdrawn if the company buys back its debt or if it is acquired or merged and the debt is therefore withdrawn from the market. Because each row must sum to one in order for the matrix to be a valid Markov transition matrix, these withdrawn ratings must be 'put back' into the Markov matrix by making some assumptions. Given the typical causes of a withdrawn rating, we make the expedient assumption that the distribution of rating migrations for those counterparties with withdrawn ratings is identical to that for those which did not have their rating withdrawn.

Unconditional credit migration probabilities

Beginning Rating		AAA	AA	A	BBB	BB	B	CCC	Default
		Ending rating							
AAA		0.894	0.098	0.006	0.002	0.001	0	0	0
AA		0.009	0.909	0.071	0.008	0.001	0.002	0	0
A		0.001	0.026	0.901	0.06	0.008	0.004	0	0.001
BBB		0.001	0.003	0.063	0.851	0.063	0.015	0.002	0.003
BB	$\phi M =$	0	0.002	0.006	0.074	0.789	0.102	0.012	0.015
B		0	0.001	0.004	0.006	0.061	0.83	0.038	0.061
CCC		0.002	0	0.002	0.01	0.015	0.12	0.66	0.191
Default		0	0	0	0	0	0	0	1

Source: Standard & Poor's (95)

Fig. 15. Unconditional credit migration probabilities

Default and migration probabilities

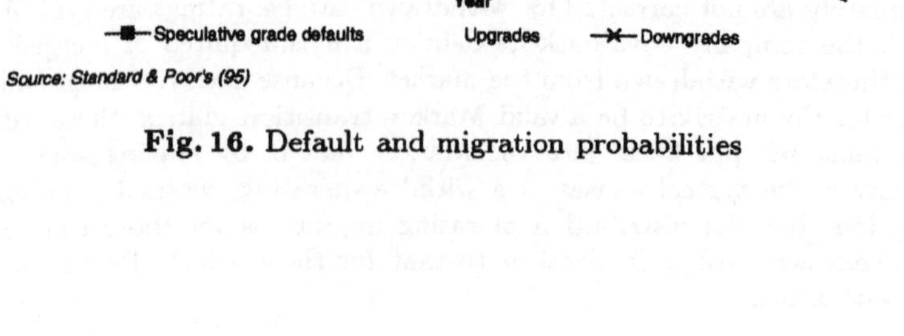

Source: Standard & Poor's (95)

Fig. 16. Default and migration probabilities

the average speculative default rate: If the speculative default rate is higher than average, then downward migrations increase and upward migrations decrease; these observations are confirmed by the correlations between the two given in the Figure 17.

Rating migration and default correlations

	Correlations			
	Investment	Speculative	Downgrades	Upgrades
Investment grade	1,00	0,18	0,23	0,32
Speculative grade	0,18	1,00	0,55	-0,14
Downgrades	0,23	0,55	1,00	0,47
Upgrades	0,32	-0,14	0,47	1,00

Fig. 17. Ration migration and default correlations

The empirical relationships between the speculative default rates and the credit migration as well as default rates for other rating classes discussed earlier can be used to make the *un*conditional Markov rating transition matrix *conditional* on the current speculative default rate. Using these relationships and defining $M(SDP_t/\phi SDP)$ as the one period Markov transition matrix *conditional* on the level of the current speculative default rate relative to its long term average, the figure below demonstrates how the probability mass can be redistributed for different levels of the speculative default rate relative to its unconditional average level[14]:

If the ratio of the actual speculative default rate to its unconditional mean $(SDPt/\phi SDP)$ is equal to one, then the conditional transition matrix is equal to the unconditional matrix; if the ratio of current speculative defaults to average speculative defaults is greater than one (e.g. more defaults than average), then more of the probability mass is shifted into downgraded and default states and vice versa if the ratio is less than one. This is demonstrated in the figure, which illustrates the shift in probability for a ratio equal to 1.2 and 0.8, respectively.[15]

[14] Further refinements regarding the modelling of the relationship between speculative default rates and rating migrations by rating class should be followed up; for example, for those categories where the relationship is less strong, what is the percent of default volatility which is explained by these empirical relationships?

[15] An alternative starting point might have been to model the one year probability of default for each rating class rather than the one year Markov rating transition matrix. We chose this approach instead because we are interested in measuring

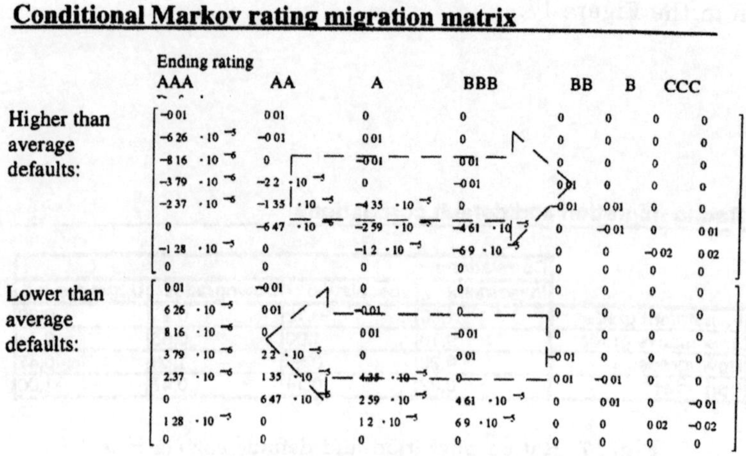

Fig. 18. Conditional Markov ration migration matrix

Based on these conditional one year Markov transition matrices, we can calculate the cumulative conditional probabilities of migrating from one rating class to another over any (yearly) time frame using the following formula:

$$M_{,t} = \Pi_{i=1..t} M(p_{,t}/\phi p) \tag{4}$$

where M_t represents the t-year conditional cumulative rating distribution for a given future path of the speculative default rates, $M(p_{,}t/\phi p)$ is the conditional one year rating transition matrix dependent upon the speculative default rate, p_t as the realised speculative default rate at time t, ϕp is the average speculative default rate and $\Pi_{i=1..t}$ is the multiplication operator. Graphs of these cumulative transition probabilities for starting ratings of AAA and A are given in Figure 19 below.

3.6 Multi-factor, systemic risk model

Using these as well as other historical relationships, we can construct a multi-factor model for the evolution of default and credit migration probabilities. In the remainder of this section, we describe in more detail how these simple, intuitive concepts are implemented in a technically rigorous manner. We do this

the credit risk of multi-year engagements; for such engagements, credit migrations leading to later defaults can be just as important as immediate defaults, implying that modelling one year default rates would have been inappropriate.

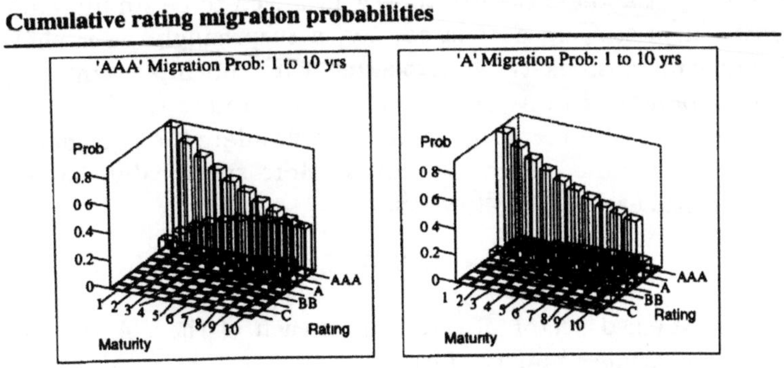

Fig. 19. Cummulative rating migration probabilities

in three steps: first, describing the evolution of the global macro-economy; second, mapping this into country/industry-specific one-year speculative default rates or default indices; and, finally, mapping these into country/industry-specific cumulative rating migration probabilities.

Defining the state of the economy At the beginning of any period t, the state of the economy is drawn from a (conditional) distribution; given our Logit models for the determination of the country/industry-specific speculative default rates, the state of the economy is defined by the various macro-economic variables listed in previous tables. For simplicity, we have chosen to use a set of univariate, auto-regressive of order two (AR(2)) equations to model the development of the individual macro-economic time series which describe the state of the economy[16]. More precisely, we assume that the evolution of each of the macro-economic series is governed by the following functional form:

$$X_{i,t} = k_{i,0} + k_{i,1} * X_{i,t-1} + k_{i,2} * X_{i,t-2} + \epsilon_{i,t} \tag{5}$$

where $X_{i,t}$ is the value of the ith macro-economic variable at time t, $k_{i,j}$, $j = 1..3$, are the three constants to be estimated for each of the i macro-economic variables and $\epsilon_{i,t}$ is the error term, assumed to be distributed $iidN(0, \sigma_i)$.

[16] While it is clear that a better modelling strategy might have been to estimate a vector auto regressive moving average (VARMA(p,q)) model or even a multi-country structural equation model, we chose the independently estimated AR(2) representation because of its simplicity. Further work in the area of improving the macro-economic models should be undertaken.

The constant parameters (k) are estimated using historical data, e.g. data for which $t < 0$. Once the constant parameters have been estimated, the $\epsilon_{i,t}$ terms for $t \geq 0$ can be considered as the forecast errors or 'surprises' for the t4-step ahead forecast and are the variables which need to be simulated. Because the forecast 'surprises' are correlated, e.g. a surprisingly poor showing for GDP growth will most likely be associated with an unpleasant surprise in terms of unemployment rates, we also need to estimate the joint covariance matrix for the forecast error terms in order to simulate the joint development of the different macro-economic variables. More specifically, based on the assumptions made earlier, it follows that

$$\epsilon \sim iidN(0, \Sigma_\epsilon) \tag{6}$$

where ϵ is the stacked vector of errors from each of the i AR(2) equations and Σ_ϵ is the covariance matrix of ϵ's.

Defining country/industry segment default indices Based on this state of the economy, we determine the average speculative default rate for each country/industry segment. As in previous Figures, intuitively, an urn designated for each segment is filled up with the appropriate number of black and white balls: the worse the state, the more black balls, and the better the state, the more white balls. These urns are 'filled up' by using the macro-economic forecast and simulated forecast error terms to drive the Logit functions described in Observations 3 and 4; these are, in turn, used to construct the t-step ahead cumulative rating distributions using the empirical regularities specified in Observation 5.

By combining the earlier equations, we derive that the speculative default rates for all of the j country/industry segments, given a particular realisation of the various macro-economic innovations and innovations to the Logit functions, is given by the following system of equations:

$$P(X, \nu, \epsilon, \eta, t; k, \beta) = \begin{pmatrix} p_t \\ p_{other,t} \end{pmatrix} \tag{7}$$

where p_t is a $jx1$ vector of country/industry-segment average speculative default probabilities. Because we may not have historical speculative default time series for every counterparty in our portfolio, it is necessary to define a catch-all default index, $p_{other,t}$, which is defined as the speculative default probability for all 'other' segments, assumed to be governed by a geometric Brownian motion process with mean equal to ϕp, the unconditional average speculative default rate based on Moody's data from 1970-1994 ($\phi p = 4.6\%$) and volatility, $\phi\sigma$, chosen so $p_{other}(\eta_t)$ has the same unconditional proportional volatility of the one year speculative default rate as the Moody's data[17].

[17] Note that other formulations provide a better fit with the historical data such as a Geometric Brownian Motion with or without mean reversion, etc.

More specifically, we define a system of equations governing the joint evolution of the country/industry specific speculative default rates and associated macro-economic variables:

$$
\begin{aligned}
p_j(y_{j,t}) &= \frac{1}{1+exp(y_{j,t})} \quad and \\
p_{other}(\eta_t) &= \phi p * exp(-\sigma^2/2 + \sigma * \eta_t) \\
y_{j,t} &= \beta_{j,0} + \beta j, 1 * X_{j,1,t} + \beta_{j,2} * X_{j,s,t} + \beta_{j,3} * X_{j,3,t} + \nu_{j,t} \qquad (8) \\
X_{i,t} &= k_{i,0} + k_{i,1} * X_{i,t-1} + k_{i,2} * X_{i,t-2} + \epsilon_{i,t} \\
E &= \boldsymbol{u}, \epsilon, \boldsymbol{\nu} * iidN(0,\Sigma), \Sigma \equiv \begin{bmatrix} \Sigma_\nu & \Sigma_{\nu,\epsilon} & 0 \\ \Sigma\epsilon, \nu & \Sigma_\epsilon & 0 \\ 0 & 0 & 1 \end{bmatrix}
\end{aligned}
$$

where E is the $(j+i+1)x1$ vector of innovations affecting the system of equations given above and Σ is the $(j+i+1)x(j+i+1)$ covariance matrix of macro-economic variable forecast errors (ν), country/industry-specific speculative default rate 'shocks' (ϵ) and the random innovation to the unconditional speculative default probability.

Defining A as the nxn Cholesky decomposition of Σ such that $\Sigma = AA'$, we are now in a position to simulate the joint speculative default rates across all country/industry-segments over some time horizon, T. This is done by, first, drawing a sequence of realisations for z_t, $t = 1$ to T, of $(j+i+1)$ $iidN(0, I)$ random variables, where I is the $(j+i+1)x(j+i+1)$ identity matrix; second, calculating the realisations E_t, incorporating the correlations between the various macroeconomic and segment specific default innovations, using the relationship $E_t = A * z_t$; third, calculating $P(t)$ using the above system of equations.

Figure 20 gives an example of five simulations of the speculative default rates for Germany over a ten year horizon. For example, we can interpret the uppermost simulation run as representing an immediate recession followed by a gradual recovery and the lowermost simulation run as being an immediate and sustained improvement in the global economy.

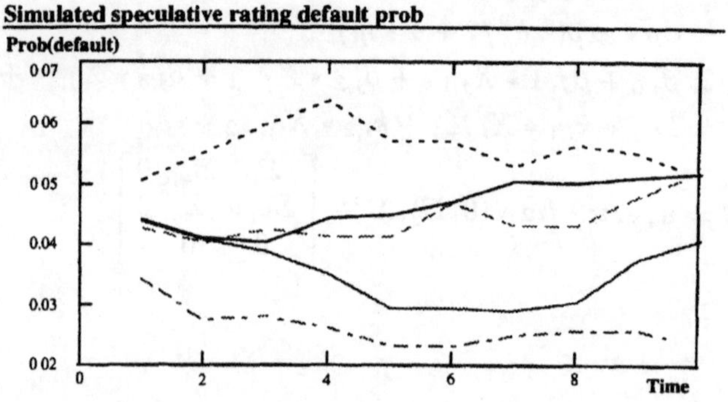

Fig. 20. Simulated speculative rating default probability

Tabulating the cumulative migration distributions Once the speculative default rates for each country/industry segment have been simulated, we then calculate the unique Markov transition matrix for each country/industry segment for any time horizon by using equation (5). Using the relationships outlined in Observation 5 and the simulated speculative default rates, we can calculate the conditional cumulative rating distribution over any arbitrary time horizon initial rating classes where the conditioning is on the simulated path of the speculative default rates. Figure 21 gives, as an example, the histogram of simulated cumulative default probabilities for a 5 year single-A counterparty in Germany based on 500 simulations.

Similar histograms can be tabulated for both cumulative default probabilities as well as cumulative credit rating migration probabilities for any maturity, rating class, country and industry combination which has been modelled. It is interesting to note that the histogram in the previous figure does not look 'normal': first, it is truncated at zero (a natural phenomenon since probabilities should not in general be negative) and, because of this it is skewed. In general, this phenomenon is most prevalent for higher probabilities on average. The distributions for lower rated, longer maturity portfolios

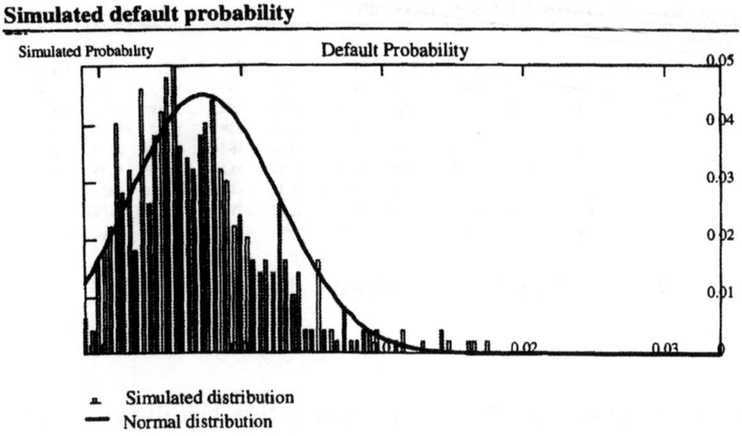

Fig. 21. Simulated default probability

tend to be more symmetric in shape (since the mean is quite a bit higher), although they also feature skewness and kurtosis.

Systematic risks Given the simulated default probability distribution for each of the country/industry/rating segments, we can tabulate several interesting distribution parameters such as the expected conditional default probability and maximum possible default probability for each segment. The figure below tabulates the conditional expected cumulative default rates for the German index portfolio for exposures of any maturity up to 10 years and any initial rating from triple-A to default, conditional on the macro-economic environment in Germany in 1995. This table *could* in principle represent the expected portfolio loss rates *if and only if* four conditions are met: first, the portfolio was diversified to the extent that it mimicked the German index portfolio (e.g. a large number of small German exposures in all industries); second, if recovery rates were zero; third, if the exposure for each individual counterparty were constant over its entire life; and, finally, if the losses are to be treated on a non-discounted basis. Since these assumptions are restrictive in practice for an arbitrary portfolio of credit exposures, we present the cumulative expected default probabilities here only to illustrate the properties of the systematic risk model; *we relax each of these restrictive assumptions in the next section*, demonstrating there how to tabulate the actual loss distributions for any arbitrary portfolio.

Cumulative conditional default probability

Germany, in %

Year	1	2	3	4	5	6	7	8	9	10	Rating
E_Prob =	0	0.008	0 035	0.093	0.193	0 345	0.562	0 863	1.252	1.741	AAA
	0	0.055	0 178	0 382	0 684	1.094	1 628	2.315	3.143	4.122	AA
	0 121	0 364	0.78	1 385	2 206	3 241	4 493	5 999	7.697	9 579	A
	0 575	1 61	3 153	5.121	7.483	10 135	13.013	16 148	19 367	22 633	BBB
	2 872	7 087	12 16	17 51	22 941	28 209	33.227	38.094	42.599	46 75	BB
	11 798	22 17	31.473	39 438	46 34	52 253	57.333	61.858	65 755	69.12	B
	26.999	44.02	55.545	63 445	69 169	73.446	76 76	79.494	81.726	83 582	CCC
	100	100	100	100	100	100	100	100	100	100	D

Fig. 22. Cumulative conditional default probability

According to this table, the average cumulative default rates for a 3 year, single-A diversified German index position is 0.755%, conditional on the current macro-economic environment. The last row of this table is equal to 100 since this row represents counterparties which are currently in default and the model assumes that default is an absorbing state. In a similar manner, we can also tabulate other critical values of the index loss distribution. In the figure, we have tabulated the additional probability of default within a 99% confidence interval, above and beyond the expected default probability, for the German index portfolio conditional on the current macro-economic environment. As with the figure, this table *could* be interpreted as the Credit Risk Capital (CRC) for a portfolio of German exposures *if and only if* the portfolio mimicked the German index, had no recovery potential in the case of default, had constant exposures over the life of the commitments and if the losses were to be treated on a non-discounted basis. Again, we will demonstrate how to calculate the CRC from an arbitrary portfolio of exposures in the next section.

There are two interesting and related phenomenon which emerge from the figures given in Figure 23: The excess maximum default probability above their expected values is neither monotone, increasing by rating category, nor by maturity, as one might expect given its close relationship to credit risk capital; these observations are reflected most clearly in the declining numbers for the triple-C rating category for maturities longer than 2 years and for the profile by rating for 8 year exposures which peaks at the double-B class and declines thereafter.

The explanation for this phenomenon is straightforward, for lower rated exposures, the uncertainty of loss relative to the expected loss actually decreases with maturity after some point, reflecting the fact that more of the exposure will be lost with increasing certainty as the maturity is extended. This interesting effect disappears, however, if one focuses on the total max-

Cumulative excess loss probability*

Germany, in %

Max_Prob = Year	1	2	3	4	5	6	7	8	9	10	Rating
	0	0.01	0.02	0.06	0.12	0.24	0.41	0.63	0.83	1.08	AAA
	0	0.04	0.1	0.22	0.36	0.64	1	1.43	1.77	2.17	AA
	0.05	0.17	0.36	0.68	1.03	1.67	2.4	3.19	3.72	4.3	A
	0.25	0.74	1.35	2.24	3.05	4.39	5.68	6.85	7.38	7.91	BBB
	1.26	3.06	4.57	6.29	7.35	9.05	10.27	11.05	10.92	10.79	BB
	5.16	7.79	8.84	9.87	9.97	10.67	10.88	10.72	9.96	9.31	B
	7.07	8.37	7.94	7.75	7.11	7.03	6.78	6.42	5.83	5.36	CCC
	0	0	0	0	0	0	0	0	0	0	D

* maximum default probability within 99% CI less expected default probability

Fig. 23. Cumulative excess loss probability

imum risk, defined as the total maximum default probability within a 99% confidence interval, which, like expected losses, is monotone increasing in maturity for all rating classes. This is demonstrated in Figure 24:

Total risk (= expected + maximum loss)

Germany, in %

E_Loss + Max Loss = Year	1	2	3	4	5	6	7	8	9	10	Rating
	0	0.01	0.06	0.15	0.31	0.58	0.97	1.49	2.08	2.82	AAA
	0	0.09	0.28	0.6	1.05	1.74	2.63	3.75	4.91	6.29	AA
	0.17	0.54	1.14	2.06	3.23	4.91	6.89	9.19	11.41	13.87	A
	0.83	2.35	4.5	7.37	10.54	14.53	18.69	23	26.75	30.54	BBB
	4.13	10.15	16.73	23.8	30.29	37.25	43.5	49.14	53.52	57.54	BB
	16.96	29.96	40.31	49.31	56.31	62.92	68.21	72.58	75.71	78.43	B
	34.07	52.39	63.49	71.19	76.28	80.48	83.54	85.91	87.55	88.94	CCC
	100	100	100	100	100	100	100	100	100	100	Default

Fig. 24. Total risk

4 Modelling portfolio loss distributions

In the previous section, we developed a multi-factor model for systemic default and credit migration risk based on historical average default rate time series for different country/industry segments. We now turn our attention to developing an efficient method for tabulating the loss distribution for any arbitrary portfolio, capable of handling portfolios with *large, undiversified positions* **and/or** *diversified* portfolios, portfolios with *non-constant exposures*, such as those found in derivatives trading books, **and/or** *constant exposures*, such as those found in commercial lending books, and portfolios comprising *liquid credit-risky positions* such as secondary market debt or loans **and/or** *illiquid exposures* which must be held to maturity, such as some commercial loans or trading lines. We demonstrate how to tabulate the loss distributions for the simplest case (e.g. constant exposures, non-discounted losses) and then build upon the simplest case to handle more complex cases (e.g. non-constant exposures, discounted losses, liquid positions, correlated recovery rates, etc.).

4.1 Numerical example

In order to illustrate how to tabulate the conditional loss distributions for an arbitrary portfolio, consider Figure 25 which illustrates an abstract time-line for the determination of credit events.

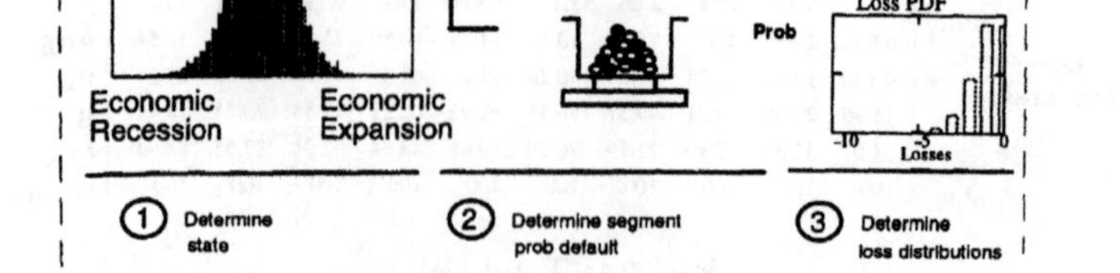

Fig. 25. Model structure

In our multi-factor model, time is divided into discrete periods, indexed by t, and during each period a sequence of three events occurs: First, the state of the economy is determined; second, the conditional cumulative probability of defaults for each country/industry segment is determined; finally, actual defaults for the portfolio are determined by sampling from the relevant segment default distributions, given the composition of the portfolio. To make the calculations more tangible through a numerical example, consider the single period, two segment model numerical example illustrated in Figure 26.

Numerical example

① Determine state

State	GDP	Prob
Expansion	+1	33.33%
Average	0	33.33%
Recession	-1	33.33%

② Determine segment prob default

State	'Low beta' Prob A	'High beta' Prob B
Expansion	2.50%	0.75%
Average	2.97%	3.45%
Recession	4.71%	5.25%

③ Determine loss distributions

Fig. 26. Numerical example

Determine the state: During any given time period, the first step is to determine the state of the world or macro-economic health of the economy. In this simple one period example, there are three possible states of the economy which can occur: an economic 'expansion' (with GDP growth of +1%), an 'average' year (with GDP growth of 0%) and an economic 'recession' (with GDP growth of -1%); each of these states can occur with equal probability (33.33%).

Determine segment prob(default): The second is to then translate the state of the world into conditional probabilities of default for each customer segment. In this simple example, there are two counterparty segments, a 'low beta' segment whose probability of default reacts less strongly to macro-economic fluctuations (with a range from 2.50% to 4.71%) and a 'high-beta' segment which reacts quite strongly to macro-economic fluctuations (with a range from 0.75% to 5.25%). These default probabilities would be driven by the systematic risk models described in the last section.

4.2 Constant exposures, non-discounted losses

Using the simple numerical example developed in the last section, we now demonstrate how to tabulate the (non-discounted) loss distribution for portfolios which are constant over their life, cannot be liquidated and have known recovery rates, including both diversified and non-diversified positions; as such, the approach outlined in this section might be used to get a reasonable first-cut approximation to potential loss distributions for portfolios of non-liquid, traditional commercial banking exposures. In later sections, we show how to relax each of these assumptions within the framework of this model in order to more accurately estimate the expected losses and risk capital from credit events.

Undiversified portfolios The conditional loss distribution in the simple two counterparty, three state numerical example is tabulated by recognising that there are three independent draws or states of the economy and that, *conditional* on each of these states, there are only four possible default scenarios (e.g. A defaults, B defaults, A+B defaults or no-one defaults), illustrated in Figure 27.

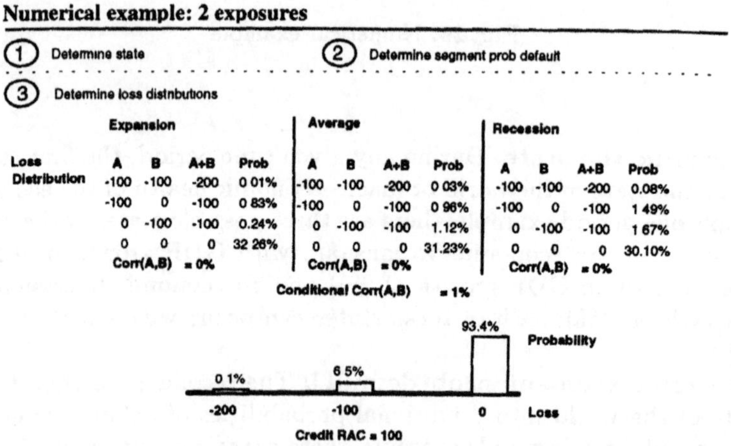

Fig. 27. Numerical example: 2 exposures

The conditional probability of each of these loss events for each state of the economy is calculated by convoluting each position's individual loss distribution for each state; thus, the conditional probability of a \$200 loss in the expansion state is 0.01%, whereas the unconditional probability of achieving the same loss given the entire distribution of future economic states (e.g. expansion, average, recession) is 0.1%. For this illustrative example, the expected portfolio losses is \$6.502 and the credit risk capital is \$100 since this is the maximum potential loss within a 99% confidence interval across all possible future states of the economy.

The implicit assumption underlying our calculation method is that *all default correlations are caused by the correlated segment default indices, i. e. that no further information beyond country, industry, rating and the state of the economy is useful in terms of predicting the default correlation between any two counterparties.* To underscore this point, suppose that management were confronted with two single-A counterparties in the German construction industry with the prospect of either a recession or an economic expansion in the near future. Using the traditional approach, which ignores the impact of the economy in determining the default probabilities for this and other segments, we would conclude that the counterparty default rates were correlated. Using our approach, we observe that, in a recession, the probability of default for both counterparties is significantly higher than in during an expansion and that their joint conditional probability of default is therefore also higher, leading to correlated defaults. However, because we assume that all idiosyncratic or non-systematic risks can be diversified away, no other information beyond the counterparties' country, industry and rating (e.g. the counterparties' segmentation criteria) are useful in determining their joint default correlation.

As an aside, if the recovery rate were known with certainty, we could adjust the conditional loss distributions appropriately to reflect the fact that a default event did not imply that the full exposure was lost; for example, if the recovery rate was 71.57% (Standard & Poor's average recovery rate for triple-B senior secured debt), then the expected losses would be \$1.85 and the Credit Risk Capital would be \$28.43. Note that the assumption of a constant recovery rate will be relaxed in a later section.

Portfolio diversification Intuitively, as we increase the number of counterparties per segment, we should be able to diversify away all idiosyncratic risk, leaving only systematic or non-diversifiable risk. More succinctly, *as we diversify our holdings within a particular segment, that segment's loss distribution will converge to the loss distribution implied by the segment index.* This logic is consistent with other single or multi-factor models in finance; for example, in the Capital Asset Pricing Model, as the portfolio of equity holdings becomes more and more diversified, all idiosyncratic risk is diversified away leaving only the residual systematic or market index (reference Copeland & Weston (83) for a more in depth discussion).

Our multi-factor model for systemic default risks is qualitatively similar: as the portfolio becomes more and more diversified, only the residual systematic or market index risk remains. In our simple numerical example, for a well diversified portfolio consisting of a large number of counterparties in each segment (the NA & NB = Infinity case), all idiosyncratic risk per segment is diversified away, leaving only the systematic risk per segment (see Figure 28).

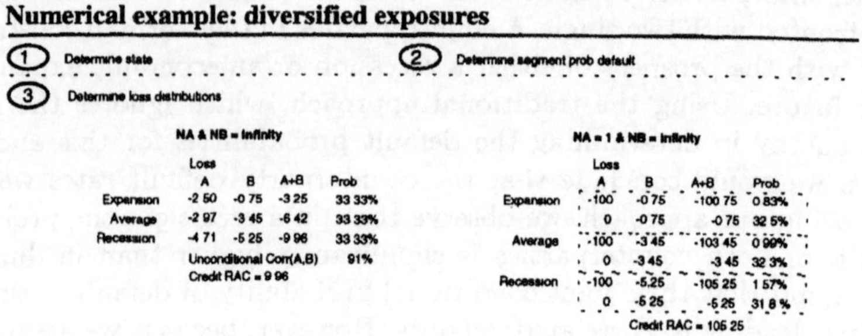

Fig. 28. Numerical example: diversified exposures

In other words, because of the law of large numbers, the actual loss distribution for the portfolio will converge to the expected loss for each state of the world, implying that the unconditional loss distribution has only three possible outcomes representing each of the three states of the world, each occurring with equal probability and each with a loss per segment consistent with the conditional probability of loss for that segment given that state of the economy. While the expected losses from the portfolio would remain constant, this remaining systematic risk would generate a CRC value of only \$9.96 for the \$200 million exposure in this simple example, demonstrating both the benefit to be derived from portfolio diversification as well as the fact that not all systematic risk can be diversified away.

In the second case (labelled NA = 1 & NB = Infinity), all of the idiosyncratic risk is diversified away within segment B leaving only the systematic risk component for segment B; the segment A position, on the other hand, still contains idiosyncratic risk since it comprises only a single risk position. Thus, for each state of the economy, two possible outcomes are possible: either the counterparty in segment A goes bankrupt or it doesn't; the unconditional probability that counterparty A will default in the economic expansion state is 0.83% (e.g. 33.33% probability that the expansion state occurs times 2.5% probability of default for a segment A counterparty given that state); regardless of whether or not counterparty A goes into default, the segment B position losses will be known with certainty, given the state of the economy, since all idiosyncratic risk within that segment has been diversified away.

To illustrate the results using our simulation model, suppose that we had equal $100, 10 year exposures to single-A counterparties in each of five country segments: Germany, France, Spain, the US, and the UK at the beginning of 1996. The aggregate loss distribution for this portfolio of diversified country positions based on sixty simulations, conditional on the then-current macroeconomic for the different countries at the end of 1995, is given in Figure 29.

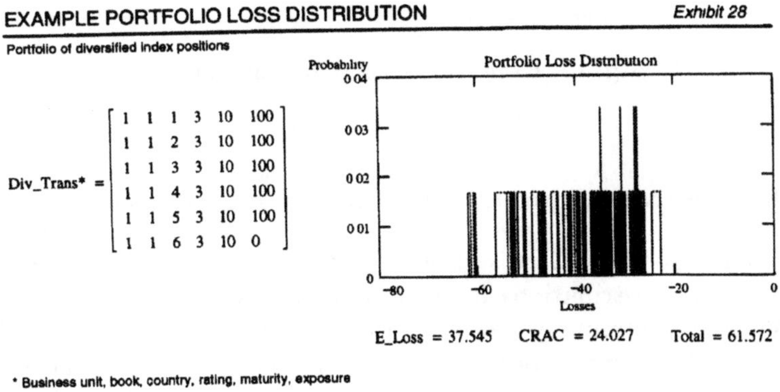

Fig. 29. Example portfolio loss distribution

The impact of introducing one large, undiversified exposure into the same portfolio is illustrated in Figure 30. In this Figure, we take the same 5 country portfolio of diversified index positions which were used in the previous figure, but add a single, large, undiversified position to the 'other' country's position.

As can be seen in the figure, the inclusion of a single, large, undiversified position has the following impact: first, the return distribution becomes bimodal, reflecting the fact that, for each state of the world, two events might occur: either the large counterparty will go bankrupt, generating a 'cloud' of portfolio loss events centered around -140, or the undiversified position will not go bankrupt, generating a similar 'cloud' of loss events centred around -40, but with higher probability. This calculation also demonstrates the large exposure risk capital premium needed to support the addition of a large, undiversified exposure.

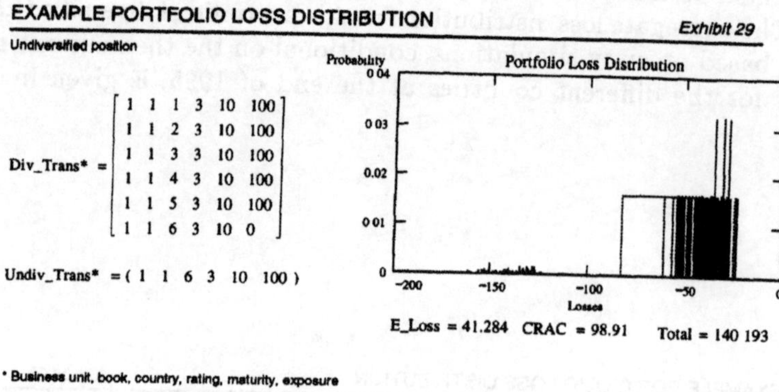

Fig. 30. Example portfolio loss distribution

4.3 Non-constant exposures and/or discounted losses

The calculations in the last section illustrated how to tabulate the (non-discounted) loss distributions for non-liquid portfolios with constant exposures. While useful in many instances, these portfolio characteristics abstract from reality along two important dimensions: first, the potential exposure profiles generated by trading products is typically not constant (as pointed out in our Risk Measurement Framework discussion) and, second, it ignores the time value of money, e.g. a potential loss in the future is somehow 'less painful' in terms of today's marked-to-market value than a loss today.

The challenge to be overcome, as can be seen in the following figure, is that the amount of potential economic loss in the event of default varies over time, either due to discounting or to non-constant exposure or both. Thus, if the counterparty were to go into default sometime during the second year, then the present value of the portfolios loss would be \$50 in the case of non-constant exposures and \$100 $* \, exp{-r_2} * 1$ in the case of discounted exposures, as opposed to \$100 and \$100 $* \, exp{-r_1} * 1$ if the counterparty had gone into default some time during the first year. As opposed to constant, non-discounted exposures where the timing of the default was inconsequential, non-constant exposures or discounting of the losses implies that the *timing of the default is critical in terms of tabulating the economic loss in the event of default.*

Addressing both of these issues requires that one work with *marginal*, as opposed to *cumulative*, default probabilities: whereas the *cumulative* default probability tells you the probability of observing a default in *any of the prior years*, the *marginal* default probability tells you the probability of observing

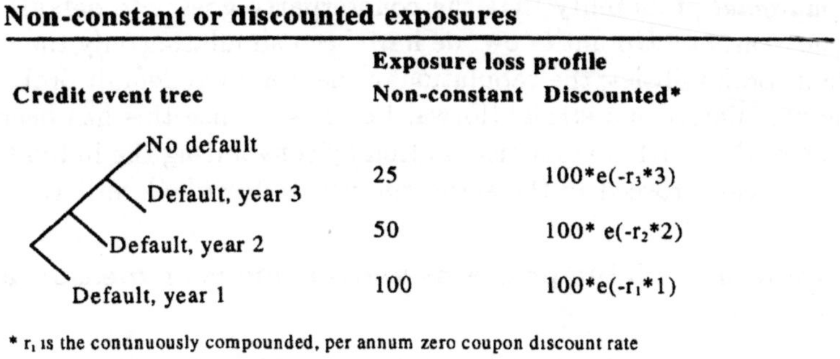

Fig. 31. Non-constant or discounted exposures

a *loss in each specific year*, assuming that the loss has not already occurred in a previous period. Figure 31 illustrates the impact of non-constant loss exposures in terms of tabulating loss distributions.

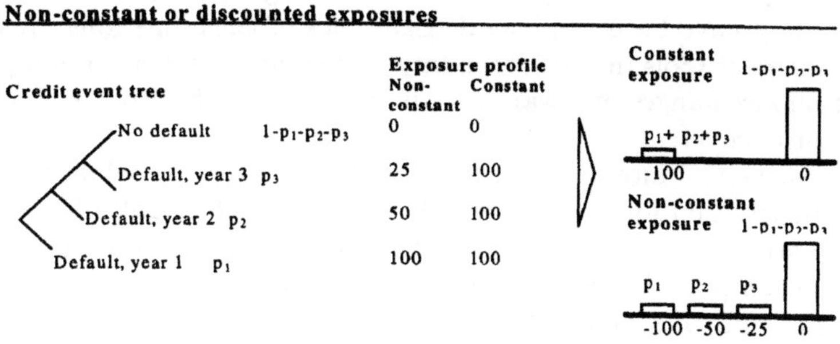

Fig. 32. Non-constant or discounted exposures

With constant exposures, the loss distribution for a single exposure is bi-modal - either it goes into default at some time during its maturity, with a *cumulative* default probability covering the entire three year period equal to $p_1 + p_2 + p_3$ in the above diagram, implying a loss of 100, or it does not; if the exposure is non-constant, however, you stand to lose a different amount depending upon the exact timing of the default event. In the above example, you would stand to lose 100 with probability p_1, the *marginal* probability that the counterparty goes into default during the first year, 50 with probability p_2, the *marginal* probability that the counterparty goes into default during the second year, etc. Up until now, we have been simulating only the cumulative default probabilities; the tabulation of the marginal default probabilities from the cumulative is a straightforward exercise. Once this has been done, tabulation of the portfolio loss distribution by convoluting the individual loss distributions can proceed in the same manner as described earlier.

4.4 Liquid or tradable positions and/or one year measurement horizons

Up until now, we have also assumed that the counterparty exposure must be held until its maturity and that it cannot be liquidated at some market price at any time prior to its maturity or expiration. This assumption is inadequate for three reasons:

- First, many financial institutions are faced with the increasing probability that a bond name will also show up in their loan portfolio; as such, they want to measure the credit risk contribution arising from their secondary bond trading operations and integrate it into an overall credit portfolio perspective.
- Second, while this assumption has been historically appropriate for many traditional asset classes, especially corporate loans and credit lines, to support off-balance sheet products, financial institutions are gaining increasingly more flexibility in terms of managing their credit exposures after they have been originated. Examples of such emerging credit risk management tools include secondary and securitized loan trading, credit portfolio exchanges or 'swaps', credit derivatives, third party guarantees or insurance, etc.
- Alternatively, many financial institutions choose to tabulate the actual loss distribution for holding period horizons which might differ from the 'liquidation' period of the commitment, choosing, for example, a capital attribution period consistent with the periodicity of their capital allocation planning and budgeting process.

In all three cases, management is presented with two specific measurement challenges. First, as is the case in measuring market risk capital or Value at Risk, management must decide on the appropriate time horizon over which to measure the potential loss distribution. In the previous examples, the relevant time horizon coincided with the maximum maturity of the exposure based on the assumption that management could not liquidate the position prior to its expiration. Relaxing this assumption is commonly done either by specifying a *common holding period* consistent with your institution's planning and budgeting process or, alternatively, specifying a *position-specific holding period* horizon coinciding with the time it would take management to recognise a problem situation and then liquidate the position in an orderly manner. This latter approach requires that management have a clear perspective regarding the market liquidity for each individual position.[18]

The second challenge arises with regards to tabulating the marked-to-market value losses should a credit event occur. Up until now we have defined the loss distribution *only* in terms of *default events*; it is clear that, if the position can be liquidated prior to its maturity, then other credit events (e.g. credit downgrades and upgrades) will affect its marked-to-market value prior to at any time prior to its expiration. For example, if you have locked in a single-A spread and the credit rating of the counterparty decreases to a triple-B rating, then you would suffer an economic loss: while the market would demand a higher, triple-B spread, your commitment only provides a lower, single-A spread.

In order to calculate the marked-to-market loss distribution[19] for positions which can be liquidated prior to their maturity, we therefore need to modify our approach along two important dimensions. First, we need to simulate not only the cumulative default probabilities for each rating class, but also the probabilities that the counterparty's rating might migrate from one class to another over some time horizon; this is a straightforward, albeit memory intensive, calculation. Complicating this calculation, however, is the fact that, if the time horizons are different for different asset classes, then a continuum of rating migration probabilities might need to be scored, one for each possible maturity or liquidation period. In order to reduce the complexity of the task, we tabulate migration probabilities for yearly intervals only and make the expedient assumption that the rating migration probabilities for any liquidation horizon which falls between years can be approximated by linear interpolation.[20]

Second, and more challenging, we need to be able to tabulate the change in marked-to-market value of the exposure for each possible change in credit rating. In the case of traded loans or debt, a pragmatic approach is to simply

---

[18] Complicating this process is the fact that, while any credit exposure can be liquidated for a price, whether or not it actually is liquidated depends upon the price offered; very often, management is far more willing to acknowledge the possibility of liquidating a credit-risky position until it is faced with the need to actually do so.

[19] It is interesting to note that, in the event of a credit upgrade, it is also possible to make a profit from a credit event by, for example, locking in a single-A spread for a counterparty which then migrates to a double-A risk class; as such, it may be a misnomer to call the resultant distribution in marked-to-market value changes a 'loss' distribution.

[20] Clearly, this assumption can be relaxed if the Markov migration probability matrix can be calculated for time horizons of higher frequency; the problem which one might face is that this calculation may introduce more noise than explanatory power due to the higher frequency of the data.

define a table of average credit spreads, in basis points per annum, as a function of rating and the maturity of the underlying exposure. The potential loss (or gain) from a credit migration can then be tabulated by calculating the change in marked-to-market value of the exposure due to the changing of the discount rate implied by the credit migration.[21]

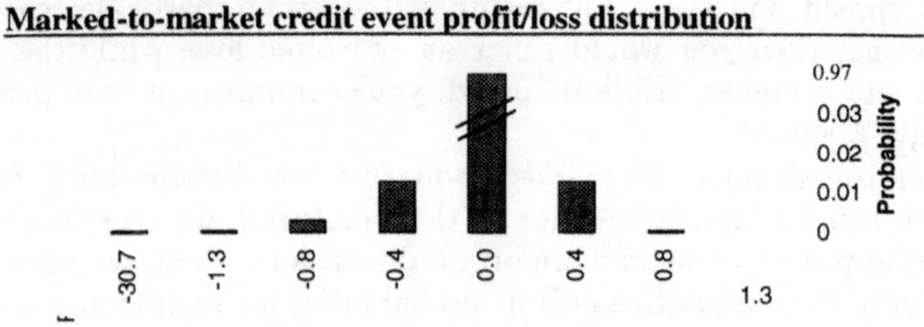

Fig. 33. Marked-to-market credit event profit/loss distribution

The results of applying this approach are illustrated in Figure 33, where we have tabulated the potential profit and loss profile from a single traded credit exposure, originally rated triple-B, which can be liquidated prior to one year. For this example, we have used a recovery rate of 71.57%, the average recovery rate for senior secured credits rated triple-B. As is clear from the figure, it is inappropriate to talk about 'loss distributions' in the context of marked-to-market loan or debt securities since a profit or gain in marked-to-market value can also be created by an improvement in the counterparty's credit standing.

While this approach allows one to capture the impact of credit migrations, holding the level of interest rates and spreads constant, it must be seen as a complement to a market risk measurement system which accurately captures the potential profit or loss impact of changing interest rate and/or average credit spread levels. If your market risk measurement system does not capture these risks, then a more complicated approach could be implemented, e.g. to jointly simulate interest rate levels, average credit spread levels and credit rating migrations; while potentially desirable, such an approach may come

[21] For credit-risky exposures such as the potential exposure from trading lines which can only be liquidated via other methods such as credit guarantees or credit derivatives, one would have to provide a cost of credit insurance table, again as a function of the exposure's maturity and credit rating.

close to the fabled 'all singing, all dancing' models that our grandmothers warned us about.

4.5 Random/correlated recovery rates

Throughout our discussions so far, we have also assumed that the amount which could be recovered in the event of default was a known constant, dependent only upon the rating of the counterparty. In reality, however, the actual amount which one can recover is neither constant nor is it dependent solely on the rating of the counterparty. Figure 34 illustrates these observations using Moody's (95) data by plotting the distribution of the bond prices upon default. Clearly, the market's expectation of potential recovery from defaulted securities depends upon the individual security.

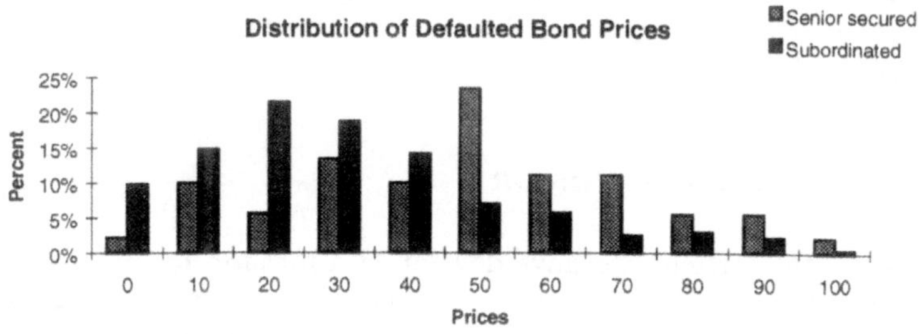

Fig. 34. Distribution of defaulted bond prices

One expedient way to model the impact of random recovery rates is to jointly simulate defaults and recovery rates. As is apparent from the figure, however, we must first specify a recovery distribution for each of the relevant recovery classes. We do this for various classes, e.g. senior secured, subordinated, unsecured, etc., by using Moody's or Standard & Poor's historical data. For every simulation of the systematic risk factors, we then take a random draw on this recovery distribution in order to tabulate the exposure's loss in the event of default.

This solution technique implicitly assumes that the random recovery rates can be drawn independently from one another across different macro-economic scenarios and counterparties. This assumption clearly breaks down, however, in two cases of interest:

- First, when we are tabulating the potential losses in the event of default arising from trading exposures: depending upon the compositions of their portfolios, it is quite probable that the recovery rates for two client exposures highly correlated (for example, if both are corporate counterparty's who are hedging their liabilities with a long-dated USD interest rate swap). The solution would again be to develop a model which jointly simulates the portfolio's potential exposure, driven by market rates, and credit events; in addition to its inherent modelling complexity (e.g. building an 'all singing, all dancing' model), this approach might also suffer from some concrete systems constraints: in order to implement such an approach, all of the counterparty's transactions, including netting and collateral set information and valuation methods, would have to reside in a common system. While this may be a valid end-game objective for many institutions, very few institutions are in a position to implement such a system currently.

- Second, for certain asset classes, most notably mortgages and some collateralised loans, the recovery rate is highly negatively correlated with the probability of default: very rarely will a residential homeowner default on their property when there is still positive equity in the home. In these cases, it is sometimes useful to add regional property price indices to the set of explanatory variables used in the Logit modelling of our systematic risk indices.

5 Summary and conclusion

In the course of this article, we have developed a method for explicitly tabulating the loss distributions from an arbitrary portfolio of credit exposures. Our method is based on two distinct components: a multi-factor, systematic risk model for credit risk uncertainty and a method for explicitly tabulating the discrete loss distributions arising from an arbitrary portfolio of credit exposures.

5.1 Multi-factor, systematic credit risk model

Credit risk uncertainty, in the form of default and rating migration uncertainty, was represented by a multi-factor, systematic risk model. This model, driven by macro-economic factors, was then estimated to fit the historical average default rates for various country/industry/rating segments and used to both forecast and simulate future credit risk uncertainty. This model of the systematic credit risk differs from other models in several respects:

- First, it explicitly recognises that the probability of a credit event occurring (e.g. a credit migration or default) is dependent upon the current state of the global and regional economy; more simply stated, defaults and downgrades tend to increase during a recession and decrease during an economic expansion.

- Second, it explicitly recognises the differing influence of macro-economic scenarios on individual industry segments through the estimation of 'sectoral betas' as well as the potential impact of industry-specific shocks. While these are intuitive observations for credit professionals, they are not reflected in many of the models used to represent systematic credit risk.
- Finally, the model is estimated and calibrated using average, historical default data; this implies that it can be calibrated to exactly reflect the loss experiences of any portfolio of counterparty's.

5.2 Tabulating discrete loss distributions

The model presented here then proceeds to explicitly tabulate the discrete, marked-to-market value loss distribution for any arbitrary portfolio of credit exposures. From these distributions, management can then calculate key management risk measures such as expected losses, often used to set the reserve levels for the institution, as well as the maximum loss within a known confidence interval, often interpreted as the amount of risk capital required to support a given portfolio of credit exposures.

The tabulation methods outlined are quite flexible and can be used across a wide variety of different portfolio compositions including:

- Both diversified and undiversified portfolios, leading to the ability to explicitly calculate a large exposure premium.
- Both constant and non-constant exposure profiles, allowing the model to be used equally well for standard commercial banking products as well as for exposure profiles generated by trading businesses.
- Both liquid and illiquid portfolios, allowing the model integrate trading and commercial banking portfolios as well as giving management the flexibility to define risk capital allocation horizons consistent with their own planning and budgeting process.
- Both constant and uncertain recovery rates, allowing management to capture the uncertainty surrounding the amount which might be recovered in the event of default.

The methods outlined in this paper differ from other models in that they can explicitly handle these types of portfolios; whereas many other models are content with using a normal approximation to the portfolio loss distribution for diversified, constant exposure, illiquid portfolios, the model developed here can be used in more general and practical cases by explicitly modelling the discrete loss distributions for and arbitrary portfolio of credit exposures.

References

Altmann, Edward, 1983, Aggregate Influences on Business Failure Rates, in Corporate Financial Distress: A Complete Guide to Predicting, Avoiding and Dealing with Bankruptcy, John Wiley & Sons

DPG or Derivatives Policy Group, 1995, Voluntary Oversight of OTC Derivatives

Black, F. and M. Scholes, 1973, The Pricing of Options and Corporate Liabilities, Journal of Political Economy, May-June

Cumming, C., and K. Saini, 1981, The Macroeconomic Determinants of Corporate Bankruptcies in Japan and the UK., Federal Reserve Bank of New York, Research Paper, December

Gollinger, T. L., and J. B. Morgan, 1993, Calculation of an Efficient Frontier for a Commercial Loan Portfolio, Journal of Portfolio Management, Winter, Volume 19, Number 2

G30 or Group of Thirty, 1993, Derivatives: Practices and Principles, Global Derivatives Study Group, Washington DC, July

Hull, J. and A. White, 1991, The Impact of Default Risk on Option Prices, Working Paper, University of Toronto

Jarrow, R. A. and S. M. Turnbull, 1995, Pricing Options on Derivative Securities Subject to Credit Risk, Journal of Finance 50(1)

Kealhofer, Stephen, 1995a, Managing Default Risk in Portfolios of Derivatives, in Derivative Credit Risk: Advances in Measurement and Management, Risk Publications, London, 1995

Kealhofer, Stephen, 1995b, Portfolio Management of Default Risk, proprietary documentation, KMV Corporation, San Francisco

Langstaff, F. and E. Schwartz, 1992, Valuing Risky Debt: A New Approach, Working Paper, UCLA

Lawrence, D. 1995, Aggregating Credit Exposures: The Simulation Approach, in Derivative Credit Risk: Advances in Measurement and Management, Risk Publications, London

Moody's Investor Service, 1994, Corporate Bond Defaults and Default Rates, 1970-1993, Moody's Investor Service

Noto, N. A. and D. Zimmerman, 1981, A Comparison of Failure and Liability Trends and Implications for Business Assistance, Report No. 81-36E, Congressional Research Service, January

Rowe, D., 1995, Aggregating Credit Exposures: The Primary Risk Source Approach, in Derivative Credit Risk: Advances in Measurement and Management, Risk Publications, London

Thomas, L. C., Crook, J. N., and D. B. Edelman, 1992, Credit Scoring and Credit Control, Clarendon Press, Oxford

Wilson, T., 1996, Calculating Risk Capital, in Handbook of Risk Management edited by C. Alexander, J. Wiley & Sons

Contributions to Economics

Hagen Bobzin
Indivisibilities
1998. ISBN 3-7908-1123-8

Helmut Wagner (Ed.)
Current Issues in Monetary Economics
1998. ISBN 3-7908-1127-0

Peter Michaelis/Frank Stähler (Eds.)
Recent Policy Issues in Environmental and Resource Economics
1998. ISBN 3-7908-1137-8

Jessica de Wolff
The Political Economy of Fiscal Decisions
1998. ISBN 3-7908-1130-0

GPSR Compliance

The European Union's (EU) General Product Safety Regulation (GPSR) is a set of rules that requires consumer products to be safe and our obligations to ensure this.

If you have any concerns about our products, you can contact us on ProductSafety@springernature.com

In case Publisher is established outside the EU, the EU authorized representative is:

Springer Nature Customer Service Center GmbH
Europaplatz 3
69115 Heidelberg, Germany

Batch number: 10725973

Printed by Printforce, the Netherlands